Stochastic Dynamic Programming and the Control of Queueing Systems

Stochastic Dynamic Programming and the Control of Queueing Systems

LINN I. SENNOTT
Illinois State University

A Wiley-Interscience Publication
JOHN WILEY & SONS, INC.
New York • Chichester • Weinheim • Brisbane • Singapore • Toronto

This book is printed on acid-free paper. ♾

Published simultaneously in Canada.

Library of Congress Cataloging-in-Publication Data:

Sennott, Linn I., 1943–
Stochastic dynamic programming and the control of queueing systems
/Linn I. Sennott.
p. cm.
"A Wiley-Interscience publication."
Includes bibliographical references and index.
ISBN 0-471-16120-9 (cloth : alk. paper)
1. Stochastic programming. 2. Dynamic programming. 3. Queueing theory. I. Title.
T57.79.S47 1999
519.7′03—dc21 98-2777

Printed in the United States of America

10 9 8 7 6 5 4 3 2 1

To my husband Jim,
for his unfailing encouragement and support

Contents

Preface

The subject of stochastic dynamic programming, also known as stochastic optimal control, Markov decision processes, or Markov decision chains, encompasses a wide variety of interest areas and is an important part of the curriculum in operations research, management science, engineering, and applied mathematics departments.

This book is unique in its total integration of theory and computation, and these two strands are interleaved throughout. First the theory underlying a particular optimization criterion (goal for system operation) is developed, and it is proved that optimal policies (rules for system operation that achieve an optimization criterion) exist. Then a computational method is given so that these policies may be numerically determined.

Stochastic dynamic programming encompasses many application areas. We have chosen to illustrate the theory and computation with examples mostly drawn from the control of queueing systems. Inventory models and a machine replacement model are also treated. An advantage in focusing the examples largely in one area is that it enables us to develop these important applications in depth and to concomitantly expand the subject of control of queueing systems. However, the theory presented here is general and has applications in diverse subject areas.

A total of nine numerical programs are fully discussed in the text. Text problems give suggestions for further exploration of these programs.

It is intended that the book can be successfully used by an audience ranging from advanced undergraduates to researchers. This may be done as follows:

- For advanced undergraduates, omit all proofs. Focus on motivation of the concepts, and exploration and extension of the nine programs.
- For first- and second-year graduate students, accompany the motivation of the concepts by reading selected proofs under the direction of a professor.
- For advanced graduate students, professionals, and researchers, read a selection of proofs as desired. The more difficult proofs are starred, and it is suggested that these be deferred to a second reading.

- For the reader whose primary interest is in applications and computation, omit the proofs as desired and concentrate on the material relating to computation.

The important background material is given in the appendixes. The appendixes are intended to be used as references, to be dipped into as needed. Some of the appendix material includes proofs. These are for the convenience of the interested reader and are not requisite to understanding the text.

The mathematical background necessary for comprehension of the text would be encompassed by a semester course on basic probability and stochastic processes, especially on the theory of Markov chains. However, since all the necessary background results are reviewed in the appendixes, the number of specific results the reader is expected to bring to the table is minimal. Perhaps most important for the reader is a bit of that famous ever-vague "mathematical maturity," which is always helpful in understanding certain logical ideas that recur in many of the arguments. The prospective student of this text should keep in mind that understanding the basic arguments in stochastic dynamic programming is a skill that is developed and refined with practice. It definitely gets easier as one progresses!

The chapter dependencies are shown in the flowchart (Fig. P.1). Chapter 1 is an introduction, and Chapter 2 gives the definitions of the optimization criteria. Chapter 3 presents theory and computation for the finite horizon optimization

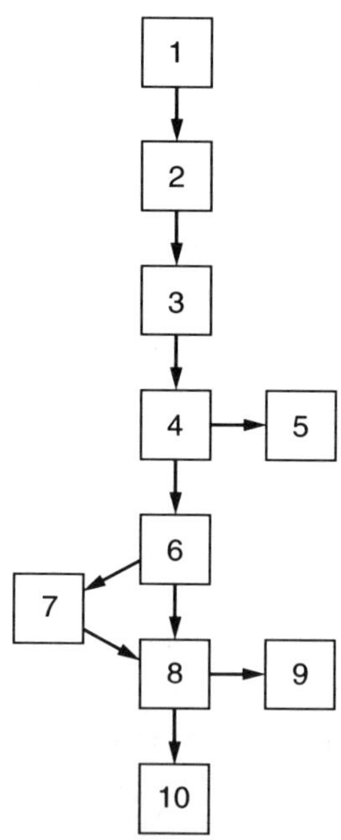

Figure P.1

criterion, and Chapter 4 presents theory and computation for the infinite horizon discounted optimization criterion. Chapter 5 presents an inventory model under the infinite horizon discounted cost criterion. This model is not a prerequisite to any other material.

Chapter 6 presents theory and computation for the average cost optimization criterion, when the state space of the process is a finite set. For computation, a thorough and very general treatment of value iteration is developed. Chapter 6 sets the stage for Chapter 8, which deals with the *computation* of average cost optimal policies when the state space of the process is an infinite set. Most of the material in Chapter 8 refers directly to results in Chapter 6. Chapter 7 deals with the *existence* theory of average cost optimal policies when the state space is infinite. The bulk of this material is not requisite to the computational results in Chapter 8 and may be omitted or referred to as desired.

Chapter 9 deals with (discrete time) models in which actions may only be taken at selected epochs. It is shown that the theory for this situation reduces to the general theory previously given. The computational examples focus on the average cost criterion. This material is not requisite to understanding Chapter 10.

Chapter 10 deals with the average cost optimization of certain continuous time systems. Again the theory here is reduced to that previously given.

The text is unique in combining theory and programs. The computational output from nine programs is presented and fully discussed. Numerous problems, both theoretical and computational, illuminate the text and give the reader practice in applying the ideas. Some of the problems involve explorations of the programs and include ideas for modifying them to obtain further insight.

I would like to express my profound thanks to Kevin Rock, a former student and now an instructor at Illinois State University, who read the complete manuscript carefully for understanding and identification of errors. He also helped set up a web site for the book.

I would like to express my deep gratitude to Professor Apostolos Burnetas, his Ph.D. student E. Lerzan Ormeci, and Professor Ulrich Rieder. Each of them read large portions of the manuscript, identified some errors, and made many helpful suggestions that improved the presentation. Any remaining mistakes, of course, are my responsibility. I would like to express my thanks to Professor Bob Cooper, who provided crucial references for Section 10.6 and checked the formula in Proposition 10.6.1.

I would also like to thank other professional colleagues who have provided insights, support, and encouragement over the years, including Rolando Cavazos-Cadena, Emmanuel Fernandez-Gaucherand, Onesimo Hernandez-Lerma, Mary Johnson, Marty Puterman, Sheldon Ross, Henk Tijms, and Mark VanOyen. I would like to thank friends and colleagues at Illinois State University for their support and wish particularly to mention Ken and Laura Berk, Lotus Hershberger, Jan Neuleib, Virginia Owen, the late Bob Ritt, Paul Schollaert, and Jane Swafford. My thanks also go to Steve Quigley, my editor at John Wiley & Sons, for his support and encouragement; his assistant Alison Bory;

and associate managing editor Angioline Loredo, for her gracious and careful attention to the book's production.

Finally, I wish to express my appreciation to my husband Jim, my son Kyle, and my new daughter-in-law Anastasia, for their love, understanding, and belief in me.

This book is for the reader to delve into, to study, and ultimately to take off from. Perhaps it will suggest new avenues for the reader's exploration and development and give impetus to the growth of this exciting and ever-developing field.

Comments are invited at sennott@math.ilstu.edu. The Pascal source code for the programs is available for viewing and downloading on the Wiley web site at http://www.wiley.com/products/subject/mathematics. The site contains a link to the author's own web site and is also a place where readers may discuss developments on the programs or other aspects of the material. The source files are also available via ftp at ftp://ftp.wiley.com/public/sci_tech_med/stochastic.

Stochastic Dynamic Programming and the Control of Queueing Systems

CHAPTER 1

Introduction

We are considering systems, evolving in time, that have chance or random aspects to their behavior. Such a system may evolve either in discrete or in continuous time. In the discrete setting the time axis is partitioned into fixed equal length segments, called *slots* or *periods*. Events affecting the system may take place during a slot, but typically they are registered by the system at the beginning of the following slot. In contrast, in a continuous time system events can occur at any instant of time and are registered as they occur. Many interesting models occur naturally in discrete time, while others occur naturally in continuous time. In this book attention is focused on discrete time systems, with the exception of Chapter 10 which treats a class of continuous time systems.

Our focus is the *sequential control* (also known as *dynamic* or *real time* control) of discrete time systems with random aspects. Such systems are called discrete time controlled *stochastic* systems. With the advent of computer-controlled processes, it is the case that control will often be applied at discrete time steps, even if the system under control occurs in continuous time. Therefore the control of an inherently continuous time model with random aspects is often well treated as a discrete time controlled stochastic system.

To control such systems, various actions may be taken, at various times, to affect the future behavior of the system. In the discrete time setting we assume that control can only be exercised at the beginning of a given slot and not at any other time during the slot. A fundamental dichotomy exists: Either control is exercised at the beginning of *every* slot or only at the beginning of certain selected slots, called *epochs*.

It turns out that the case of control available in every slot is more fundamental than the case of control available at selected epochs. The theory for control available in every slot is developed in Chapters 2 through 8. The theory for control available at selected epochs turns out to be a special case of this, and no new theoretical results are needed. This topic is treated in Chapter 9.

A certain class of continuous time control processes may be treated within the discrete time framework; this class consists of processes governed by exponential distributions. This development is in Chapter 10.

The next section illustrates control problems of the type that are covered in the text. Let us emphasize that the theory developed in the book is of a general nature and its applications are not limited to the particular models chosen for illustration.

1.1 EXAMPLES

The theory of discrete time controlled stochastic systems is motivated by systems that arise in applications. We are especially interested in using these results to gain an understanding of discrete time controlled queueing systems. A *queueing system* includes servers, customers and, usually, waiting lines or queues for the customers awaiting service. In the discrete time setting the servers may, for example, be transmitters, computers, or communication lines, or they may be stations on a production line. The customers may be messages, or fixed length groups of bits known as packets, or objects in the process of being manufactured. The queues are often called buffers, and we usually employ this terminology.

An inventory system is another example of a queueing system. In inventory systems the "customers" are items in the inventory, and the "servers" are external demands that remove these customers from the system. This example illustrates the importance of being very open in our thinking about what constitutes a queueing system. As we develop this flexibility, we will begin to see queueing systems everywhere.

In order to understand the types of control mechanisms that are of interest, let us now examine some common queueing systems. In Section 1.2 we will revisit some of these examples with more specificity.

Example 1.1.1. Single-Server Queue (Fig. 1.1). Packets (customers) enter the buffer, wait their turn, are served by the single server, and depart the system. Service is usually in the order of arrival, which is First Come First Served, but can be by other service disciplines such as Last Come First Served or Service in Random Order. Obviously, if the customers in a queueing system are human beings, FCFS is preferred.

Now consider the control options. We might place the controller at the entrance to the buffer to decide which packets to admit to the buffer. Or we could impose a control on the server that would adjust the rate at which packets are served. Both methods of control can be imposed simultaneously. □

Two questions arise as we begin to consider controlling queues. The first question is: To what end are we controlling the system? Clearly the control must be designed to achieve a goal. This goal is known as our *optimization criterion*, on which we will have more to say in the next section. A *policy* is a rule of operation that tells the controller which actions to choose, and an *optimal policy* is one that realizes the goal of the particular optimization criterion.

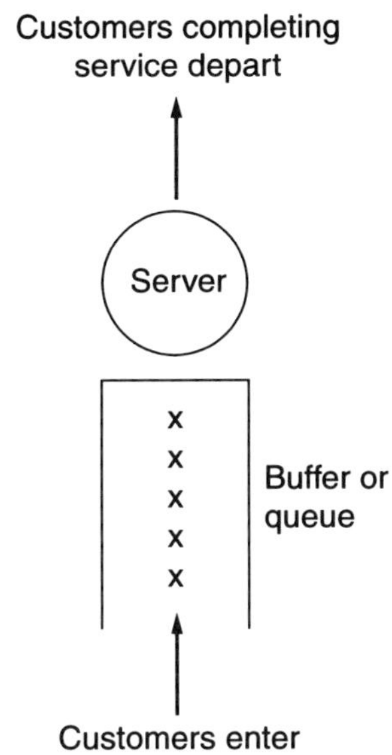

Figure 1.1 Single-server single-buffer system.

The second question is: What information about the current system condition, or *state*, is known to the controller? In this example, it would certainly be helpful to know how many packets are in the buffer before making a decision on admission or service rate. Throughout the book it is assumed that the controller has *full information* about the system. This may be contrasted with situations where the controller has only partial information, information corrupted by observation errors, or information received with a time delay. The full information case is fundamental to the development of a comprehensive theory and needs to be well understood before the case of partial or delayed information is treated.

Example 1.1.2. An Inventory Model. The demand for the product follows a known probability distribution. The demand for a particular period is assumed to be fully revealed just at the end of that period and is satisfied, as much as possible, from existing inventory and/or items produced during that period. Unfilled demand is *backlogged*; that is to say, these orders are registered to be filled in the future as inventory comes available. For example, if 5 items are on hand at the beginning of a period, 7 items are produced during that period, and 10 items are demanded during that period, then at the beginning of the next period the inventory level is 2. If 15 items are demanded, then the level is a backlog of 3 items.

The inventory level is observed at the beginning of each period. The control actions that may be taken relate to the number of items that may be produced during that period. □

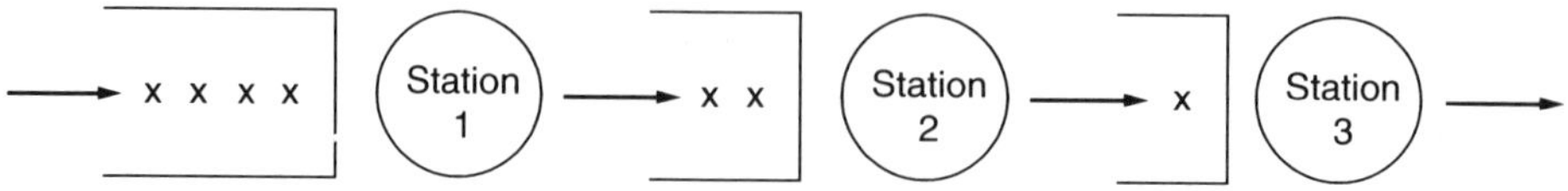

Figure 1.2 Tandem system.

Example 1.1.3. Tandem Queueing System (Fig. 1.2). Here we have a number of stations (servers) in series. There may or may not be buffers before each station. Customers enter the system at the buffer of station 1, receive service at station 1, enter the buffer at station 2, receive service at station 2, and so on, until they pass through all stations and leave the system.

Control may be exercised by restricting entry to the system, by adjusting the service rate of each of the servers, or by combinations of both. □

Example 1.1.4. Routing to Parallel Queues (Fig. 1.3). Here there are a number of servers with individual buffers. Customers arrive at the router and are sent to one of the buffers. It is assumed that once the routing has taken place, the customer cannot switch from one queue to another (called *jockeying*) and must therefore remain in the buffer to which it was routed until it receives service and leaves the system.

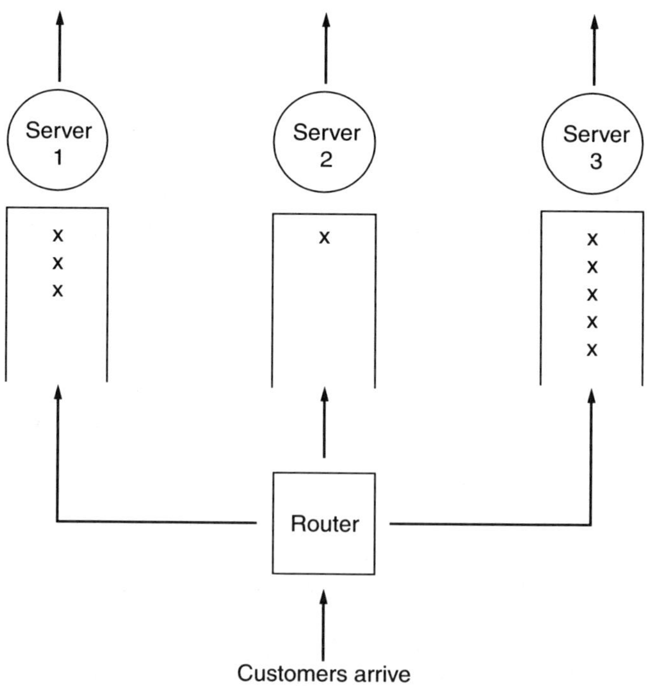

Figure 1.3 Routing to parallel buffers.

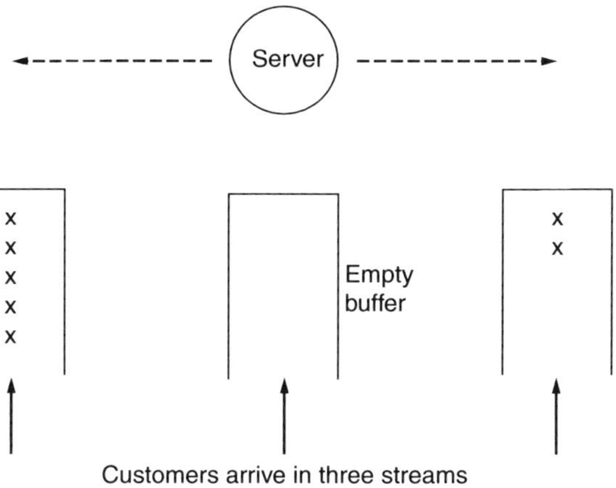

Figure 1.4 Single-server serving multiple buffers/classes.

In this example we assume that the service rates of the servers are constant. The control mechanism is invoked through the routing decision for each arriving customer (or batch of customers). □

Example 1.1.5. A Single-Server, Multiple-Buffer Model (Fig. 1.4). A single server is responsible for serving multiple queues. The server/controller may be considered the same mechanism. The server must decide which buffer to serve at a decision epoch and (possibly) how fast to serve.

It is important to note that the buffers need not be physically distinct. For example, the buffers might represent priority classes. The customers might all reside in the same location but be identified (or *tagged*) by their priority class. The decision of which buffer to serve is then the decision of which priority class to serve. The control options might also include the rate at which a given class is served. □

Example 1.1.6. A Single-Buffer, Multiple-Server Model (Fig. 1.5). In this model the service rates are fixed, but they may vary server to server. At most one customer can receive service from any server at any time. All customers not receiving service are queued in the single buffer. If there is a customer awaiting service and at least one server is free, then the control options include sending the customer to a selected free server or letting the customer remain in the queue. □

Example 1.1.7. Queueing Network (Fig. 1.6). Here a number of stations provide service to customers, and each station has its own buffer. Customers arrive from outside (*exogenous* customers) to each buffer. In addition, when a customer finishes service at a station, the customer may be routed to another station (or back to the one it just left) according to known routing probabilities.

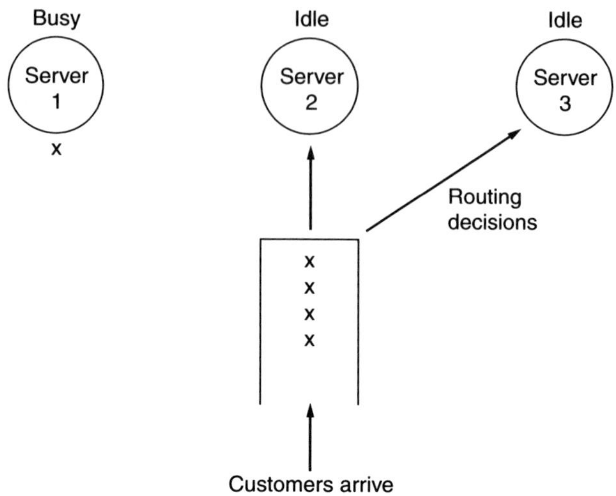

Figure 1.5 Single buffer served by multiple servers.

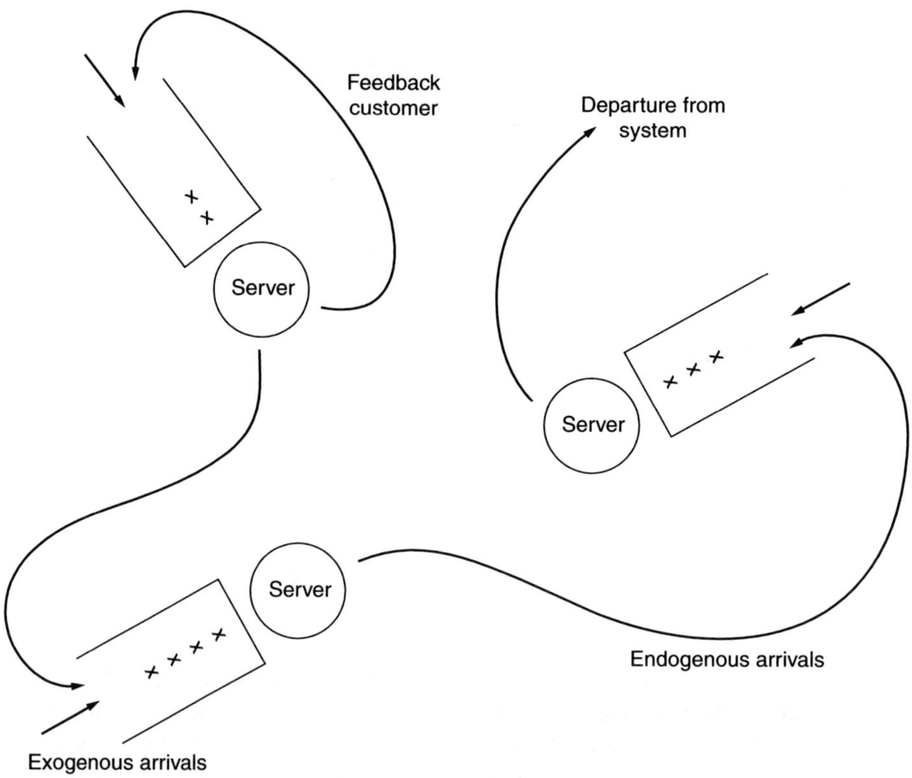

Figure 1.6 Network.

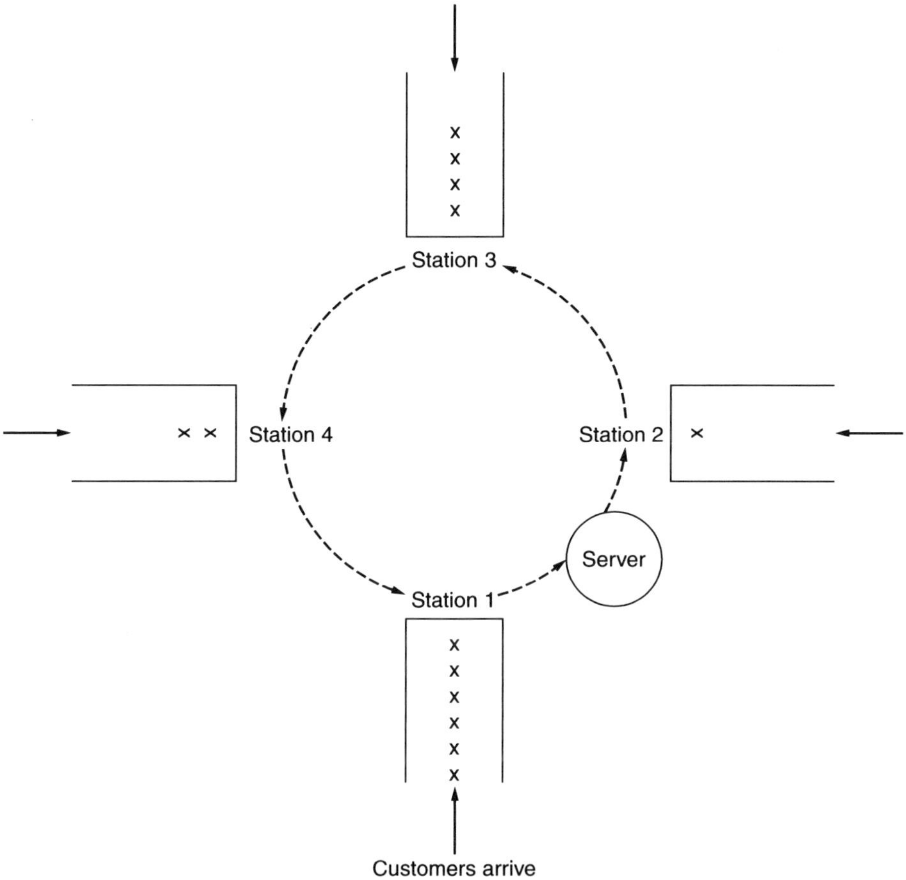

Figure 1.7 Polling system.

These are *endogenous* customers. It is assumed that every customer eventually leaves the system. Such a structure is called an *open* network. Control options include admission control and/or service control. □

Example 1.1.8. Cyclic Polling System (Fig. 1.7). Here a number of stations, each with its own buffer, are arranged in a ring. A server travels around the ring, say counterclockwise, from station 1 to station 2, then to station 3, and so on. When at station k the server has the control option of remaining there (idling if the buffer is empty, or serving packets in k's buffer if it is nonempty) or of moving to the next station. It is usually desirable to model a nonnegligible transit time in moving from one station to another. □

Example 1.1.9. Machine Replacement. A machine may be in one of various conditions, with condition 0 corresponding to a perfect machine and other conditions corresponding to various levels of wear. If the machine is in a state of wear, then we may make the decision to replace it with a perfect machine or to do nothing. A machine in a state of wear will continue to deteriorate according to a certain probability distribution. A perfect machine may be assumed to remain perfect for one slot and then begin to deteriorate. A more complex model would also allow the option of repairing a worn machine to bring it into a better condition than its present state. □

1.2 ASPECTS OF CONTROL

This section introduces the framework that will be employed for the control of the models in Section 1.1 and other systems as well. The discussion is on a general level with more precise definitions developed in Chapter 2.

The framework in which we will work is known as *stochastic dynamic programming*. A stochastic dynamic program is also called a Markov decision process. When time is discrete the process is (usually) called a *Markov decision chain*.

A Markov decision chain consists of

1. States
2. Actions
3. Costs
4. State equations (optional)
5. Transition probability distributions
6. An optimization criterion

The *state* of the system is the relevant information needed to describe the current condition of the system. The *state space* is the set of all system states. Because we are treating the full information case, we need to include in the state description all the relevant information concerning the current situation.

In Example 1.1.1 the relevant state information is the number i of packets in the buffer at the beginning of a slot (this includes the packet being served, if any). In this case the state space $S = \{0, 1, 2, \ldots\}$.

In this example we are allowing the buffer to be of *infinite capacity*. This is a useful modeling device, even though it is not physically realizable. One can simply imagine that when a new batch of packets arrives, the capacity of the physical buffer is increased to accommodate it.

Models involving infinite capacity buffers occur frequently. They may be contrasted with models assuming *finite capacity* buffers. In the latter there is a fixed capacity K for the buffer content, and no more than K customers can reside in the buffer. In some models the assumption of finite capacity buffers is appropriate, whereas in other models the assumption of infinite capacity buffers is more desirable. The following are possible reasons for allowing buffer capacities to be infinite: (1) We may not have information on the actual buffer capacity. (2) We may not want to lose customers and may prefer to assume that buffer capacity can be expanded as customers arrive. (3) We may wish to use the model with infinite capacity buffers to gain information on the appropriate sizing of buffer capacity.

The theoretical results developed in the text are general and apply to models with either finite or infinite capacity buffers.

We are also interested in obtaining computational results, both for models with finite buffer capacities and for those with infinite capacities. In the literature computational results have largely been confined to the case of finite capacity buffers. Here an approach called the *approximating sequence method* is developed that allows rigorous computation in the presence of infinite capacity buffers. The idea is to replace the infinite capacity model with a sequence of finite capacity models so that, as the capacities increase, the computations for these models are guaranteed to converge to the correct result for the original model. Nine computational examples are given in the book, and a program for each example is available on the companion web page.

Various *actions* are available to the controller and the available actions may depend on the current state of the system. Take Example 1.1.1. Let us assume that the actions are the available service rates. Note that when the system is empty, the server is idle and has only the "idle" action available.

There is a nonnegative *cost* associated with each state and available action in that state. The cost is associated with the state-action pair. The subject of stochastic dynamic programming has two major developmental strands. One can seek to minimize costs or to maximize rewards. We choose to deal with cost minimization because it is more congenial for the types of control problems we are most interested in treating.

However, all is not lost for those who wish to maximize their rewards! A reward associated with a particular state-action pair can be incorporated into this framework as a negative cost (under some nonrestrictive conditions). Chapter 5 treats an inventory model in which costs are imposed for holding and/or producing inventory and rewards are earned when inventory is sold. This model shows, in detail, how to work with rewards.

In Example 1.1.1, where actions are service rate adjustments, we might assume a cost for storing a packet in the buffer (related to delay in being served) and a cost for serving a packet (faster service costing more).

Now consider Example 1.1.4. An appropriate system state is the vector $\mathbf{i} = (i_1, i_2, \ldots, i_K)$ of buffer contents, where i_k is the number of packets in buffer k. The cost is then a function of the pair $(\mathbf{i}, k)$, where k is the action chosen (i.e., the server to which the customer is routed). This cost could consist of a holding cost reflecting the number of customers in the system and a cost of routing to server k.

Suppose that the packets that arrived in slot t were routed to buffer 1 but that at the beginning of slot $t + 1$ the controller wishes to route the newly arriving packets in the current slot to buffer 2. Under this circumstance it might be assumed that a cost is incurred for *switching the routing*, namely for changing the routing from one slot to the next. To handle this situation, we would enlarge the state description to be $(\mathbf{i}, 1)$, where the current buffer content vector is augmented by the previous routing decision. The cost is then a function of the state-action pair $[(\mathbf{i}, 1), 2]$. By means of this augmenting device, additional information can be built into the state description.

In this same example, let us now assume that there is no cost for switching the routing and a cost of 1 unit for each slot of time a packet resides in one of the buffers. Notice that the total cost over a time interval of T slots is the same as the total amount of delay suffered by the packets in the system.

The *delay* incurred by a communication system is an important measure of its performance. The minimization of delay may generally be modeled directly in our framework. Another important measure is system *throughput*, a measure of the number of packets successfully served. The maximization of throughput may generally be modeled using the device for incorporating rewards into our framework.

Assume that the system is in a given state and that the controller has decided on an action. Then, as discussed above, there is a cost incurred. The state of the system at the beginning of the following slot is governed by a *transition probability distribution*. This distribution will generally depend both on the current state and on the chosen action. A representation of the evolution of the system may be given by a *state equation*. The state equation, which is optional in specifying the system, can be helpful in picturing how the system evolves in time.

Consider Example 1.1.1 with the actions being service rate adjustments. The state of the system at time t may be represented by a random variable X_t (since it will be a random quantity rather than a deterministic quantity). Let the random variable Y_t represent the number of new packets arriving in slot t. Let Z_t be an indicator random variable that equals 0 if the buffer is empty or if a service is not completed in slot t, and equals 1 if there is a service completion. Then the evolution of the system is given by the state equation

$$X_{t+1} = X_t + Y_t - Z_t, \qquad t \geq 0. \tag{1.1}$$

This follows since the buffer content at time $t + 1$ is determined by the number in the buffer at time t plus any new arrivals during that slot minus 1 if there is a service completion during that slot. Note that X_0 is the initial buffer content.

Let us assume that the distribution of Y_t is independent of time. For example, suppose that P(no packets arrive) = 0.5, P(a single packet arrives) = 0.3, and P(two packets arrive) = 0.2. The distribution of Z_t depends on whether or not the buffer is empty and, if it is nonempty, on the service rate chosen by the controller.

Assume that the service rate adjustment takes place through a choice of the probability of a successful service. For purposes of this example, let us assume that the controller chooses the action of successfully serving a packet with probability 0.9. Finally let us assume that $X_t = 5$. Then the probability distribution of X_{t+1} is $P(X_{t+1} = 4) = (0.5)(0.9) = 0.45$, $P(X_{t+1} = 5) = (0.3)(0.9) + (0.5)(0.1) = 0.32$, $P(X_{t+1} = 6) = (0.3)(0.1) + (0.2)(0.9) = 0.21$, and $P(X_{t+1} = 7) = (0.2)(0.1) = 0.02$. (Check these!) These calculations are valid under the assumption that the chosen service rate does not influence the arrival process.

This information can be imparted in another way, namely by specifying the transition probability distributions. This is the probability distribution of the next state, given that the current state $X_t = i$, and is given by

$$\begin{aligned}
P_{0j} &= P(j \text{ arrive}), \quad j = 0, 1, 2, \\
P_{ii-1} &= (0.9)P(0 \text{ arrives}), \\
P_{ii} &= (0.1)P(0 \text{ arrives}) + (0.9)P(1 \text{ arrives}), \\
P_{ii+1} &= (0.1)P(1 \text{ arrives}) + (0.9)P(2 \text{ arrive}), \\
P_{ii+2} &= (0.1)P(2 \text{ arrive}), \quad i \geq 1.
\end{aligned} \tag{1.2}$$

The reader should realize that (1.2) contains a complete specification of how the system probabilistically evolves. The state equation (1.1) is helpful but optional.

We require the transition probability distributions to be independent of time (*time homogeneous*). This means that they cannot depend on the time slot number, undoubtedly a limitation in modeling some actual systems that do exhibit time-varying transition behavior. However, one approach to overcome this limitation is to build time-varying behavior into the state space description, at the cost of increased complexity of the state space. Another approach is to argue that if the system is slowly time varying, then we can analyze the system piecewise over those portions of time for which it is approximately time homogeneous. Considering the piecewise analyses together yields valuable information about controlling the original system.

Finally we come to the *optimization criterion*, our goal in controlling the system. The criteria are described here in general terms and more precisely in Chapter 2.

We may be interested in optimizing system behavior over the *finite horizon*. In this case the behavior of the system is considered for slots $t = 0$ to $t = K$ for a fixed positive integer K.

Or, we may be interested in allowing the system to operate for an infinite number of slots $t = 0, 1, 2, \ldots$. This is the *infinite horizon* and is appropriate if the system is to operate for a lengthy period and there is no a priori cutoff time. One approach for optimizing operation over the infinite horizon is to consider the total accumulation of costs where future costs have been discounted. Discounting reflects the economic principle that costs incurred in the future have a smaller present value.

Another approach to working with the infinite horizon is by means of averaging. We may look at the average cost incurred per slot over a fixed time horizon and then let the time horizon become ever longer. In this way we obtain a limit that reflects what happens on average far into the future.

There are other popular optimization criteria. However, these three are arguably the most important and are the ones treated in the book.

1.3 GOALS AND SUMMARY OF CHAPTERS

The goals of the book are as follows:

Goal 1. To develop the theory for optimization under the finite horizon, infinite horizon with discounting, and infinite horizon average cost criteria.

Goal 2. To show how optimization may be performed computationally, both when buffers are finite and when they are infinite.

Goal 3. To illustrate the theory and computational method with a rich set of examples drawn largely from the field of queueing control.

This text is unique in its total integration of *theoretical development* and *computational method*. For each optimization criterion, the theoretical development yields conditions for the existence of a particularly desirable type of policy that is optimal for that criterion. The approach to computation, known as the approximating sequence method, is a flexible and rigorous method for the computation of optimal policies in the presence of models with infinite buffers (more generally, models with infinite state spaces). To carry out the method, the original problem is replaced with a sequence of finite state approximation problems for which the convergence to the true value in the original problem is guaranteed. One may then compute optimal policies for a few members of the sequence (usually 2 or 3 suffice) and be confident that a close approximation to the optimal policy for the original problem has been attained.

The ability to compute optimal policies, while extremely valuable in itself, has two important corollaries. First, it allows us to examine sensitivity issues. This is done by varying the parameters of a problem to see whether the optimal policy is affected and, if so, to what degree. Second, it allows us to compare system performance under the optimal policy (which requires full information about the system state) with system performance under various suboptimal policies that do not require full state information. There is usually some cost involved in designing a system so that the controller has knowledge of the system state. If, for example, there exists a suboptimal policy with a performance within 5% of the optimal policy, this might be an acceptable level of performance. Having this type of knowledge is valuable when designing a system.

For the convenience of the interested reader, a brief summary of the chapter contents is given here. The Preface contains a discussion of chapter interdependencies and the reader is particularly advised to look at the flowchart given there.

In Chapter 2 notation and definitions are given. In Chapter 3 optimization over the finite horizon is treated. The computational model is Example 1.1.1 with control exercised through the acceptance or rejection of arriving packets.

In Chapter 4 optimization over the infinite horizon with discounting is treated. Chapter 5 illustrates the computational method under this criterion with a detailed treatment of the inventory model Example 1.1.2, showing the computation of optimal production levels.

In Chapter 6 optimization over the infinite horizon with averaging is treated for systems with finite state spaces. In Chapter 7 the theory of optimization over the infinite horizon with averaging is treated for systems with infinite state spaces. Chapter 8 develops the approximating sequence method for this criterion and illustrates it with two computational examples. The first is Example 1.1.1 with service rate control, and the second is Example 1.1.4.

In Chapter 9 we show how to treat the situation of control exercised only at selected epochs. This idea is illustrated with computations involving Example 1.1.1. Here the service time of a customer follows a general discrete time probability distribution, and service may be adjusted only when one service is completed and a new service is ready to commence.

Chapter 10 treats a class of continuous time systems. Three computational examples are given. The first is the service rate control of an M/M/1 queueing system. This is the continuous time analog of Example 1.1.1. The second example assumes that there is a pool of servers available, and that servers can be turned on or off. The problem is to determine the policy for dynamically adjusting the number of servers turned on. The third example is a continuous time version of the polling system in Example 1.1.8.

Our hope is that this material will be both interesting in its own right and an impetus to further development of the theoretical and computational aspects of stochastic dynamic programming. It is especially important to expand our knowledge (both theoretical and practical) concerning effective computational methods, and it is hoped that the work presented here will contribute to enthusiastic efforts in this direction.

BIBLIOGRAPHIC NOTES

Bellman (1957) is credited with founding the subject of stochastic dynamic programming. A second important early researcher is Howard (1960). However, the historical roots of the subject go deeper than Bellman's work. See Puterman (1994, p. 16) for interesting historical background.

PROBLEMS

1.1. Identify at least eight queueing situations that one might meet in everyday life. These systems have humans as the customers and/or the servers. For each situation, discuss the nature of the customers, the servers, and the queues.

1.2. For each of your examples in Problem 1.1 discuss whether and how it is feasible to control the system.

1.3. Discuss the aspects of control as they might be applied to Examples 1.1.3 and 1.1.5.

CHAPTER 2

Optimization Criteria

The mathematical structure we consider is known as a Markov decision chain. The Markov decision chain (also known as a discrete time Markov decision process or as a stochastic dynamic program) is a flexible construct for analyzing the control of discrete time systems involving random aspects. This chapter sets up the basic notation for a Markov decision chain and defines the important concepts.

The efficiency of system operation is measured by a suitable optimization criterion. The optimization criteria treated in the book are defined. A policy is a rule for the operation of the Markov decision chain. The various types of policies are discussed. An optimal policy is the best rule of operation for the system under the chosen criterion. Our goal is to show that optimal policies exist and to compute them. To this end, the notation of an approximating sequence is introduced. The approximating sequence method is the approach employed to compute optimal policies when the state space of the system is infinite.

2.1 BASIC NOTATION

Recall that time is divided into distinct equal portions, called slots or periods. A state represents the condition of the system at the beginning of a slot, and the state space S is the collection of all states. We assume that S is a countable set, which means that it is either a finite set or a denumerably infinite set. (A set is *denumerably infinite* if its elements can be enumerated, i.e., put into one-to-one correspondence with the natural numbers 1, 2, 3, ...)

When the system is in state $i \in S$, the controller has available various actions. These actions comprise a finite (and nonempty) set A_i. For any system whose control is digitally implemented, the assumption of finite action sets will suffice. Besides being adequate for the majority of applications, this assumption also has the advantage of simplifying the theory. When modeling a system whose control is implemented by an analog device, one may desire the flexibility of allowing the action sets to be intervals of real numbers. We do not treat this case.

Suppose that the system is currently in state i and that the action $a \in A_i$ is selected by the controller. Then a nonnegative (finite) cost $C(i, a)$ is incurred.

Under some state-action pairs we may wish to assume, additionally, that a nonnegative reward is earned. Rewards can be incorporated into the structure under certain conditions. Suppose that a cost of $C(i, a)$ is incurred and a reward of $R(i, a)$ is earned. Then the net cost is $C(i, a) - R(i, a)$ which may be negative. Assume that there exists a nonnegative number B such that the net costs are uniformly bounded below by $-B$, for all state-action pairs. Define a new cost structure by $C^*(i, a) = C(i, a) - R(i, a) + B \geq 0$. We can then determine the optimal policy for system operation under the C^* cost structure. Because our optimization criteria are not affected by the addition of a constant to all costs, the optimal rule of operation just determined is also optimal for the system operating under the original net cost structure. For this reason, if rewards are present, then we can assume that they have been incorporated into the system as negative costs and that the resulting (net) costs are nonnegative. Certain models with unbounded rewards are not treatable within this framework. Chapter 5 contains an inventory example that shows the treatment of rewards in detail.

If the system is in state i and action a is chosen, then the state at the beginning of the next slot is j with probability $P_{ij}(a)$, where $\sum_{j \in S} P_{ij}(a) = 1$. This means that the next state is determined according to a probability distribution that may depend on the current state-action pair. Since the transition probabilities sum to one, exit from S is not possible. In the future a summation over j will be understood to mean all states $j \in S$. It may also be helpful (but is optional) to indicate the evolution of the system by means of state equations.

The structure introduced above comprises a *Markov decision chain* (MDC) which is denoted by the symbol Δ. Keep in mind that to define an MDC requires the specification of four things: countable state space, finite action sets, nonnegative costs, and transition probability distributions.

Now we show how to model some of the examples from Chapter 1 as Markov decision chains.

Example 2.1.1. This is Example 1.1.1 with arrival control. (See Fig. 2.1.) The state of the system is the number of packets in the buffer at the beginning of a slot, and thus $S = \{0, 1, 2, \ldots\}$. At the beginning of each slot a batch of packets arrives and $p_j = P$(a batch containing j packets arrives), where $\sum_{j \geq 0} p_j = 1$. In every state there are two actions available: a = accept the incoming batch, or r = reject the incoming batch. The action must be chosen before the size of the batch is observed.

There is a nonnegative holding cost $H(i)$ incurred when there are i packets in the buffer, and we assume that $H(0) = 0$. The holding cost may be regarded as a cost of delaying those packets. For example, if $H(i) = i$, then for every slot in which a packet resides in the buffer, a delay cost of 1 unit is charged for that packet. In addition there is a positive rejection cost R incurred whenever a batch is rejected. The cost structure is $C(i, a) = H(i)$ and $C(i, r) = H(i) + R$.

Service occurs according to a geometric distribution with fixed rate μ, where

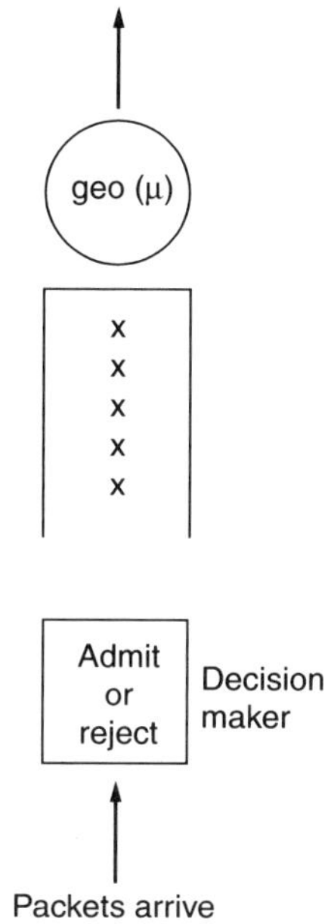

Figure 2.1 Example 2.1.1.

$0 < \mu < 1$. This means that the probability of a successful service in any slot is μ. If the service is unsuccessful, then another try is made in the next slot with the same probability of success, and this continues until the packet has been successfully served. If a batch arrives to an empty buffer and is accepted, then its packets are available for service at the beginning of the following slot.

If X_t, Y_t, and Z_t are as in (1.1), then the state equation is

$$X_{t+1} = X_t + I(a \text{ chosen})Y_t - Z_t, \qquad t \geq 0. \tag{2.1}$$

The indicator random variable I is 1 if a is chosen (and hence the new batch is admitted) and 0 if it is rejected.

The transition probability distributions are given by

$$\begin{aligned} P_{00}(r) &= 1, \\ P_{0j}(a) &= p_j, \qquad j \geq 0, \end{aligned}$$

$$\begin{cases} P_{i\,i-1}(r) = \mu, \\ P_{i\,i}(r) = 1 - \mu, \qquad i \geq 1, \end{cases}$$

$$\begin{cases} P_{i\,i-1}(a) = \mu p_0, \\ P_{i\,i+j}(a) = \mu p_{j+1} + (1-\mu)p_j, \qquad i \geq 1,\ j \geq 0. \end{cases} \tag{2.2}$$

This specifies the states, actions, costs, and transition probability distributions, and hence this example has been modeled as an MDC. □

Example 2.1.2. This is Example 1.1.1 with service rate control. The state space is as in Example 2.1.1. In state 0 there is no control action available since there are no packets to serve. We may think of this as the availability of a single action, namely in this case "take no service action." In any situation where there is a single action available in a given state (which just means that the controller has no choice), we refer to the single action as a *null* action. In the case of a null action we omit the notation a when specifying the costs and transition probabilities. In state $i \geq 1$ the actions consist of the allowable service rates $a_1 < a_2 < \ldots < a_M$, where $0 < a_1$ and $a_M < 1$. The conditions mean that the server must serve if the buffer is nonempty and that perfect service is unavailable.

The holding cost is as in Example 2.1.1. There is a nonnegative cost $C(a)$ of choosing to serve at rate a during a particular slot. The cost in state 0 is then 0, and for $i \geq 1$ we have $C(i,a) = H(i) + C(a)$. Notice that we have the opportunity to choose a new service rate at the beginning of each slot (if the buffer is nonempty).

The state equation is given in (1.1). In state 0 the transition probabilities are $P_{0j} = p_j$. For service rate choice a, the transition probability distributions are given by

$$\begin{cases} P_{i\,i-1}(a) = a p_0, \\ P_{i\,i+j}(a) = a p_{j+1} + (1-a)p_j, \qquad i \geq 1,\ j \geq 0. \end{cases} \tag{2.3}$$

□

Example 2.1.3. This is similar to Example 2.1.1 except that the size of the incoming batch may be observed before making a decision to accept or reject it. At the beginning of a slot the state is (i, k), where i denotes the number of packets in the buffer and k the number of packets in the incoming batch. The state space $S = \{(i,k) | i = 0, 1, 2, \ldots, k = 0, 1, 2, \ldots\}$ is a denumerable set.

The holding cost is as in Example 2.1.1. There is a positive rejection cost $R(k)$ incurred whenever a batch of size $k \geq 1$ is rejected. If a batch of size zero is observed, then there is no action taken and so $C(i,0) = H(i)$. A holding cost is not incurred on newly accepted packets until the slot following their arrival. Hence for $k \geq 1$ the cost structure is $C[(i,k),a] = H(i)$ and $C[(i,k),r] = H(i) + R(k)$.

The transition probabilities are somewhat more involved than when the batch size is unobserved. If k denotes the size of the current batch and j the size of the next batch, then some of the transition probabilities are

$$P_{(0,k)(0,j)}(r) = P_{(0,k)(k,j)}(a) = p_j, \qquad k \geq 1,\ j \geq 0,$$

$$\begin{cases} P_{(i,0)(i-1,j)} = \mu p_j, \\ P_{(i,0)(i,j)} = (1-\mu)p_j, \qquad i \geq 1,\ j \geq 0. \end{cases} \tag{2.4}$$

(Problem 2.1 asks you to develop all the transition probabilities for this example.) □

Example 2.1.4. This is Example 1.1.4. Let us assume that the batch arrival process is as in Example 2.1.1. The problem concerns the routing of an incoming batch to one of K parallel servers. Each server maintains its own queue, and server k serves its packets at geometric rate μ_k, where $0 < \mu_k < 1$. There may or may not be a cost associated with changing the routing to which the current batch is to be sent. Let us model the system under the supposition that there is a switching cost. We also assume that the routing decision is made before the size of the incoming batch is observed. An arriving batch is not "counted" in the buffer to which it is routed until the beginning of the following slot.

The state space S for this example is discussed in Chapter 1 and consists of pairs $(\mathbf{i}, u)$, where $\mathbf{i}$ is the vector of buffer levels and $u \in \{1, 2, \ldots, K\}$ is the previous routing decision.

There is a nonnegative holding cost $H_k(i_k)$ associated with the contents of buffer k. The total holding cost is $H(\mathbf{i}) = \sum_k H_k(i_k)$. In addition there is a nonnegative cost $C(u, k)$ for changing the routing from server u to server k, where $C(k, k) = 0$. The cost structure is $C[(\mathbf{i}, u), k] = H(\mathbf{i}) + C(u, k)$.

Some thoughtful notation can facilitate the writing of the transition probabilities. Let $\mathbf{j}(k)$ be a vector with j in the kth place and 0's elsewhere. Then $P_{(\mathbf{0},u)(\mathbf{j}(k),k)}(k) = p_j$.

Now let $\mathbf{i}$ be a state vector with at least one nonzero component. Let $F = F(\mathbf{i})$ be the set of nonzero coordinates of $\mathbf{i}$, and let $E = E(\mathbf{i})$ be a (possibly empty) subset of F representing those servers who complete service during the current slot (recall they can only serve packets already in their buffer). The probability of this event is $P(E) = \Pi_{k \in E}\, \mu_k \Pi_{k \in F-E}\, (1 - \mu_k)$. Finally let $\mathbf{e}(E)$ be a vector with 1 in every coordinate $k \in E$ and 0's elsewhere. Then we claim that

$$P_{(\mathbf{i},u)(\mathbf{i}+\mathbf{j}(k)-\mathbf{e}(E),k)}(k) = p_j P(E). \tag{2.5}$$

(Problem 2.2 asks you to explain this.) □

2.2 POLICIES

Informally a policy is a rule for the operation of a Markov decision chain. Let $t = 0$ be the initial slot. The MDC may be operated in one of two modes. In the infinite horizon mode the system is operated for slots $t = 0, 1, 2, \ldots$. In the finite horizon mode a fixed integer $n \geq 1$ is specified, and the system is operated for the n slots $t = 0, 1, \ldots, n - 1$.

We first define a policy for the infinite mode of operation. The beginning reader need not be overly concerned with the details, but it is important to grasp both the general idea of how a policy governs the operation of the MDC and the definition of a stationary policy.

Assume that the initial state i of the system is known. Here is how the controller operates under a policy θ. The history at time $t = 0$ is given by $h_0 = (i)$. The initial action is chosen from A_i according to the distribution $\theta(a|i) = \theta(a|h_0)$. This is a probability distribution on the actions $a \in A_i$. Assume that action a_0 is selected.

Then the state of the system at $t = 1$ is determined by the transition probability distribution associated with i and a_0. Suppose that this state is j. The history at time $t = 1$ is then given by $h_1 = (i, a_0, j)$. The action at time $t = 1$ is chosen from A_j according to the distribution $\theta(a|i, a_0, j) = \theta(a|h_1)$.

Once this action has been chosen (say it is a_1), then the state of the system at $t = 2$ is determined by the distribution associated with j and a_1. Suppose that this state is k. The history at time $t = 2$ is then given by $h_2 = (i, a_0, j, a_1, k)$. The process continues in this fashion.

Assume that the process has been operating for slots $t = 0, 1, \ldots, n - 1$, and that the state at time $t = n$ has just been determined. A history at time n is a tuple $h_n = (i, a_0, i_1, a_1, \ldots, i_{n-1}, a_{n-1}, i_n)$ of the past states and actions and the current state. Then the action at n is chosen according to the probability distribution $\theta(a|h_n)$ on the action set associated with i_n. Once this action has been chosen, the state at $t = n + 1$ may be determined. The process continues in this way for infinitely many steps.

We see that the controller's actions under a policy can be based on the previous states visited, the actions chosen in those states, and when those visits occurred. There are several important types of policies. These are classified by how much of the history may be utilized by the controller. We start with the most restrictive (and most important) type and work toward the less restrictive.

A *stationary policy*, denoted by f, operates as follows: Associated with each state i is a distinguished action $f(i) \in A_i$. If *at any time* the controller finds the system in state i, then the controller always chooses the action $f(i)$. Thus a stationary policy depends on the history of the process only through the current state. *To implement a stationary policy, the controller need only know the current state of the system.* Past states and actions are irrelevant. The advantages for implementation of a stationary policy are clear, since it necessitates the storage of less information than required to implement a general policy. The stationary policy is by far the most important type of policy.

A slightly less restrictive type of policy is the *randomized stationary policy*, denoted by δ. Associated with each state i is a probability distribution $\delta(i)$ on A_i. If at any time the controller finds the system in state i, then the controller always chooses action a with probability $\delta(i)(a)$. As is the case for a stationary policy, a randomized stationary policy depends on the history of the process only through the current state. To implement it, the controller needs to know the current state of the system. If f is a stationary policy, then it is a "degenerate" randomized stationary policy, since we may define the distribution associated with state i to be the degenerate distribution that chooses action $f(i)$ with probability 1.

A *deterministic Markov policy* is a sequence $\theta = (f_0, f_1, f_2, \ldots)$ of stationary policies. It operates as follows: If the process is in state i at time $t = n$, then the controller chooses action $f_n(i)$. Thus a deterministic Markov policy depends on the history of the process only through the current state and the time index. To implement it, the controller needs to know the current state of the system and the time index.

A *randomized Markov policy* is a sequence $\theta = (\delta_0, \delta_1, \delta_2, \ldots)$ of randomized stationary policies. It operates as follows. If the process is in state i at time $t = n$, then the controller chooses action $a \in A_i$ with probability $\delta_n(i)(a)$. Thus a randomized Markov policy depends on the history of the process only through the current state and the time index. To implement it, the controller needs to know the current state of the system and the time index.

The following example clarifies the various types of policies:

Example 2.2.1. This is Example 2.1.2 with $M = 3$ available service rates. If the process is operating under a given policy, then a history is a list of the previous buffer levels and service rates employed up to the current time, together with the current buffer level. If the buffer is empty at time t, then a_t is the null action. Now fix integers $L < U$ with $1 \le L$, and let i be the current state.

The policy θ operates as follows: Serve at rate a_1 if $1 \le i \le L$, serve at rate a_2 if $L < i \le U$, and serve at rate a_3 if $U < i$. Then θ is a stationary policy. To implement it only requires the controller to monitor the current buffer level.

The policy ψ operates just as the policy θ except when $i = U$. In this case the server randomizes equally between rates a_2 and a_3. Then ψ is a randomized stationary policy. To implement it requires the controller to monitor the current buffer level and to perform a randomization if the level is U.

The policy χ operates as follows: If the current time is less than 50 (and the buffer is nonempty), then serve at the lowest rate. If the current time is 50 or more (and the buffer is nonempty), then randomize equally between all three service rates. Here the controller needs to monitor the current buffer level (only to see that the buffer is nonempty) and the time index. If $t \ge 50$, then a randomization must be performed. Hence χ is a randomized Markov policy.

The policy ξ operates as follows: Assume a given history at time n, and let w_n be the average buffer level for slots $t = 0$ to $t = n$. Note that the average level is a function of the history up to and including the current level. If $w_n \le U$, then serve at lowest rate, while if $w_n > U$, then serve at highest rate. Because

w_n requires knowledge of the history of the process, this is a general policy. However, it can be implemented in a more efficient way than through the histories. If the controller keeps track of the current buffer level i, the time index n, and the previous value w_{n-1}, then w_n may be computed recursively since $w_n = [i + nw_{n-1}]/(n + 1)$. □

Now consider the meaning of a policy θ for the process operating over the finite n horizon, namely over slots $t = 0, 1, \ldots, n - 1$. (This policy may be denoted by θ_n.) The policy is defined exactly as above, except that when the history $h_n = (i_0, a_0, \ldots, i_{n-1}, a_{n-1}, i_n)$ is observed, then the process stops. So the state at time $t = n$ is determined, and then the process terminates. We often assume that a *terminal cost* is incurred that is a function of the terminal state. Observe that under this situation exactly n choices of actions will be made by the controller.

Because the histories include the time index, under a general policy it is always clear to the controller what the present time is, and hence how many slots are left before termination. Because of this, the action chosen under θ at time t may also depend on the number $n - t$ of slots until termination, the *steps to go*.

2.3 CONDITIONAL COST DISTRIBUTIONS

The purpose of this section is to clarify conceptually the meaning of the expectation given in (2.6). These expectations are the building blocks of the optimization criteria to be introduced in Section 2.4.

Assume that the initial state is i and that the process operates under an arbitrary policy θ. It is clear from the discussion in the previous section that the state of the process at time t depends on various probability distributions and hence typically is not a deterministic quantity. The state at time t is a random variable, which we denote by X_t. Similarly the action chosen at time t is a random variable, which we denote by A_t. (Note that this notation is not to be confused with the action set associated with a particular state.) The joint probability distribution of (X_t, A_t) is given by $P_\theta(X_t = j, A_t = a|X_0 = i)$, where clearly we must have $a \in A_j$.

It is the case that this probability distribution is well-defined. We do not prove this but instead show how it may be calculated for $t = 0, 1, 2$. This will be sufficient to indicate the operative ideas. Now $P_\theta(X_0 = i, A_0 = a|X_0 = i) = \theta(a|i)$. For $t = 1$ we have

$$P_\theta(X_1 = j, A_1 = a|X_0 = i) = \sum_{b \in A_i} \theta(b|i)P_{ij}(b)\theta(a|i, b, j).$$

Here a term is the probability of originally choosing action b, then transitioning to j, and then choosing action a; the terms are summed over the action $b \in A_i$. For $t = 2$ we have

$$P_\theta(X_2 = j, A_2 = a | X_0 = i) = \sum_{b \in A_i} \theta(b|i)$$

$$\cdot \left(\sum_k P_{ik}(b) \sum_{d \in A_k} \theta(d|i,b,k) P_{kj}(d) \theta(a|i,b,k,d,j) \right).$$

This calculates the probability of a selection of actions and states leading to the pair (j, a) and sums over all such selections.

Associated with the random pair (X_t, A_t) is the cost $C(X_t, A_t)$. This is the cost incurred at time t when the controller operates under θ. Because $C(X_t, A_t)$ is also a random variable, one effective way to assess it is to employ its expectation. This is given by

$$E_\theta[C(X_t, A_t)|X_0 = i] = \sum_j \sum_{a \in A_j} C(j,a) P_\theta(X_t = j, A_t = a | X_0 = i), \qquad (2.6)$$

and represents the *statistical average cost* at time t. Because the costs are nonnegative, it is the case that the expectation in (2.6) is well-defined. In some examples, it may have value $+\infty$.

Let us consider the important situation when θ is a stationary policy f. In this case we employ some special notation. The cost associated with state i is denoted $C(i,f)$, where this is understood to be $C(i,f(i))$. Similarly the transition probabilities are denoted $P_{ij}(f)$, where this is understood to be $P_{ij}(f(i))$.

Then $P_f(X_t = j,\ A_t = a|X_0 = i)$ is zero unless $a = f(j)$, and we have

$$\begin{aligned} &P_f(X_t = j, A_t = f(j) | X_0 = i) \\ &= P_f(X_t = j | X_0 = i) \\ &= \sum_{k_1 \in S} P_{ik_1}(f) \sum_{k_2 \in S} P_{k_1 k_2}(f) \ldots \sum_{k_{t-1} \in S} P_{k_{t-1} j}(f) \\ &=: P_{ij}^{(t)}(f). \end{aligned} \qquad (2.7)$$

Then (2.6) becomes

$$E_f[C(X_t, A_t)|X_0 = i] = \sum_j C(j,f)P_f(X_t = j|X_0 = i)$$

$$= \sum_j C(j,f)P_{ij}^{(t)}(f). \tag{2.8}$$

This is the expected cost at time t under the stationary policy f.

For most models it is impossible to obtain a closed form expression for the quantity in (2.6) (or even for the quantity in (2.8)). The reader can relax—we will not typically be calculating these quantities! What is crucial is a conceptual understanding of (2.6) and (2.8) rather than facility in calculation.

2.4 OPTIMIZATION CRITERIA

Four optimization criteria will be treated in the book:

1. The finite horizon expected discounted cost criterion.
2. The finite horizon expected cost criterion.
3. The infinite horizon expected discounted cost criterion.
4. The long-run expected average cost criterion.

It will be seen shortly that each criterion is based on the fundamental building block of the statistical average cost $E_\theta[C(X_t, A_t)]$ at time t, as defined in (2.6). These basic building blocks are put together in different ways under each criterion.

We first discuss the concept of discounted costs. A discount factor is a number α satisfying $0 < \alpha \le 1$ such that future costs are discounted at rate α. What this means is that a cost of 3 units incurred at time 0 is considered to be a cost of 3α when incurred at time 1, of $3\alpha^2$ when incurred at time 2, and in general, of $3\alpha^t$ when incurred at time $t \ge 0$. This embodies the economic idea that a cost to be incurred in the future is discounted in today's money. (It is certainly possible to have $\alpha = 0$, but this case is uninteresting.) Note that $\alpha = 1$ corresponds to no discounting.

To define the criterion in 1, assume that the process operates over the finite horizon n and that there is a nonnegative terminal cost $F(k)$ incurred whenever the process halts in state k. Let the initial state i, the horizon n, and the policy θ be given. The n horizon expected (total) discounted cost under θ is denoted by $v_{\theta,\alpha,n}(i)$.

In defining this quantity, it is helpful to allow the possibility of $n = 0$. For $n = 0$ we assume that the initial state is observed and the terminal cost assessed, but that no action is taken. Hence $v_{\theta,\alpha,0}(i) = F(i)$. For $n \ge 1$ we define

$$v_{\theta,\alpha,n}(i) = E_\theta\left[\sum_{t=0}^{n-1} \alpha^t C(X_t, A_t) + \alpha^n F(X_n) | X_0 = i\right]$$

$$= \sum_{t=0}^{n-1} \alpha^t E_\theta[C(X_t, A_t)|X_0 = i] + \alpha^n E_\theta[F(X_n)|X_0 = i]. \tag{2.9}$$

The second line follows from the linearity of the expectation. The function $v_{\theta,\alpha,n}$ is well-defined but may be $+\infty$.

The n horizon expected discounted value function is defined as

$$v_{\alpha,n}(i) = \inf_\theta v_{\theta,\alpha,n}(i), \tag{2.10}$$

where the infimum is taken over all policies for the n horizon. The quantity $v_{\alpha,n}$ is the greatest lower bound on all the n horizon expected discounted costs and is the best result that could be desired. Here is the definition of an optimal policy under this criterion.

Definition 2.4.1. Let θ be a policy for the n horizon. Then θ is *optimal for the expected discounted cost criterion for the n horizon* if $v_{\theta,\alpha,n}(i) = v_{\alpha,n}(i)$ for $i \in S$. □

Remark 2.4.2. The quantities in (2.9–10) and others to be defined shortly in (2.11–16) may equal $+\infty$, and we denote $+\infty$ by ∞. The approach of allowing these quantities to be infinite (unless stated otherwise) gives us the greatest degree of flexibility in our theoretical development, since we need not be concerned with imposing potentially complicated conditions to make these quantities finite. However, quantities introduced in a model such as a holding cost or cost for service are always assumed to be finite quantities. This convention is used without further mention. □

To define the criterion in 2, assume that the process operates as in 1 but that future costs are undiscounted. This corresponds to criterion 1, with $\alpha = 1$. The n horizon expected cost under θ is denoted by $v_{\theta,n}$, where $v_{\theta,0}(i) = F(i)$. From (2.9–10) we have

$$v_{\theta,n}(i) = \sum_{t=0}^{n-1} E_\theta[C(X_t, A_t)|X_0 = i] + E_\theta[F(X_n)|X_0 = i] \tag{2.11}$$

and

$$v_n(i) = \inf_\theta v_{\theta,n}(i), \tag{2.12}$$

where the infimum is taken over all policies for the n horizon. Here is the definition of an optimal policy under this criterion.

Definition 2.4.3. Let θ be a policy for the n horizon. Then θ is *optimal for the expected cost criterion for the n horizon* if $v_{\theta,n}(i) = v_n(i)$ for $i \in S$. □

The remaining two criteria deal with the infinite horizon case. Now assume that θ is a policy for the infinite horizon. To define the criterion in 3, assume that the discount factor α, initial state i, and policy θ are given. Here we must have $\alpha < 1$. The expected (total) discounted cost under θ is denoted by $V_{\theta,\alpha}$ and defined as

$$\begin{aligned} V_{\theta,\alpha}(i) &= E_\theta\left[\sum_{t=0}^{\infty} \alpha^t C(X_t, A_t) | X_0 = i\right] \\ &= \sum_{t=0}^{\infty} \alpha^t E_\theta[C(X_t, A_t) | X_0 = i]. \end{aligned} \tag{2.13}$$

The expected discounted value function (*discounted value function*, for short) is defined as

$$V_\alpha(i) = \inf_\theta V_{\theta,\alpha}(i), \tag{2.14}$$

where the infimum is taken over all policies for the infinite horizon. The quantity V_α is the greatest lower bound on all the expected discounted costs, over the infinite horizon, and is the best result that could be desired. Here is the definition of an optimal policy under this criterion.

Definition 2.4.4. Let θ be a policy for the infinite horizon. Then θ is *optimal for the expected discounted cost criterion* if $V_{\theta,\alpha}(i) = V_\alpha(i)$ for $i \in S$. □

Analogously to the finite horizon case one is tempted to set $\alpha = 1$ in (2.13) to obtain an undiscounted criterion for the infinite horizon. The problem with this approach is that for the systems we desire to model, the resulting expected total cost would be ∞ for all policies. Instead, we employ the idea of averaging the expected (total) cost over n steps and then use a limiting procedure.

Given initial state i, the long-run expected average cost under policy θ is denoted by $J_\theta(i)$ and defined by

$$J_\theta(i) = \limsup_{n \to \infty} \frac{1}{n} E_\theta \left[\sum_{t=0}^{n-1} C(X_t, A_t) | X_0 = i \right]$$

$$= \limsup_{n \to \infty} \frac{v_{\theta,n}(i)}{n}. \tag{2.15}$$

The limit supremum concept is reviewed in Appendix A. The reader who is uncomfortable with the limit supremum should feel free to think of this as a limit until more experience with this concept is gained. There will be no loss of understanding. The limit supremum is taken since the limit sometimes fails to exist (see Example 6.2.1). The limit supremum is the largest limit point of the expected average costs, and hence is the worst case situation.

We define the long-run expected average cost function (*average cost*, for short) by

$$J(i) = \inf_\theta J_\theta(i), \tag{2.16}$$

where the infimum is taken over all policies for the infinite horizon. The quantity $J(\cdot)$ is the greatest lower bound on the average costs, and is the best result that could be desired. Here is the definition of an optimal policy under this criterion.

Definition 2.4.5. Let θ be a policy for the infinite horizon. Then θ is *optimal for the average cost criterion* if $J_\theta(i) = J(i)$ for $i \in S$. □

We very occasionally need the average cost concept but with the limit supremum replaced by the limit infimum. The quantity $J_\theta^*(i)$ is defined as in (2.15) but with the limit supremum replaced by the limit infimum. This is the smallest limit point of the expected average costs and hence is the best case situation. The quantity $J^*(i)$ is defined analogously to (2.16).

This completes the definition of the optimization criteria. The deeper meaning of each criterion will be revealed in subsequent chapters.

2.5 APPROXIMATING SEQUENCE METHOD

The approximating sequence method is a general framework for the computation of optimal policies when the state space is denumerably infinite. In this section we define an approximating sequence and discuss some important ways of constructing such sequences.

Now assume that the Markov decision chain (MDC) Δ is given, and recall that it consists of four items: the state space S (assumed in this section to be denumerably infinite), the action sets A_i, the costs $C(i, a)$, and the transition probabilities $P_{ij}(a)$.

We define a sequence (Δ_N) of MDCs that *approximates* Δ. The state space of each Δ_N is finite, and because of this the computation of an optimal policy may be carried out in Δ_N. Then under certain conditions the results of these computations will converge to an optimal policy for Δ. It will generally be sufficient to compute for only two or three members of the sequence to get a good approximation to an optimal policy for Δ.

Definition 2.5.1. Let N_0 be a nonnegative integer. The sequence $(\Delta_N)_{N \geq N_0}$ is an *approximating sequence* (AS) for Δ if there exists an increasing sequence $(S_N)_{N \geq N_0}$ of nonempty finite subsets of S such that $\cup S_N = S$. Each Δ_N is an MDC with state space S_N satisfying two conditions:

(i) For $i \in S_N$ the action set is A_i and the cost at a is $C(i, a)$.

(ii) For each $i \in S_N$ and $a \in A_i$, $P_{i-}(a; N)$ is a probability distribution on S_N such that

$$\lim_{N \to \infty} P_{ij}(a; N) = P_{ij}(a), \qquad j \in S. \tag{2.17}$$

□

If we are dealing with the finite horizon case with a terminal cost F, then this same terminal cost applies to Δ_N. The integer N is said to be the *approximation level.*

At first glance this definition seems formidable, but in reality it is quite simple. The MDC Δ_N has as its state space a finite subset of S. On this finite subset the action sets and costs for Δ_N are exactly the same as for Δ. Only the transition probabilities are different. These form distributions on the finite subset that converge pointwise to the original distributions on Δ. The distributions in Definition 2.5.1(ii) are called *approximating distributions.* Keep in mind that only two items are required to specify an AS: the finite subsets and the approximating distributions.

Example 2.5.2 Consider the state space $S = \{0, 1, 2, \ldots\}$; there is one action in each state. We have $P_{0j} = (1/2)^{j+1}$ for $j \geq 0$, and $P_{ii-1} = 1$ for $i \geq 1$.

Let $N_0 = 2$ and $S_N = \{0, 1, \ldots, N\}$. Let $P_{ii-1}(N) = P_{ii-1}$ for $1 \leq i \leq N$. The distribution for 0 is given by

$$P_{00}(N) = \frac{1}{2} - \frac{1}{N},$$

$$P_{0j}(N) = \frac{1}{2^{j+1}}, \qquad 1 \le j \le N-1,$$

$$P_{0N}(N) = \sum_{j=N}^{\infty} \frac{1}{2^{j+1}} + \frac{1}{N} = \frac{1}{2^N} + \frac{1}{N}. \tag{2.18}$$

This satisfies (2.17) and hence is an approximating distribution. □

The most important way to define the approximating distributions is by means of an *augmentation procedure*. This procedure is useful when the state space is multidimensional as well as when it is one-dimensional. Informally the idea is as follows:

Suppose that the process is in state $i \in S_N$ and action a is chosen. For $j \in S_N$ the probability $P_{ij}(a)$ is left unchanged. Now assume that $P_{ir}(a) > 0$ for some $r \notin S_N$. This means that under this probability the original process would transition to state r outside of S_N. This is said to be *excess probability* associated with (i, a, r, N), and something must be done with this excess probability. It is redistributed (i.e., given or sent) to the states of S_N according to a specified distribution. In full generality this distribution may depend on i, a, r, and N; it is called the augmentation distribution associated with (i, a, r, N). Moreover it is no loss of generality to require it to be defined even if $P_{ir}(a) = 0$. The formal definition of an augmentation procedure is now given.

Definition 2.5.3. The approximating sequence (Δ_N) is an *augmentation type approximating sequence* (ATAS) if the approximating distributions are defined as follows: Given $i \in S_N$ and $a \in A_i$, for each $r \notin S_N$ there exists a probability distribution $(q_j(i, a, r, N))_{j \in S_N}$, called the *augmentation distribution* associated with (i, a, r, N), such that

$$P_{ij}(a; N) = P_{ij}(a) + \sum_{r \in S - S_N} P_{ir}(a) q_j(i, a, r, N), \qquad j \in S_N. \tag{2.19}$$

□

Under an augmentation procedure the original probabilities on S_N are never decreased, but they may be augmented by the addition of portions of excess probability (see Fig. 2.2). Note that Example 2.5.2 is not an ATAS.

To help the reader become comfortable with the concept of an ATAS, we now give some terminology and examples. After these are completed, we prove that (2.19) does indeed define an approximating probability distribution.

Here are some ATASs that arise frequently and the informal terminology used to describe them. Suppose that there exists a finite subset G of S such that it is always the case that $\sum_{j \in G} q_j(i, a, r, N) = 1$. This means that all excess probability is given, in some way, to the elements of G. We say that this ATAS

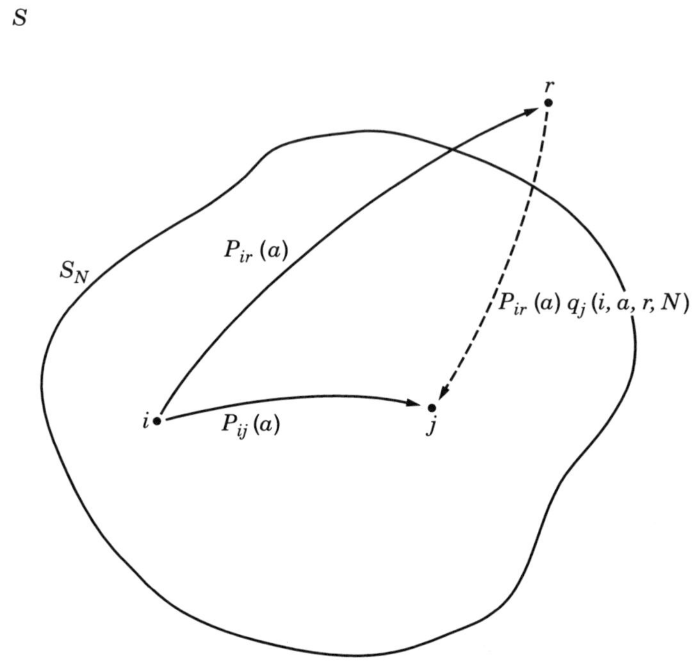

Figure 2.2 Augmentation type approximating sequence.

sends excess probability to a finite subset, or that it *sends excess probability to G*. If $G = \{x\}$, we say that it *sends excess probability to x*. (In defining such an ATAS there is no loss in generality in assuming that N_0 is so large that S_N contains G.)

If $S = \{0, 1, \ldots\}$, $S_N = \{0, 1, \ldots, N\}$, and $q_N(i, a, r, N) \equiv 1$, then we say that this ATAS *sends excess probability to N*.

The following examples illustrate the idea of an ATAS:

Example 2.5.4. The MDC Δ and S_N are as in Example 2.5.2. There is excess probability associated only with 0. Hence for an ATAS we must have $P_{ii-1}(N) = P_{ii-1}$, for $1 \le i \le N$. Let $Y(N) = \sum_{r=N+1}^{\infty} P_{0r}$. Here are four ways of defining the approximating distribution $(P_{0j}(N))_{0 \le j \le N}$:

1. Let $P_{0j}(N) = P_{0j}$ for $1 \le j \le N$, and let $P_{00}(N) = P_{00} + Y(N)$. This defines an ATAS that sends excess probability to 0. Formally we have $q_0(0, r, N) = 1$ for $r \ge N + 1$. This ATAS is shown in Fig. 2.3.
2. Let $P_{0j}(N) = P_{0j}$ for $2 \le j \le N$. Let $P_{00}(N) = P_{00} + (0.5)\, Y(N)$ and $P_{01}(N) = P_{01} + (0.5)\, Y(N)$. This defines an ATAS that sends excess probability to $\{0, 1\}$. Formally we have $q_0(0, r, N) = q_1(0, r, N) = 0.5$ for $r \ge N + 1$.
3. Let $P_{0j}(N) = P_{0j}$ for $2 \le j \le N$. Let $P_{00}(N) = P_{00} + P_{0N+1}$ and $P_{01}(N) =$

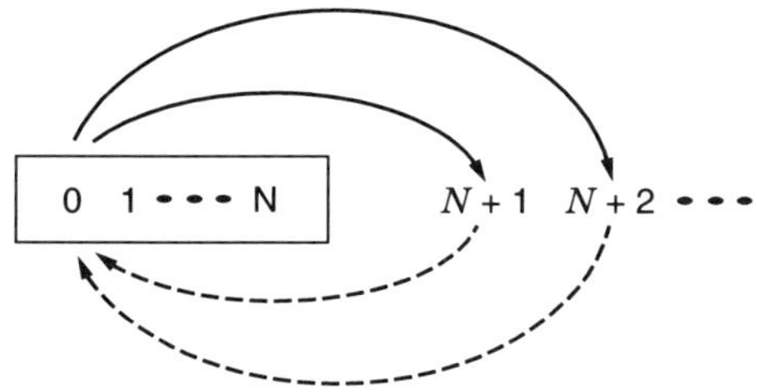

Figure 2.3 Excess probability in state 0 sent to 0.

$P_{01} + \sum_{r=N+2}^{\infty} P_{0r}$. This also defines an ATAS that sends excess probability to $\{0, 1\}$. In this case we have $q_0(0, N+1, N) = 1$ and $q_1(0, r, N) = 1$ for $r \geq N + 2$.

4. Let $P_{0j}(N) = P_{0j}$ for $0 \leq j \leq N - 1$, and let $P_{0N}(N) = P_{0N} + Y(N)$. This defines an ATAS that sends excess probability to N. We have $q_N(0, r, N) = 1$ for $r \geq N + 1$. This ATAS is shown in Fig. 2.4. □

Example 2.5.5. This is Example 2.1.4. Recall that this example concerns the routing of batches of packets to one of K parallel servers. The state space S consists of all pairs $(\mathbf{i}, u)$, where $\mathbf{i}$ is the vector of buffer levels and u is the server to which the previous batch was routed.

Let S_N be the set of pairs $(\mathbf{i}, u)$, where $\mathbf{i}$ satisfies $i_k \leq N$ for $1 \leq k \leq K$. This means that in the approximating sequence no buffer is allowed to contain more than N packets. To simplify the definition of the ATAS, let us assume that $K = 2$. How to approach the general situation will be clear from this case.

Let us first discuss a numerical example with $N = 10$. Assume that the current state is [(8, 4), u], that action 1 is chosen, and that a batch of size 5 arrives. The following states outside of S_{10} may be reached on the next transition: [(12, 3), 1], [(13, 3), 1], [(12, 4), 1], [(13, 4), 1]. The first state corresponds to service completions at both buffers, the second state corresponds to a service completion only at the second buffer, and so on. The probability associated with the states [(12, 3), 1] and [(13, 3), 1] is given to state [(10, 3), 1] $\in S_{10}$. Similarly the probability associated with the states [(12, 4), 1] and [(13, 4), 1] is given to state [(10, 4), 1].

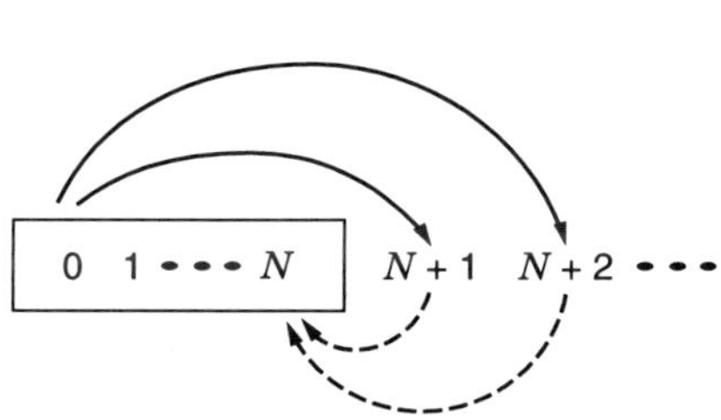

Figure 2.4 Excess probability in state 0 sent to N.

Let us now describe generally how this operates. Suppose that the system is in state $[(i_1, i_2), u] \in S_N$ and that action 1 is chosen. Only the level of buffer 1 may increase after this decision. The level of buffer 2 will either stay the same or decrease by 1 (if there is a service completion). We see that the excess probability associated with this state-action pair involves states of the form $[(r, x), 1]$, where $r > N$. Here $x = i_2$ if $i_2 = 0$ or if there is no service completion at buffer 2, and $x = i_2 - 1$ if there is a service completion at buffer 2. The excess probability involving $[(r, x), 1]$ is sent to $[(N, x), 1]$. Formally the augmentation distribution is given by

$$q_{[(N,x),1]}([(i_1, i_2), u], 1, [(r,x), 1], N) = 1, \qquad r > N,\ x = i_2 \text{ or } i_2 - 1. \tag{2.20}$$

The augmentation distribution, if decision 2 is made, is defined similarly. □

Finally we show that (2.19) does indeed define an approximating probability distribution.

Proposition 2.5.6. Equation (2.19) defines an approximating probability distribution on S_N.

Proof: To show that (2.19) defines a probability distribution, note that

$$\begin{aligned}
\sum_{j \in S_N} P_{ij}(a; N) &= \sum_{j \in S_N} P_{ij}(a) + \sum_{j \in S_N} \sum_{r \notin S_N} P_{ir}(a) q_j(i, a, r, N) \\
&= \sum_{j \in S_N} P_{ij}(a) + \sum_{r \notin S_N} P_{ir}(a) \left(\sum_{j \in S_N} q_j(i, a, r, N) \right) \\
&= \sum_{j \in S_N} P_{ij}(a) + \sum_{r \notin S_N} P_{ir}(a) \\
&= \sum_{j \in S} P_{ij}(a) \\
&= 1.
\end{aligned} \tag{2.21}$$

The interchange of the order of summation in the second line is valid since all terms are nonnegative. The third line follows since the probabilities in an augmentation distribution sum to 1. The remaining lines are clear.

We now show that the distribution in (2.19) satisfies (2.17). First observe that

$$1 = \sum_{j \in S_N} P_{ij}(a) + \sum_{r \notin S_N} P_{ir}(a). \tag{2.22}$$

Since the finite sets S_N increase to S, it is the case that the first term on the right of (2.22) approaches 1 as $N \to \infty$. Hence we have

$$\lim_{N \to \infty} \sum_{r \notin S_N} P_{ir}(a) = 0. \tag{2.23}$$

Now fix $j \in S$, and assume that N is so large that $j \in S_N$. Since the terms in (2.19) are nonnegative and $q_j \leq 1$, it follows that

$$P_{ij}(a) \leq P_{ij}(a; N) \leq P_{ij}(a) + \sum_{r \notin S_N} P_{ir}(a). \tag{2.24}$$

We take the limit in (2.24) as $N \to \infty$, and the result follows from (2.23). □

BIBLIOGRAPHIC NOTES

The foundational works of Bellman (1957) and Howard (1960) were mentioned in the Notes to Chapter 1. Derman has a series of papers culminating in (1970). Hinderer (1970) has foundational material. Fundamental early work was done by Blackwell (1965), for example. Also see Strauch (1966).

Denardo has a series of papers culminating in (1982). See also Ross (1970, 1983) and Bertsekas (1987). White and White (1989) give an interesting survey. Recent books include Puterman (1994) and Bertsekas (1995).

Much work on theoretical aspects of policies has been done by Feinberg (1991), for example, and Feinberg and Park (1994).

The approximating sequence method was introduced in Sennott (1997a). A somewhat similar approximating technique was introduced in Langen (1981) in a much more theoretical setting.

PROBLEMS

2.1. Develop the transition probabilities for Example 2.1.3.

2.2. Explain the transition probabilities in Example 2.1.4.

2.3. Consider a single server queue with batch packet arrivals. There is a probability p_j that a batch of size $j \geq 0$ will arrive at the beginning of any slot. The state $i \geq 0$ denotes the number of packets currently in the buffer. In state $i \geq 1$ the action set is $\{0, 1, \ldots, i\}$, where action $k \in \{0, 1, \ldots, i\}$ means that in the next slot a perfect (batch) service of k packets will occur. (If $k = 0$, then the server is idle during the next slot.) There are costs $H(i)$ and $C(k)$ as usual. Model this as an MDC.

2.4. Formulate Example 1.1.3 as an MDC.

2.5. Formulate the priority queueing system in Example 1.1.5 as an MDC.

2.6. Consider the MDC in Example 2.1.1. Let the current state of the system be i. For each policy specified below decide whether it is stationary, randomized stationary, deterministic Markov, randomized Markov, or general. Discuss what information is required to implement each policy.

(a) If $i \leq 25$, then accept incoming batches, while if $i > 25$, then reject incoming batches. If $i = 25$, then accept them with probability 0.25 and reject them with probability 0.75.

(b) At $t = 0$ accept the incoming batch with probability 0.5 and reject it with probability 0.5. At time $t \geq 1$, if the previous decision was to accept, then reject the next batch, and vice versa if the previous decision was to reject.

(c) If $i < 100$, then accept, while if $i \geq 100$, then reject.

(d) If the proportion of slots in which the batch was rejected does not exceed 0.2, then reject the incoming batch. Otherwise, accept it.

(e) Assume that $i \leq 100$. If the time t is even, then accept, while if it is odd, then reject. If $i > 100$, then reject.

2.7. Consider an MDC with $S = \{0, 1, 2\}$. There is a single action in each state, and we have $P_{01} = P_{12} = P_{21} = P_{10} = 1$ (deterministic transitions from 0 to 1 to 2 to 1 and back to 0). The costs are given by $C(i) = i + 1$. Calculate $v_{14}(0)$ (assume a terminal cost of zero), $V_\alpha(0)$, $J(0)$, and $\lim_{\alpha \to 1}(1 - \alpha)V_\alpha(0)$. Compare the last two quantities.

2.8. For the priority queueing system modeled in Problem 2.5, discuss three different ways to set up an ATAS for this MDC.

CHAPTER 3

Finite Horizon Optimization

In this chapter we derive an equation for the finite horizon expected discounted (or undiscounted) value functions. Necessary and sufficient conditions for a policy to be optimal for the finite horizon criteria are given. It is shown that an optimal deterministic Markov policy exists.

The remainder of the chapter is devoted to the topic of computation of optimal policies when the state space is infinite. Conditions are given so that the finite horizon expected value functions in an approximating sequence converge to the analogous quantities in the original MDC and likewise for the optimal policies. These ideas are illustrated with the development of an approximating sequence for Example 2.1.1. A specific case of this is ProgramOne. Computational output is discussed for several scenarios. Suggestions for further exploration of this model are in the chapter problems.

3.1 FINITE HORIZON OPTIMALITY EQUATION

Let θ be an arbitrary policy for the n horizon. Recall that the n horizon expected cost under θ, defined in (2.11), may be obtained from the n horizon expected discounted cost, defined in (2.9), by setting $\alpha = 1$. With $0 < \alpha \leq 1$ we may develop the theory for the expected discounted and undiscounted cost criteria at the same time. We speak in general of the n horizon expected value function. Let it also be understood that α is *fixed*, and this convention will hold throughout the chapter.

The expected value function $v_{\alpha,n}$ defined in (2.10) represents the smallest expected discounted cost that can possibly be achieved when the process is operated over the n horizon, namely over n time slots. We may think of an n horizon as beginning from an arbitrary slot and continuing for n slots. The goals of this section are as follows: First, we want to provide an equation satisfied by the value function. This is the *finite horizon optimality equation*. Second, we want to give necessary and sufficient conditions for an n horizon policy to be optimal. Third, we want to show that there exists an optimal policy of deterministic Markov form.

It is helpful to introduce the auxiliary function

$$u_{\alpha,n}(i,a) =: C(i,a) + \alpha \sum_j P_{ij}(a) v_{\alpha,n-1}(j), \qquad n \geq 1. \tag{3.1}$$

Let $B_i(\alpha, n) = \{b \in A_i | u_{\alpha,n}(i,b) = \min_{a \in A_i} \{u_{\alpha,n}(i,a)\}\}$. These actions are said to achieve or realize the minimum. In most cases $B_i(\alpha, n)$ is a singleton, but it is possible for it to contain more than one action.

Remark 3.1.1. Quantities similar to the minimization above occur frequently throughout the book. If the terms being minimized involve the state i, then the minimization is understood to be over actions $a \in A_i$ unless otherwise specified. Subsequently these minimizations are denoted by $\min_a$. □

Recall that $v_{\theta,\alpha,0}(i)$ equals the terminal cost $F(i)$. This implies that $v_{\alpha,0} = F$. There are no actions to take for a horizon length of 0. Now assume that $n \geq 1$, so that the process will be operated for at least one step and actions will be taken. We engage in some informal reasoning, both to gain insight and to suggest the statement of the major theorem. So suppose that the horizon n is given and that the process is initially in state i. It is desired to operate the system as close to optimality as possible. Some initial action must be taken (or a randomization among actions performed, on the basis of which one of them is chosen). Let us assume that the controller tentatively selects action $a \in A_i$. Then a cost of $C(i,a)$ is incurred and the process transitions to state j with probability $P_{ij}(a)$. It is clear that to obtain the best overall result, the controller should act optimally for the $n - 1$ horizon problem with initial state j. But this means that the n horizon expected discounted cost is given by $C(i,a) + \alpha \Sigma_j P_{ij}(a) v_{\alpha,n-1}(j)$. In reconsidering what to do initially, the controller sees that the action that realizes the minimum of these quantities should be chosen. This suggests that $v_{\alpha,n}(i) = \min_a \{u_{\alpha,n}(i,a)\}$ and that, at time $t = 0$, an optimal policy for the n horizon problem should choose an action in $B_i(\alpha, n)$.

This insight leads directly to the major theorem of this chapter. This result gives a recursive equation satisfied by the finite horizon value functions. Parts (i–ii) give necessary and sufficient conditions for an arbitrary policy to be optimal for the n horizon optimization criterion.

The proof makes use of Proposition A.1.1 in Appendix A and the reader interested in this proof should examine this result before proceeding.

Theorem 3.1.2. The finite horizon expected value function satisfies the finite horizon optimality equation

$$v_{\alpha,n}(i) = \min_a \left\{ C(i,a) + \alpha \sum_j P_{ij}(a) v_{\alpha,n-1}(j) \right\}, \qquad i \in S, n \geq 1. \quad (3.2)$$

(i) A policy θ is optimal for the 1 horizon if and only if given initial state i, the distribution $\theta(a|i)$ is concentrated on the set $B_i(\alpha, 1)$ (that is, equals zero outside this set).

(ii) A policy θ is optimal for the $n \geq 2$ horizon if and only if

(1) Given initial state i, the distribution $\theta(a|i)$ is concentrated on the set $B_i(\alpha, n)$.

(2) Given the process moves to state j at $t = 1$, then θ follows an $n - 1$ horizon optimal policy with initial state j.

Proof: The proof is accomplished by induction on the horizon.

First assume that $n = 1$. In the one-period case the controller acts at $t = 0$, then observes the state at $t = 1$ and incurs the terminal cost. Let the initial state be i, and let θ be an arbitrary policy for the 1 horizon. Then

$$\begin{aligned}
v_{\theta,\alpha,1}(i) &= \sum_a \theta(a|i)\, E_\theta[C(X_0, A_0) + \alpha F(X_1) | X_0 = i, A_0 = a] \\
&= \sum_a \theta(a|i) \left\{ C(i,a) + \alpha \sum_j P_{ij}(a) v_{\alpha,0}(j) \right\} \\
&= \sum_a \theta(a|i) u_{\alpha,1}(i,a) \\
&\geq \min_a \{u_{\alpha,1}(i,a)\}. \qquad (3.3)
\end{aligned}$$

The first line follows from (2.9) by conditioning on the initial action chosen. The other lines follow easily.

Since (3.3) holds for all θ, it follows that $\inf_\theta v_{\theta,\alpha,1}(i) \geq \min_a \{u_{\alpha,1}(i,a)\}$. Then from (2.10) it follows that

$$v_{\theta,\alpha,1}(i) \geq v_{\alpha,1}(i) \geq \min_a \{u_\alpha(i,a)\}. \quad (3.4)$$

Observe that the last expression in (3.4) is the right side of (3.2).

Now let θ be a policy with $\theta(a|i)$ concentrated on $B_i(\alpha, 1)$. Proposition A.1.1 of Appendix A tells us that any such policy satisfies $v_{\theta,\alpha,1}(i) = \min_a \{u_{\alpha,1}(i,a)\}$. That means that the terms in (3.4) are all equal. This implies that (3.2) holds for $n = 1$ and that θ is optimal.

This proves the sufficiency of the condition in (i). To prove the necessity, let θ be an arbitrary policy. It again follows from Proposition A.1.1 and (3.4) that θ is optimal only if $\theta(a|i)$ is concentrated on $B_i(\alpha, 1)$. This completes the proof for $n = 1$.

Now assume the truth of the statements for $n - 1$. The only assumption that will be used in carrying out the induction is the existence of an $n - 1$ horizon optimal policy θ^*. Using this, we will show the truth of the statements for n.

Let the initial state be i, and let θ be an arbitrary policy for the n horizon problem. If the initial action is a and the state at $t = 1$ is j, let $\psi(i,a,j)$ be the policy rule for the $n - 1$ horizon under θ, starting at time $t = 1$. Then

$$
\begin{aligned}
v_{\theta,\alpha,n}(i) &= \sum_a \theta(a|i) E_\theta [C(X_0, A_0) \\
&\quad + \alpha \sum_{t=1}^{n-1} \alpha^{t-1} C(X_t, A_t) + \alpha^n F(X_n) | X_0 = i, A_0 = a] \\
&= \sum_a \theta(a|i) \left\{ C(i,a) + \alpha \sum_j P_{ij}(a) v_{\psi(i,a,j),\alpha,n-1}(j) \right\} \\
&\geq \sum_a \theta(a|i) \left\{ C(i,a) + \alpha \sum_j P_{ij}(a) v_{\theta*,\alpha,n-1}(j) \right\} \\
&= \sum_a \theta(a|i) \left\{ C(i,a) + \alpha \sum_j P_{ij}(a) v_{\alpha,n-1}(j) \right\} \\
&= \sum_a \theta(a|i) u_{\alpha,n}(i,a) \\
&\geq \min_a \{u_{\alpha,n}(i,a)\}.
\end{aligned}
\tag{3.5}
$$

From (3.5) it follows that $v_{\theta,\alpha,n}(i) \geq v_{\alpha,n}(i) \geq \min_a \{u_{\alpha,n}(i,a)\}$, where the last term is the right side of (3.2).

Now assume that $\theta(a|i)$ is concentrated on $B_i(\alpha, n)$ and then follows the policy θ^*. From Proposition A.1.1 it follows that the last line in (3.5) is an equality. Since $\psi(i,a,j) = \theta^*$, it follows that the third line is an equality. Hence (3.2) holds for n, and there exists an optimal policy of the claimed form. This proves the sufficiency of (ii).

It remains to show the necessity of (ii). First assume that $\theta(a|i) > 0$, for some $a \notin B_i(\alpha, n)$. Then from Proposition A.1.1 it follows that the last inequality in

(3.5) is strict, and hence θ cannot be optimal. Now assume that $\theta(a|i)$ is concentrated on $B_i(\alpha, n)$. Assume that there exists $b \in B_i(\alpha, n)$ such that $\theta(b|i) > 0$ and j such that $P_{ij}(b) > 0$. Note that this means that state j may be reached at time $t = 1$ under the policy θ. Suppose that θ does not act optimally for the $n-1$ horizon at j. This implies that $v_{\psi(i,a,j),\alpha,n-1}(j) > v_{\alpha,n-1}(j)$. But this means that the first inequality in (3.5) is strict, and hence θ cannot be optimal. □

The form of an n horizon optimal policy embodies a famous result known as *Bellman's principle of optimality*. The principle says that if a policy is to have a chance of being optimal and certain actions have been taken for periods $0, \ldots, t-1$, then the remaining actions must constitute an optimal policy for the $n - t$ horizon.

The implication of Theorem 3.1.2 is that both the finite horizon value function and a finite horizon optimal policy can be built up inductively for $n = 1$, then $n = 2$, and so on. The following example illustrates this procedure in a simple setting. The reader may find it useful to work through the calculations in detail.

Example 3.1.3. Consider an MDC with $S = \{0, 1\}$. State 0 has actions a and a^*, while state 1 has actions b and b^*. We have $C(0,a) = 1$, $C(0,a^*) = 0.75$, $C(1,b) = 4$, and $C(1,b^*) = 3$. The terminal costs are $F(0) = 1$ and $F(1) = 2$. The transition probabilities are completely specified by the conditions $P_{00}(a) = 0.5$, $P_{00}(a^*) = 0.25$, $P_{11}(b) = 0$, and $P_{11}(b^*) = 0.5$. See Fig. 3.1.

In this example let us assume that $\alpha = 1$. In this case the set defined following (3.1) is denoted by $B_i(n)$. Our aim is to construct an optimal policy for the $n = 2$ horizon. Observe that

$$\begin{aligned} v_1(0) &= \min\{C(0,a) + 0.5\, F(0) + 0.5\, F(1), C(0,a^*) + 0.25\, F(0) + 0.75\, F(1)\} \\ &= \min\{2.5, 2.5\}. \end{aligned}$$

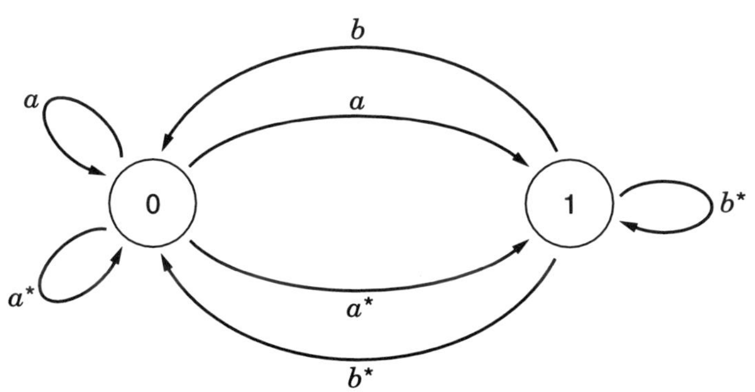

Figure 3.1 Example 3.1.3.

Hence $v_1(0) = 2.5$ and $B_0(1) = \{a, a^*\}$. Similarly we find that $v_1(1) = \min\{5, 4.5\} = 4.5$ and $B_1(1) = \{b^*\}$.

For $n = 2$ we have

$$\begin{aligned} v_2(0) &= \min\{C(0,a) + 0.5\, v_1(0) + 0.5\, v_1(1), C(0,a^*) + 0.25\, v_1(0) + 0.75\, v_1(1)\} \\ &= \min\{4.5, 4.75\}. \end{aligned}$$

Hence $v_2(0) = 4.5$ and $B_0(2) = \{a\}$. Similarly we find that $v_2(1) = \min\{6.5, 6.5\} = 6.5$ and $B_1(2) = \{b, b^*\}$.

Let's build up an optimal policy for the 2 horizon. The policy will be of deterministic Markov form. Define the stationary policies f_1 and f_2 by $f_1(0) = a^*$, $f_1(1) = b^*$, $f_2(0) = a$, and $f_2(1) = b$. According to Theorem 3.1.2 the deterministic Markov policy $\theta = (f_2,f_1)$ is optimal.

To check this out, note that

$$\begin{aligned} v_{\theta,1}(0) &= v_{f_1,1}(0) \\ &= C(0,f_1) + 0.25\, F(0) + 0.75\, F(1) \\ &= 2.5, \end{aligned}$$

and indeed $v_{\theta,1}(0) = v_1(0)$. Similarly we can show that $v_{\theta,1}(1) = 4.5 = v_1(1)$. Then $v_{\theta,2}(0) = C(0,f_2) + 0.5\, v_{\theta,1}(0) + 0.5\, v_{\theta,1}(1) = 4.5 = v_2(0)$. And finally $v_{\theta,2}(1) = C(1,f_2) + v_{\theta,1}(0) = 6.5 = v_2(1)$.

Let us define a history dependent optimal policy ψ under the assumption that the process is in state 1 at $t = 0$. The policy chooses action b with probability 0.2 and action b^* with probability 0.8. If the process is in state 1 at time $t = 1$, then it chooses b^*. If the process is in state 0 at this time and action b was chosen at $t = 0$, then action a is chosen, whereas if action b^* was chosen at $t = 0$, then action a^* is chosen. It follows from Theorem 3.1.2 that ψ is an optimal 2 horizon policy for initial state 1. (Problem 3.1 asks you to verify this.) □

Example 3.1.3 shows that there may be more than one optimal policy and that an optimal policy may be history dependent. However, the most important type of policy for the n horizon is a deterministic Markov policy. Theorem 3.1.2 tells us how to define such a policy to ensure that it is optimal. At time $t = 0$ we choose and fix an action in $B_i(\alpha, n)$ for each i, and this defines the stationary policy f_n. At time $t = 1$ (since the policy from that point on must be optimal for the $n - 1$ horizon) we choose and fix an action in $B_i(\alpha, n - 1)$ for each i, and this defines the stationary policy f_{n-1}. We continue in this way until time $t = n - 1$ (the last time to make a decision). At this time we choose and fix an action in $B_i(\alpha, 1)$ for each i, and this defines the stationary policy f_1. Then the policy $\theta = (f_n, f_{n-1}, \ldots, f_1)$ is optimal for the n horizon. The following result formalizes this argument:

Corollary 3.1.4. Let the deterministic Markov policy $\theta = (f_n, f_{n-1}, \ldots, f_1)$

be defined as follows: The stationary policy f_{n-t} satisfies $f_{n-t}(i) \in B_i(\alpha, n-t)$ for $0 \le t \le n-1$. Then θ is optimal for the n horizon.

Proof: The notation is set up so that t represents the time period. So at time $t = 0$ we employ f_n, at time $t = 1$ we employ f_{n-1}, and so on.

The result is proved by induction on n. First let $n = 1$. Then $\theta = f_1$, where $f_1(i) \in B_i(\alpha, 1)$. By Theorem 3.1.2 this is optimal.

Now assume that the result is true for $n - 1$. Let us prove it for n. Let $\theta = (f_n, f_{n-1}, \dots, f_1)$ be as above. Let $\theta^* = (f_{n-1}, \dots, f_1)$, and observe that $\theta = (f_n, \theta^*)$. According to the definition of θ^* and the induction hypothesis, θ^* is optimal for the $n-1$ horizon. But it then follows from Theorem 3.1.2 that θ is optimal for the n horizon. □

Our goals for this section have been achieved. In particular, we may use (3.2) to recursively calculate $v_{\alpha,n}$, and then Corollary 3.1.4 may be used to identify an optimal deterministic Markov policy.

3.2 ASM FOR THE FINITE HORIZON

The approximating sequence method (ASM) is used to calculate both the finite horizon expected value function and a finite horizon optimal policy for the case when the state space is denumerably infinite. At this point the reader may want to review Definition 2.5.1.

Throughout this section let Δ be an MDC with a denumerable state space and terminal cost F, and let (Δ_N) be an approximating sequence for Δ. Then $(v^N_{\alpha,n}(i))_{i \in S_N}$ is the expected value function in Δ_N. (In general, quantities occurring in Δ_N are superscripted with N.) An optimal policy for the n horizon in Δ_N is given by $\theta^N_n = (e^N_n, e^N_{n-1}, \dots, e^N_1)$, where e^N_t is a stationary policy that is optimal for time $n - t$.

As we let $N \to \infty$ (with the horizon length n fixed), the questions of interest are as follows:

QUESTION 1. When does $v^N_{\alpha,n} \to v_{\alpha,n} < \infty$?

QUESTION 2. When does θ^N_n converge to an n horizon optimal policy in Δ?

We want to ensure both the finiteness of the value function in Δ and the convergence. The next example shows that the desired convergence may not hold.

Example 3.2.1. This is Example 2.5.2 with $C(i) = i$ and zero terminal cost. We claim that $v^N_2(0)$ does not converge to $v_2(0)$. Observe that $v_1(j) = j$. Then $v_2(0) = \sum_{j=1}^{\infty} j/2^{j+1} = 1$. (This follows by factoring out $\frac{1}{2}$ and applying (A.25).)

Now $v_1^N(j) = j$ for $0 \le j \le N$, and hence

$$v_2^N(0) = \sum_{j=0}^{N} P_{0j}(N)j$$

$$= \sum_{j=1}^{N-1} \frac{j}{2^{j+1}} + \frac{N}{2^N} + 1.$$

As $N \to \infty$ the first term approaches $1 = v_2(0)$ and the second term approaches 0. Hence $\lim_{N\to\infty} v_2^N(0) = 2 > v_2(0)$. □

The following result shows what can be proved without further assumptions. The reader may wish to review the concept of the limit infimum (supremum) of a sequence as discussed in Section A.1 of Appendix A. Recall that the limit of a sequence exists if and only if the limit infimum of the sequence equals its limit supremum (and the limit is then this common quantity).

Lemma 3.2.2. We have $\lim_{N\to\infty} v_{\alpha,0}^N = v_{\alpha,0}$. For $n \ge 1$ we have $\liminf_{N\to\infty} v_{\alpha,n}^N \ge v_{\alpha,n}$.

Proof: This is proved by induction on n. Let $n = 0$ and $i \in S$. There exists N^* such that $N \ge N^*$ implies that $i \in S_N$. Then $v_{\alpha,0}^N(i) = F(i) = v_{\alpha,0}(i)$ for $N \ge N^*$, which proves the first statement. Observe that if the limit exists, then the limit infimum is equal to the limit. Hence the second statement is true for $n = 0$ and may be used to start the induction.

Now assume that the result is true for $n - 1$. We show that it holds for n. The n horizon optimality equation in Δ_N is

$$v_{\alpha,n}^N(i) = \min_a \left\{ C(i,a) + \alpha \sum_{j\in S_N} P_{ij}(a;N) v_{\alpha,n-1}^N(j) \right\}, \qquad i \in S_N. \tag{3.6}$$

Take the limit infimum of both sides of (3.6) to obtain

$$\liminf_N v_{\alpha,n}^N(i) = \min_a \left\{ C(i,a) + \alpha \liminf_N \sum_{j\in S_N} P_{ij}(a;N) v_{\alpha,n-1}^N(j) \right\}$$

$$\geq \min_a \left\{ C(i,a) + \alpha \sum_j P_{ij}(a)(\liminf_N v_{\alpha,n-1}^N(j)) \right\}$$

$$\geq \min_a \left\{ C(i,a) + \alpha \sum_j P_{ij}(a) v_{\alpha,n-1}(j) \right\}$$

$$= v_{\alpha,n}(i). \tag{3.7}$$

(Notice that "$\to \infty$" has been suppressed and is understood in the limit infimum. Recall our convention that "$\sum_j$" indicates a summation over $j \in S$.) Here the first line follows from Proposition A.1.3(i), which says that a limit infimum can be "passed through" a minimization over a finite set. The second line follows from the generalized Fatou's lemma (Proposition A.2.5). Since the costs are nonnegative, it is the case that the value function is nonnegative. The third line follows from the induction hypothesis, and the fourth line follows from (3.2). This completes the induction. Hence the result holds for $n \geq 0$. □

The following *finite horizon* assumption, for fixed α and fixed $n \geq 1$, is the key to answering the questions.

Assumption FH(α, n). For $i \in S$ we have $\limsup_{N \to \infty} v_{\alpha,n}^N(i) =: w_{\alpha,n}(i) < \infty$ and $w_{\alpha,n}(i) \leq v_{\alpha,n}(i)$. □

The answer to Question 2 requires the concept of a stationary policy for Δ that is a limit point of a sequence of stationary policies in (Δ_N). This is given in Definition B.4 in Appendix B, and the reader may wish to refer to this definition now. There is also some background information on sequences of stationary policies.

Theorem 3.2.3. Let $n \geq 1$ be fixed. The following are equivalent:

(i) $\text{Lim}_{N \to \infty} v_{\alpha,n}^N = v_{\alpha,n} < \infty$.
(ii) Assumption FH(α, n) holds.

Assume that either (then both) of these holds, and let e_n^N be a stationary policy for Δ_N that is optimal for the n horizon at time $t = 0$. Then any limit point of the sequence $(e_n^N)_{N \geq N_0}$ is optimal in Δ for the n horizon at $t = 0$.

Proof: If (i) holds, then $\limsup_N v_{\alpha,n}^N = \lim_N v_{\alpha,n}^N = v_{\alpha,n} < \infty$, and then clearly (ii) holds.

Now assume that (ii) holds. Then $\limsup_N v_{\alpha,n}^N \leq v_{\alpha,n} \leq \liminf_N v_{\alpha,n}^N$, where the last inequality follows from Lemma 3.2.2. Moreover the first term is finite.

But this implies that all the terms are equal and finite, and thus (i) holds. This proves the equivalence of (i) and (ii).

Now assume that (i) holds. By Proposition B.5 there exists a limit point e_n of the sequence $(e_n^N)_{N \geq N_0}$. Recall from Definition B.4 that there exists a subsequence N_r such that given $i \in S$, we have $e_n^{N_r}(i) = e_n(i)$ for N_r sufficiently large (how large may depend on i).

For a fixed state i and N_r sufficiently large, (3.6) may be written

$$v_{\alpha,n}^{N_r}(i) = C(i, e_n) + \alpha \sum_{j \in S_N} P_{ij}(e_n; N_r) v_{\alpha,n-1}^{N_r}(j). \tag{3.8}$$

This follows since $e_n^{N_r}$ is n horizon optimal at $t = 0$ and chooses the same action at i as e_n for N_r sufficiently large.

We now take the limit infimum of both sides of (3.8) as $r \to \infty$ (i.e., take the smallest limit point relative to the subsequence determined by N_r). This yields

$$\begin{aligned}
\liminf_{r \to \infty} v_{\alpha,n}^{N_r}(i) &\geq C(i, e_n) + \alpha \sum_j P_{ij}(e_n)(\liminf_{r \to \infty} v_{\alpha,n-1}^{N_r}(j)) \\
&\geq C(i, e_n) + \alpha \sum_j P_{ij}(e_n)(\liminf_N v_{\alpha,n-1}^{N}(j)) \\
&\geq C(i, e_n) + \alpha \sum_j P_{ij}(e_n) v_{\alpha,n-1}(j) \\
&\geq \min_a \left\{ C(i, a) + \alpha \sum_j P_{ij}(a) v_{\alpha,n-1}(j) \right\} \\
&= v_{\alpha,n}(i).
\end{aligned} \tag{3.9}$$

Here the first line follows from Proposition A.2.5. The second line follows since the limit infimum over N is the smallest limit point. The third line follows from Lemma 3.2.2. The fourth line is clear and the last line follows from (3.2).

But by (i) we have $\liminf_r v_{\alpha,n}^{N_r}(i) = \lim_N v_{\alpha,n}^{N}(i) = v_{\alpha,n}(i)$, and hence all the terms of (3.9) are equal. This implies that $e_n(i)$ realizes the minimum in (3.2). This argument may be carried out for each i, and hence by Theorem 3.1.2, e_n is optimal for the n horizon at time $t = 0$. □

3.3 WHEN DOES FH(α, n) HOLD?

In this section we give some sufficient conditions for FH(α, n) to hold. The first result states that if the costs are bounded in Δ, then FH(α, n) holds.

Proposition 3.3.1. Assume that there exists a (finite) constant B such that $C(i, a) \leq B$ and $F(i) \leq B$, for all state-action pairs. Then FH(α, n) holds for all α and $n \geq 1$.

Proof: First observe that $v_{\alpha, n} \leq B(n+1)$, and hence the value functions in Δ are finite. We show that Theorem 3.2.3(i) holds. The proof is by induction on n. It holds for $n = 0$ by Lemma 3.2.2. Now assume that it is true for $n - 1$. Observe that $0 \leq v^N_{\alpha, n-1} \leq Bn$, and hence this is a bounded function in N for n fixed.

Consider (3.6). For a fixed action we wish to apply Corollary A.2.7 to the summation using the bounding constant Bn. By the induction hypothesis it is the case that $\lim_N v^N_{\alpha, n-1} = v_{\alpha, n-1}$. It then follows from Corollary A.2.7 that $\lim_N \sum_{j \in S_N} P_{ij}(a; N) v^N_{\alpha, n-1}(j) = \sum_j P_{ij}(a) v_{\alpha, n-1}(j)$. Using this and Proposition A.1.3(ii) yields

$$\lim_N v^N_{\alpha, n}(i) = \min_a \left\{ C(i, a) + \alpha \sum_j P_{ij}(a) v_{\alpha, n-1}(j) \right\}$$
$$= v_{\alpha, n}(i) < \infty. \tag{3.10}$$

This completes the induction, and hence Theorem 3.2.3(i) holds for $n \geq 0$. □

Proposition 3.3.1 provides a complete answer to Questions 1 and 2 in the case of bounded costs. The remainder of this section is of interest only when the costs in Δ are unbounded. We develop two situations in which FH(α, n) holds.

Proposition 3.3.2. Assume that $v_{\alpha, n} < \infty$, for $n \geq 1$. Let (Δ_N) be an ATAS that sends excess probability to a finite set. Then FH(α, n) holds for all α and $n \geq 1$.

**Proof:* For $n \geq 1$ consider the statement Z(α, n):

1. There exists a nonnegative function $z^N_{\alpha, n}(i)$, of $i \in S$ and $N \geq N_0$, bounded above by a (finite) constant $Z_{\alpha, n}$.
2. $v^N_{\alpha, n}(i) \leq v_{\alpha, n}(i) + z^N_{\alpha, n}(i)$, for $i \in S_N$ and $N \geq N_0$.
3. $\text{Lim}_{N \to \infty} z^N_{\alpha, n} = 0$.

Suppose that we could prove that $Z(\alpha, n)$ holds. Then from (2) and (3) it follows that $\limsup_N v_{\alpha,n}^N \le v_{\alpha,n} < \infty$, where the finiteness follows by assumption. Thus FH(α, n) holds. So we will show that $Z(\alpha, n)$ holds. (Note that assertion 1 was not used, but it is needed later in the argument.)

Let us obtain some preliminary results. Let f_n be a stationary policy that is optimal in Δ for the n horizon at $t = 0$. Then observe that

$$C(i,f_n) + \alpha \sum_{j \in S_N} P_{ij}(f_n) v_{\alpha,n-1}(j) \le C(i,f_n) + \alpha \sum_{j} P_{ij}(f_n) v_{\alpha,n-1}(j)$$
$$= v_{\alpha,n}(i). \tag{3.11}$$

Notice that on the left side we simply restrict the summation to states in S_N and that the AS is not involved. The first line follows since the value function is nonnegative. The second line follows from (3.2) and the optimality of f_n.

Let

$$Y_i^N(f_n) =: \sum_{r \in S - S_N} P_{ir}(f_n), \tag{3.12}$$

and note that $\lim_N Y_i^N(f_n) = 0$.

Let G be the finite set to which the excess probability is sent. We may assume that S_N contains G for $N \ge N_0$. It is now shown by induction on $n \ge 1$, that $Z(\alpha, n)$ holds. For $n = 1$ we have

$$v_{\alpha,1}^N(i) \le C(i,f_1) + \alpha \sum_{j \in S_N} P_{ij}(f_1; N) F(j)$$
$$= C(i,f_1) + \alpha \sum_{j \in S_N} P_{ij}(f_1) F(j)$$
$$+ \alpha \sum_{r \in S - S_N} P_{ir}(f_1) \left(\sum_{j \in G} q_j(i, f_1, r, N) F(j) \right)$$
$$\le v_{\alpha,1}(i) + \alpha Y_i^N(f_1) \left(\sum_{j \in G} F(j) \right). \tag{3.13}$$

The first line follows from (3.6), and the second line follows from the definition of the ATAS (Definition 2.5.3). The third line follows from (3.11–12) and the fact that $q_j \le 1$.

Set

$$z^N_{\alpha,1}(i) = \alpha Y^N_i(f_1)\left(\sum_{j \in G} F(j)\right), \tag{3.14}$$

and let $Z_{\alpha,1} = \alpha \sum_{j \in G} F(j)$. Then Z($\alpha$, 1) clearly holds.

Now assume that Z($\alpha, n-1$) holds; we should show that Z(α, n) holds. Similar reasoning to that in (3.13) yields

$$\begin{aligned} v^N_{\alpha,n}(i) &\le C(i,f_n) + \alpha \sum_{j \in S_N} P_{ij}(f_n) v^N_{\alpha,n-1}(j) \\ &\quad + \alpha \sum_{r \in S - S_N} P_{ir}(f_n)\left(\sum_{j \in G} q_j(i,f_n,r,N) v^N_{\alpha,n-1}(j)\right). \end{aligned} \tag{3.15}$$

Apply the induction hypothesis to the last two terms of (3.15). Some manipulation and the fact that $q_j \le 1$ yields

$$\begin{aligned} v^N_{\alpha,n}(i) &\le C(i,f_n) + \alpha \sum_{j \in S_N} P_{ij}(f_n) v_{\alpha,n-1}(j) + z^N_{\alpha,n}(i) \\ &\le v_{\alpha,n}(i) + z^N_{\alpha,n}(i), \end{aligned} \tag{3.16}$$

where we have defined

$$\begin{aligned} z^N_{\alpha,n}(i) =: \alpha \sum_{j} P_{ij}(f_n) z^N_{\alpha,n-1}(j) \\ + \alpha Y^N_i(f_n)\left(\sum_{j \in G} [v_{\alpha,n-1}(j) + z^N_{\alpha,n-1}(j)]\right). \end{aligned} \tag{3.17}$$

To complete the induction, it is necessary to verify assertions (1–3) for the function $z^N_{\alpha,n}$. Clearly assertion 2 holds by (3.16).

By the induction hypothesis we have $z^N_{\alpha,n-1} \le Z_{\alpha,n-1}$. This implies that the term in parenthesis in (3.17) is bounded in N. Then (3.12) implies that the limit of the last term in (3.17) is 0. Now focus on the first term and apply Corollary A.2.4 with bounding function $Z_{\alpha,n-1}$. Then

$$\alpha \lim_N \sum_j P_{ij}(f_n) z^N_{\alpha,n-1}(j) = \alpha \sum_j P_{ij}(f_n)(\lim_N z^N_{\alpha,n-1}(j))$$
$$= 0,$$

where the second line follows from the induction hypothesis. This shows that assertion 3 holds.

Finally we verify assertion 1. It follows from (3.17) and the induction hypothesis that $z^N_{\alpha,n}$ is nonnegative. From (3.17) we see that

$$z^N_{\alpha,n}(i) \le \alpha Z_{\alpha,n-1}(1 + |G|) + \alpha \sum_{j \in G} v_{\alpha,n-1}(j) =: Z_{\alpha,n}. \tag{3.18}$$

Here |G| is the cardinality of the finite set G. This completes the induction and the proof. □

A special case of this result occurs when all the excess probability is sent to a fixed state z, known as a *distinguished* state. In this case the finite horizon optimality equation (3.6) has a simple and suggestive form.

Corollary 3.3.3. Assume that $v_{\alpha,n} < \infty$, for $n \ge 1$. Let (Δ_N) be an ATAS that sends the excess probability to a distinguished state z. Then FH(α, n) holds for all α and $n \ge 1$. If $r^N_{\alpha,n} = v^N_{\alpha,n} - v^N_{\alpha,n}(z)$ (known as a *relative value* function), then the finite horizon optimality equation in Δ_N is

$$v^N_{\alpha,n}(i) = \alpha v^N_{\alpha,n-1}(z) + \min_a \left\{ C(i,a) + \alpha \sum_{j \in S_N - \{z\}} P_{ij}(a) r^N_{\alpha,n-1}(j) \right\},$$
$$i \in S_N, n \ge 1. \tag{3.19}$$

Proof: By Proposition 3.3.2 it is only necessary to show that (3.19) holds. Equation (3.6) and the definition of an ATAS that sends the excess probability to z yield

$$v^N_{\alpha,n}(i) = \min_a \left\{ C(i,a) + \alpha \sum_{j \in S_N} P_{ij}(a) v^N_{\alpha,n-1}(j) \right.$$

$$\left. + \alpha \left(1 - \sum_{j \in S_N} P_{ij}(a) \right) v^N_{\alpha,n-1}(z) \right\},$$

which is clearly equivalent to (3.19). □

Equation (3.19) is particularly well-suited for computation. The value function increases with the horizon length. However, the relative value function is more manageable. The computation can keep track of the relative value function together with the value function at z. The value function may be recovered by adding these quantities.

Our next result is a structural condition on Δ involving the augmentation distributions.

Proposition 3.3.4. Assume that $v_{\alpha,n} < \infty$ for $n \geq 1$. Let (Δ_N) be an ATAS such that the augmentation distributions satisfy

$$\sum_{j \in S_N} q_j(i,a,r,N) v_{\alpha,n}(j) \leq v_{\alpha,n}(r),$$

$$i \in S_N, a \in A_i, r \in S - S_N, n \geq 0. \tag{3.20}$$

Then $v^N_{\alpha,n}(i) \leq v_{\alpha,n}(i)$ for $i \in S_N$, all α, and $n \geq 1$. Hence FH(α, n) holds.

(Notice what hypothesis (3.20) says. If we have excess probability $P_{ir}(a)$ and send it to the states of S_N by means of an augmentation distribution, then the resulting weighted probability sum (called a *convex combination*) of values cannot exceed the value function at r.)

Proof: The result is proved by induction on n. For $n = 1$ we have

$$
\begin{aligned}
v_{\alpha,1}^{N}(i) &= \min_{a}\left\{ C(i,a) + \alpha \sum_{j\in S_N} P_{ij}(a;N)v_{\alpha,0}^{N}(j) \right\} \\
&= \min_{a}\left\{ C(i,a) + \alpha \sum_{j\in S_N} P_{ij}(a)v_{\alpha,0}(j) \right. \\
&\qquad \left. + \alpha \sum_{r\in S-S_N} P_{ir}(a)\left(\sum_{j\in S_N} q_j(i,a,r,N)v_{\alpha,0}(j) \right) \right\} \\
&\le \min_{a}\left\{ C(i,a) + \alpha \sum_{j\in S_N} P_{ij}(a)v_{\alpha,0}(j) \right. \\
&\qquad \left. + \alpha \sum_{r\in S-S_N} P_{ir}(a)v_{\alpha,0}(r) \right\} \\
&= v_{\alpha,1}(i).
\end{aligned}
\tag{3.21}
$$

Here the first line follows from (3.6). The second line follows from the definition of an ATAS and the fact that $v_{\alpha,0}^{N} = v_{\alpha,0} = F$ on S_N. The third line follows from (3.20), and the fourth line from (3.2).

Now assume that the result holds for $n-1$. Then similarly to (3.21) we obtain

$$
\begin{aligned}
v_{\alpha,n}^{N}(i) &\le \min_{a}\left\{ C(i,a) + \alpha \sum_{j\in S_N} P_{ij}(a)v_{\alpha,n-1}(j) \right. \\
&\qquad \left. + \alpha \sum_{r\in S-S_N} P_{ir}(a)\left(\sum_{j\in S_N} q_j(i,a,r,N)v_{\alpha,n-1}(j) \right) \right\} \\
&\le \min_{a}\left\{ C(i,a) + \alpha \sum_{j\in S_N} P_{ij}(a)v_{\alpha,n-1}(j) \right. \\
&\qquad \left. + \alpha \sum_{r\in S-S_N} P_{ir}(a)v_{\alpha,n-1}(r) \right\} \\
&= v_{\alpha,n}(i).
\end{aligned}
\tag{3.22}
$$

Here the first line follows from the induction hypothesis and the second line follows from (3.20). The third line follows from (3.2). This completes the induction and the proof. □

The following corollary involves a commonly occurring situation in which Proposition 3.3.4 may be applied. (For $S = \{0, 1, 2, \dots\}$ we say that a function f on S is *increasing* in i if $i \le j$ implies that $f(i) \le f(j)$.)

Corollary 3.3.5. Assume that $S = \{0, 1, 2, \dots\}$ and that $v_{\alpha,n}$ is finite and increasing in i for $n \ge 0$. Let (Δ_N) be an ATAS with $S_N = \{0, 1, \dots, N\}$ that sends the excess probability to N. Then FH(α, n) holds for all α and $n \ge 1$. If $s^N_{\alpha,n} = v^N_{\alpha,n} - v^N_{\alpha,n}(N)$ is the relative value function, then the finite horizon optimality equation in Δ_N is

$$v^N_{\alpha,n}(i) = \alpha v^N_{\alpha,n-1}(N) + \min_a \left\{ C(i,a) + \alpha \sum_{j=0}^{N-1} P_{ij}(a) s^N_{\alpha,n-1}(j) \right\},$$

$$i \in S_N, n \ge 1. \tag{3.23}$$

Proof: This proof is assigned as Problem 3.6. □

The example in the next section illustrates a computation using this result.

3.4. A QUEUEING EXAMPLE

In this section we treat Example 2.1.1. Let us set $\alpha = 1$ and $F = H$. This is the undiscounted case, and the terminal cost is the cost of holding the number of packets left at the time the process stops. We first derive the finite horizon optimality equation in Δ.

Lemma 3.4.1. Assume that $H(i)$ is increasing in i. Then v_n is increasing in i (and finite). For $n \ge 1$ the finite horizon optimality equation is

$$v_n(0) = \min \left\{ \sum_j p_j v_{n-1}(j), R + v_{n-1}(0) \right\}$$

$$v_n(i) = H(i) + \min\left\{ \mu \sum_j p_j v_{n-1}(i-1+j) \right.$$
$$+ (1-\mu) \sum_j p_j v_{n-1}(i+j), R + \mu v_{n-1}(i-1)$$
$$\left. + (1-\mu) v_{n-1}(i) \right\}, \qquad i \geq 1. \tag{3.24}$$

Proof: Equation (3.24) follows easily from (3.2) and the transition probabilities in Example 2.1.1. The first term in the minimum corresponds to admitting the arriving batch, and the second term corresponds to rejecting it.

Let us now show that the value functions are finite. Let θ be the policy that always rejects the incoming batch. Then for a fixed $n \geq 1$ we have $v_n(i) \leq v_{\theta,n}(i) \leq Rn + H(i)(n+1) < \infty$.

It is shown by induction on $n \geq 0$ that the value function is increasing in i for each fixed n. For $n = 0$ we have $v_0 = H$ which is increasing by assumption. Now assume that the result holds for $n - 1$. Observe that each term in the minimum for $v_n(0)$ is bounded above by the corresponding term for $v_n(1)$, and hence $v_n(0) \leq v_n(1)$. Now consider the right side of the second equation of (3.24). The $H(i)$ term is increasing. Suppose that the optimal decision is to reject. By the induction hypothesis both $v_{n-1}(i-1)$ and $v_{n-1}(i)$ are increasing in i. Hence $R + \mu v_{n-1}(i-1) + (1-\mu)v_{n-1}(i)$ is increasing in i (prove it!). Now suppose that the optimal decision is to accept. For each fixed j we have $v_{n-1}(i-1+j)$ increasing in i. The term $\sum_j p_j v_{n-1}(i-1+j)$ is a convex combination of increasing functions and it is easy to see that it is increasing (prove it!). The other sum is also increasing and hence so is $\mu \sum_j p_j v_{n-1}(i-1+j) + (1-\mu)\sum_j p_j v_{n-1}(i+j)$. Thus both terms in the minimum are increasing. Since the minimum of increasing functions is increasing (prove it!), this proves that v_n is increasing. □

Remark 3.4.2. It seems reasonable to hypothesize that the optimal n horizon policy is of *critical number* form, namely that there exists $0 \leq i^* \leq \infty$ such that it is optimal to accept when the buffer level is below i^* but optimal to reject when the level is at or above i^*. (Note that $i^* = 0$ means that it is optimal to always reject, and $i^* = \infty$ means that it is optimal always to accept. Hence these two extreme policies are also of critical number form.) We will not attempt to prove that the optimal policy is of critical number form. Even if this structural result were obtained, we would not have the optimal policy (since the cutoff i^* would be unknown). Here we concentrate on numerically calculating an optimal policy. □

Now let us define an AS for this model. Since v_n is increasing, this suggests that we employ Corollary 3.3.5, letting $S_N = \{0, 1, \ldots, N\}$ and sending the excess probability to N. Recall that $s_n^N(i) = v_n^N(i) - v_n^N(N)$ for $0 \le i \le N-1$. As an aid in deriving the optimality equation (3.23) for the AS, we introduce the auxiliary function

$$w_n^N(i) = \sum_{j=0}^{N-i-1} p_j s_n^N(i+j), \qquad 0 \le i \le N-1, \tag{3.25}$$

and note that (3.25) implies that $i + j \le N - 1$. Then (3.23) becomes

$$\begin{aligned} v_n^N(0) &= v_{n-1}^N(N) + \min\{w_{n-1}^N(0), R + s_{n-1}^N(0)\} \\ v_n^N(i) &= v_{n-1}^N(N) + H(i) + \min\{\mu w_{n-1}^N(i-1) + (1-\mu)w_{n-1}^N(i), \\ &\qquad R + \mu s_{n-1}^N(i-1) + (1-\mu)s_{n-1}^N(i)\}, \qquad 1 \le i \le N-1. \end{aligned} \tag{3.26}$$

Consider (3.23) for $i = N$. Observe that $w_{n-1}^N(N-1) = p_0 s_{n-1}^N(N-1)$, and hence $\mu w_{n-1}^N(N-1) < R + \mu s_{n-1}^N(N-1)$. Then (3.23) becomes

$$\begin{aligned} v_n^N(N) &= v_{n-1}^N(N) + H(N) + \min\{\mu w_{n-1}^N(N-1), R + \mu s_{n-1}^N(N-1)\} \\ &= v_{n-1}^N(N) + H(N) + \mu w_{n-1}^N(N-1), \end{aligned} \tag{3.27}$$

and it is always optimal to accept in N. (Can you explain intuitively why this is so?)

We employ (3.26–27) to compute an optimal policy when $H(i) = Hi$, for a positive constant H, and assuming that the batch size follows a Poisson distribution with mean λ packets/batch. This is ProgramOne. The user is prompted for the values H, R, λ, and μ. The approximation level N and the horizon length are constants that may be changed in subsequent runs of the program.

In the following discussion of the structure of the program the superscript N and subscript n are dropped for notational simplicity. The program carries along three arrays. One array is for the current value of s and one is for the current value of w. (The third will be discussed shortly.) The arrays are initialized by $s_0(i) = v_0(i) - v_0(N) = H(i - N)$ and $w_0(i) = He^{-\lambda} \sum_{j=0}^{N-i-1} \lambda^j(i + j - N)/j!$. The current value of $v(N)$, called v, and the next value, called v_{new}, are also maintained. These are constants.

Here is how the updating occurs. Given the current values v, s, and w, the value v_{new} is obtained from (3.27). Then s_{update} is obtained by calculating the right side of (3.26) and subtracting v_{new}. Finally w_{update} is obtained from (3.25) using s_{update}. For each iteration the optimal decision in a given state is maintained in the third array. At each iteration the optimal decision and current value of v and s are printed out. The value function can be obtained by adding the value of v to that of s.

Remark 3.4.3. Suppose that both H and R are multiplied by a positive constant U. Then in (3.26–27) the effect is to multiply the value function by U; an optimal policy remains the same. (You are asked to show this in Problem 3.7.) Hence it is only the value of R relative to that of H that is important in the computation. For this reason there is no loss of generality in assuming that $H = 1$ and considering various values for R and the other parameters. In all the scenarios we set $H = 1$. (This effect also holds in the original optimality equation (3.24).) □

Let us also consider the effect of the mean batch size λ. Recall that at most one packet can be served in any slot. If $\lambda > 1$, then, on average, more than one packet arrives in each slot. Hence the queue will rapidly build up, and we would expect an optimal policy to invoke the reject option at small horizons and for small buffer content levels. This is indeed what occurs. If $\lambda < 1$, then, on average, less than one packet arrives to the server in each slot. The queue will not build up as rapidly, and we would expect an optimal policy to employ the reject option at larger horizons and for larger buffer content levels. (Intuitively the uncontrolled queue is *stable* if $\lambda < \mu$.) Now consider the case $\lambda = 1$. For $n = R$ a special situation obtains. Hand calculations show that for $n = R = 1$ it is optimal either to accept or reject in all states. For $n = R = 2$ it is optimal to accept in states 0 or 1 and to accept or reject in the other states. In this special situation the solution is not unique, and the program gives an ambiguous result.

Scenario 3.4.4. Let $R = 10$, $\lambda = 3$, and $\mu = 0.7$. The objective is to determine an optimal policy θ_{10}. Because this is our first scenario, we discuss the reasoning in detail. Table 3.1 gives the pertinent results in summary form. Runs were made for approximation levels $N = 20$, 30, and 50. The entries are the optimal policies e_n^N. For example, the entry under $N = 30$ and $n = 5$ is e_5^{30}, which is the optimal policy for horizon 5 in Δ_{30}. This says that $e_5^{30}(i) = r$, for $0 \leq i \leq 22$, and $e_5^{30}(i) = a$, for $23 \leq i \leq 30$.

Remember that any limit point of e_n^N as $N \rightarrow \infty$ is an n horizon optimal policy for Δ. So we examine these policies to see if they are "settling down" to a policy e_n. With a high degree of confidence, it can then be asserted that e_n is n horizon optimal for Δ.

For horizons 1, 2, and 3, it is always optimal to accept in the AS (only horizon 3 is shown in Table 3.1). So we may assert that $e_1 = e_2 = e_3 \equiv a$.

At horizon 4 a change is noted. For $N = 20$ it is optimal to accept in state 0, to reject in states 1 through 11, and to accept in the rest of the states. For $N = 30$ and $N = 50$, it remains optimal to accept in 0, but the rejection region expands. It is plausible that an optimal policy for Δ is of critical number form, and this seems to be confirmed by the computations. Hence we assert that $e_4(0) = a$ and $e_4(i) = r$ for $i \geq 1$. If we wish to gain a greater degree of confidence, we can take an even larger approximation level. The run time for this program is not a significant factor.

Applying the same reasoning to horizons 5 through 10, we may assert that

Table 3.1 Results for Scenario 3.4.4

	n			
N	3	4	5	10
20	[0, 20] a	{0} a [1, 11] r [12, 20] a	[0, 12] r [13, 20] a	[0, 14] r [15, 20] a
30	[0, 30] a	{0} a [1, 21] r [22, 30] a	[0, 22] r [23, 30] a	[0, 24] r [25, 30] a
50	[0, 50] a	{0} a [1, 41] r [42, 50] a	[0, 42] r [43, 50] a	[0, 44] r [45, 50] a

$e_n \equiv r$, for $5 \le n \le 10$. (The output for horizons 6 through 9 is not shown in Table 3.1.) We can even be quite confident that $e_n \equiv r$ for $n \ge 5$. By this reasoning we see that the optimal finite horizon policy in Δ has been determined for all horizon lengths.

As a sample of the calculation of a value, this program output yields $v_{10}^{50}(50) = 549.592$ and $s_{10}^{50}(0) = -463.676$. Hence $v_{10}(0) \approx v_{10}^{50}(0) = s_{10}^{50}(0) + v_{10}^{50}(50) = 112.92$. □

Remark 3.4.5. In subsequent scenarios for this program, the reasoning process discussed above is omitted. We give the values of N and n that are used in making the inference of an optimal policy. The convergence of an optimal n horizon policy for the AS to an optimal policy for Δ follows from Theorem 3.2.3. It is desirable to have a rate of convergence result, but this issue is not discussed here. Although a rate of convergence result is of undeniable importance, employing such a result might well have some drawbacks. First, it might greatly overestimate the approximation level necessary to have confidence in the results. Second, the bound itself might involve tedious calculations. In any case, the reader may note that the convergence is rigorously guaranteed by Theorem 3.2.3; it is only the rate of convergence that is uncertain. Inferences drawn from program output must be done with care and attention, and the determination of an optimal policy requires a bit of art. But this being said, the reader can appreciate the power and elegance of this computational method. □

Scenarios 3.4.6. Additional output is given in Table 3.2. The detailed reasoning is omitted. Each column represents a different scenario. The parameter values are in the first box. In the second box are the values (or value) of (n, N) that were considered. The third box contains the optimal policy for Δ. Each optimal policy is of critical number form. As a shorthand it is denoted by a single interval representing the buffer content levels in which it is optimal to accept. At the other levels it is optimal to reject. For example, $e:[0, \infty)$ means

Table 3.2 Results for Scenarios 3.4.6

Scenario	1	2	3	4	5	6	7	8
Parameters	$R = 10$ $\lambda = 1.5$ $\mu = 0.8$	$R = 10$ $\lambda = 2$ $\mu = 0.8$	$R = 10$ $\lambda = 1.5$ $\mu = 0.4$	$R = 25$ $\lambda = 8$ $\mu = 0.9$	$R = 25$ $\lambda = 4$ $\mu = 0.9$	$R = 25$ $\lambda = 0.75$ $\mu = 0.6$	$R = 12.5$ $\lambda = 0.375$ $\mu = 0.6$	$R = 8$ $\lambda = 1.0$ $\mu = 0.7$
n N	20 30 25 50	20 30	20 30	25 50	25 50	40 70 50 50	50 50 60 50	50 50 30 70
Optimal policy	$1 \le n \le 6$ $[0, \infty)$ 7: $[0, 2]$ $n \ge 8$ $[0, 1]$	$1 \le n \le 5$ $[0, \infty)$ $n \ge 6$ $[0, 1]$	$1 \le n \le 6$ $[0, \infty)$ 7: $[0, 1]$ $n \ge 8$ $\{0\}$	$1 \le n \le 3$ $[0, \infty)$ $n \ge 4$ $\varnothing$	$1 \le n \le 6$ $[0, \infty)$ $n \ge 7$ $[0, 1]$	$1 \le n \le 33$ $[0, \infty)$ 34: $[0, 6]$ 35: $[0, 5]$ $36 \le n \le 39$ $[0, 4]$ $n \ge 40$ $[0, 3]$	$1 \le n \le 33$ $[0, \infty)$ 34: $[0, 13]$ 35, 36: $[0, 11]$ 37, 38: $[0, 10]$ $39 \le n \le 45$ $[0, 9]$ $n \ge 46$ $[0, 8]$	$1 \le n \le 7$ $[0, \infty)$ 8: Nonunique 9: $[0, 2]$ $n \ge 10$ $[0, 1]$

that it is always optimal to accept, whereas $e:\varnothing$ means that it is always optimal to reject. The policy $e:[0, 3]$ means that it is optimal to accept when the buffer content level is 3 or less and to reject when the level is above 3.

Scenario 1 has a mean batch size modestly greater than 1. In Scenario 2 this mean is slightly increased. The change in policy is modest and occurs only at horizons 6 and 7. By horizon 8 there is no longer a difference. Scenario 3 is as in the first scenario but with a service rate half as much. There is a change in the optimal policy in the conservative direction as would be expected.

In Scenario 4 both the mean batch size and the rejection cost are large. The optimal policy is very conservative and rejects all batches for horizons of 4 or more. Scenario 5 is as in 4 but with the mean batch size cut in half. Note that for both of these scenarios there is an abrupt change from always accepting to rejecting almost always.

In Scenario 6 we have $\mu < \lambda < 1$. Note that $p_0 = e^{-0.75} = 0.47$. This means that 47% of the time no batches arrive, and hence this is a fairly lightly loaded system. In this case the policy accepts all batches until horizon 34. At this point it gradually reduces the acceptance region until horizon 40. From that point on it accepts when there are three or less packets in the buffer. It is still the case that for a long horizon the policy acts quite conservatively. See Fig. 3.2.

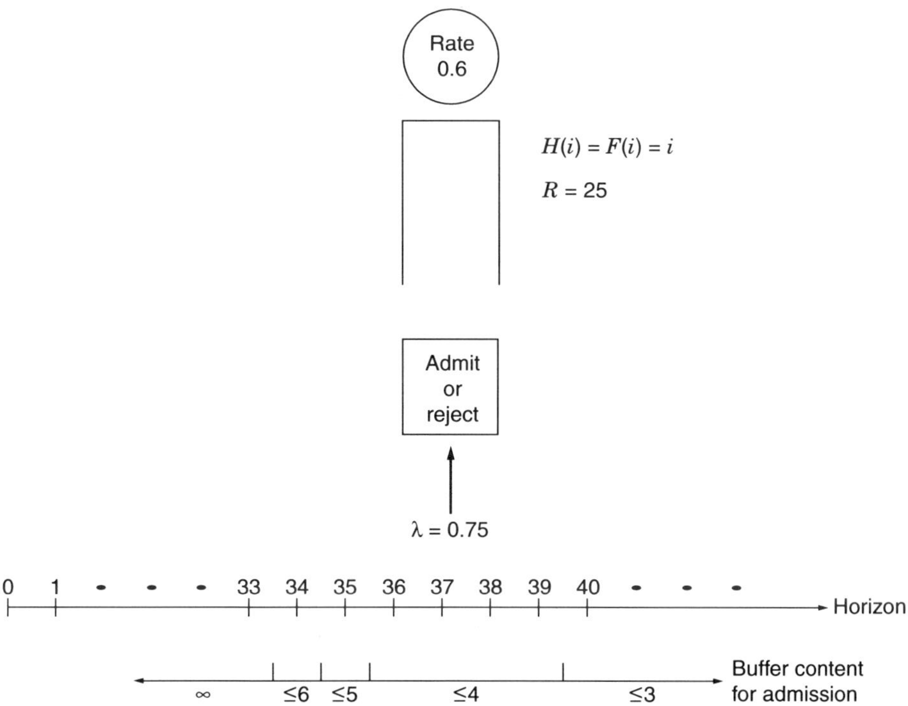

Figure 3.2 Scenario 6 from Table 3.2.

Scenario 7 examines the outcome if both the mean batch size and the rejection cost are halved. We also have $\lambda < \mu$. Note that $p_0 = e^{-0.375} = 0.69$. This means that 69% of the time no batches arrive to the controller, and hence this is a very lightly loaded system. It is quite interesting that the horizons at which changes occur are similar to the previous scenario, but the acceptance levels are modestly expanded.

Scenario 8 considers the case in which $\lambda = 1$. Since $R = 8$, an ambiguous situation occurs at $n = 8$. □

BIBLIOGRAPHIC NOTES

The material in Section 3.1 was developed primarily in Bellman (1957), Karlin (1955), Hinderer (1970), Derman (1970), and Schal (1975), with a theoretical emphasis in the latter.

The material in Sections 3.2 through 3.4 is new.

PROBLEMS

3.1. In Example 3.1.3 verify the claim made about the policy ψ.

3.2. Develop the finite horizon optimality equation (3.2) for Example 2.1.2.

3.3. Develop the finite horizon optimality equation for Example 2.1.3.

3.4. Develop the finite horizon optimality equation for Example 2.1.4.

3.5. Develop the finite horizon optimality equation for the MDC in Problem 2.4. Assume two stations.

3.6. Prove Corollary 3.3.5.

Problems 3.7–10 have to do with the model in Section 3.4.

3.7. Consider the optimality equation for Δ_N given in (3.26–27). Prove that if H is replaced by UH and R by UR, then the value function is multiplied by U and the optimal policy is unchanged. *Hint:* Prove this by induction on n. Introduce some appropriate notation.

WARNING! When running any of the programs, you should change only the constants at the top of the program. If you wish to modify the program itself, as called for in Problem 3.9, then copy the original program and give it a new name before making any modifications.

3.8. Run ProgramOne for the following scenarios:

(a) $H = 0.5$, $R = 5$, $\lambda = 1.5$, $\mu = 0.8$.

(b) $H = 1$, $R = 5$, $\lambda = 1$, $\mu = 0.9$.

(c) $H = 1$, $R = 5$, $\lambda = 0.8$, $\mu = 0.99$.

(d) $H = 1$, $R = 5$, $\lambda = 2$, $\mu = 0.45$.

(e) $H = 1$, $R = 15$, $\lambda = 0.5$, $\mu = 0.45$.

(There are three constants to be chosen: "*UB*" $= N$, "*B*" $= N - 1$, and "Horizon" $= n$. The program prompts you for the parameter values.) For each scenario determine an optimal policy and discuss your conclusions.

3.9. In this problem you are asked to modify ProgramOne. Read the Warning above before proceeding. Examine the code. Decide what needs to be done to modify it for a holding cost of the form $H(i) = Hi^2$. Carry out the modification. Make some runs for the same parameter values as in Section 3.4. Compare the results and discuss them.

***3.10.** Suppose that the distribution governing the batch sizes is bounded. For example, assume that $p_j = P(\text{batch size} = j) > 0$ for $0 \le j \le 5$. In this case at most five packets can enter the system in any slot. Write a program to determine an optimal policy.

3.11. Consider the model in Problem 3.2 with a terminal cost of zero.

(a) If $H(i)$ is increasing, prove that the expected value function is increasing in i.

(b) Develop an ATAS that sends the excess probability to N. Write the optimality equation for Δ_N. Consider the cases $i = 0$, $1 \le i \le N - 1$, and $i = N$. Employ (3.25).

(c) Prove that the expected value function in Δ_N is increasing in i.

3.12. Consider an ATAS for Problem 3.5 that sends the excess probability to the zero state. Write the optimality equation for Δ_N.

CHAPTER 4

Infinite Horizon Discounted Cost Optimization

In Section 4.1 we derive an equation for the infinite horizon expected discounted value function and prove that there exists an optimal stationary policy for the expected discounted cost criterion. In Section 4.2 it is shown that the solution to the optimality equation is not unique and various results relating to this are given. In Section 4.3 the relationship between the finite horizon and infinite horizon discounted value functions is treated. In Section 4.4 a characterization of optimal policies for the discounted cost criterion is given. In Section 4.5 we examine the behavior of the value function $V_\alpha(i)$ considered as a function of the discount factor α with the initial state i fixed.

Sections 4.6 and 4.7 consider the computation of an optimal policy when the state space is infinite. Conditions are given so that the value functions (respectively, optimal stationary policies) in an approximating sequence converge to the value function (respectively, optimal stationary policy) in the original MDC. These ideas are illustrated in an inventory model presented in Chapter 5.

4.1 INFINITE HORIZON DISCOUNTED COST OPTIMALITY EQUATION

With the exception of Section 4.5 the discount factor $\alpha \in (0, 1)$ is considered to be fixed throughout this chapter, and this is understood in our results. Notice that we do not allow $\alpha = 1$. The expected discounted value function V_α, defined in (2.14), represents the smallest expected discounted cost that can possibly be achieved when the process is operated over the infinite horizon. Recall that we refer to V_α as the discounted value function.

In this section we first derive an equation satisfied by V_α. This is the *discount optimality equation*. Second, we show that there exists an optimal stationary policy. These results form the centerpiece of the chapter.

Let us develop some preliminary results. Let θ be an arbitrary policy for the infinite horizon. We may operate the system under θ for n steps, with a terminal cost of zero. Under this condition θ becomes a policy for the n horizon, and $v_{\theta,\alpha,n}$ is its value function. The next result relates this quantity to the infinite horizon discounted value function under θ, defined in (2.13).

Lemma 4.1.1. The quantity $v_{\theta,\alpha,n}$ is increasing in n and $\lim_{n\to\infty} v_{\theta,\alpha,n} = V_{\theta,\alpha}$.

Proof: This result follows immediately from (2.13). The sum of an infinite series is defined as the limit of the sequence of its partial sums, if that limit exists. Since all costs are nonnegative, it is the case that the partial sums are increasing in n. Hence the partial sums form an increasing sequence. Such a sequence has a limit (it may be ∞). Hence it follows that

$$\begin{aligned} V_{\theta,\alpha}(i) &= \lim_{n\to\infty} \sum_{t=0}^{n-1} \alpha^t E_\theta[C(X_t, A_t)|X_0 = i] \\ &= \lim_{n\to\infty} v_{\theta,\alpha,n}(i), \qquad i \in S, \end{aligned} \tag{4.1}$$

and this completes the proof. □

Proposition 4.1.2. Let W be a nonnegative function and e a stationary policy such that

$$W(i) \ge C(i,e) + \alpha \sum_j P_{ij}(e)W(j), \qquad i \in S. \tag{4.2}$$

Then

$$W(i) \ge v_{e,\alpha,n}(i) + \alpha^n E_e[W(X_n)|X_0 = i], \qquad i \in S, n \ge 1, \tag{4.3}$$

and $W \ge V_{e,\alpha}$.

Proof: If (4.3) can be shown, then from the nonnegativity of W, it follows that $W \ge v_{e,\alpha,n}$. Hence from Lemma 4.1.1 it will follow that $W \ge V_{e,\alpha}$.

Equation (4.3) may be formally proved by induction. Here is the idea behind the proof. For $n = 1$ we have

$$\begin{aligned} W(i) &\ge C(i,e) + \alpha \sum_j P_{ij}(e)W(j) \\ &= v_{e,\alpha,1}(i) + \alpha E_e[W(X_1)|X_0 = i]. \end{aligned} \tag{4.4}$$

Iterating (4.2) once yields

$$
\begin{aligned}
W(i) &\geq C(i,e) + \alpha \sum_j P_{ij}(e)W(j) \\
&\geq C(i,e) + \alpha \sum_j P_{ij}(e)[C(j,e) + \alpha \sum_k P_{jk}(e)W(k)] \\
&= v_{e,\alpha,2}(i) + \alpha^2 E_e[W(X_2)|X_0 = i]. \qquad (4.5)
\end{aligned}
$$

The second line follows by applying (4.2) to each term of the summation in the first line. The third line follows from (2.9).

It is clear that this argument can be continued to yield (4.3). (Problem 4.1 asks you to give a formal induction proof.) □

Corollary 4.1.3. Let W be a nonnegative function satisfying

$$W(i) \geq \min_a \left\{ C(i,a) + \alpha \sum_j P_{ij}(a)W(j) \right\}, \qquad i \in S. \qquad (4.6)$$

Let f be a stationary policy that realizes the right side of (4.6); that is, for each state i, $f(i)$ is an action that achieves the minimum. Then $W \geq V_{f,\alpha} \geq V_\alpha$.

Proof: From (4.6) it follows that

$$W(i) \geq C(i,f) + \alpha \sum_j P_{ij}(f)W(j), \qquad i \in S. \qquad (4.7)$$

Then the result follows from Proposition 4.1.2 and (2.14). □

Let us introduce the auxillary function

$$U_\alpha(i,a) =: C(i,a) + \alpha \sum_j P_{ij}(a)V_\alpha(j). \qquad (4.8)$$

Let $B_i(\alpha) = \{b \in A_i | U_\alpha(i,b) = \min_a\{U_\alpha(i,a)\}\}$. These actions achieve or realize the minimum. In most cases $B_i(\alpha)$ is a singleton, but it may contain more than one action.

We are now ready to state the major theorem of the chapter.

Theorem 4.1.4. The discounted value function V_α is the minimum nonnegative solution of the discount optimality equation

$$V_\alpha(i) = \min_a \left\{ C(i,a) + \alpha \sum_j P_{ij}(a) V_\alpha(j) \right\}, \qquad i \in S. \tag{4.9}$$

Any stationary policy f_α that realizes the minimum in (4.9) is discount optimal.

Proof: Let θ be an arbitrary policy for the infinite horizon. Given history $h_1 = (i,a,j)$, let $\psi(i,a,j)$ be the policy followed by θ from time $t = 1$ onward. This policy may itself be considered a policy for the infinite horizon. This involves reindexing time, so that $t = 1$ becomes $t = 0$, etc. Then using reasoning similar to that employed in (3.5) we have

$$\begin{aligned}
V_{\theta,\alpha}(i) &= \sum_a \theta(a|i) E_\theta\left[C(X_0,A_0) + \alpha \sum_{t=1}^{\infty} \alpha^{t-1} C(X_t,A_t) | X_0 = i, A_0 = a\right] \\
&= \sum_a \theta(a|i) \left\{ C(i,a) + \alpha \sum_j P_{ij}(a) V_{\psi(i,a,j),\alpha}(j) \right\} \\
&\geq \sum_a \theta(a|i) \left\{ C(i,a) + \alpha \sum_j P_{ij}(a) V_\alpha(j) \right\} \\
&= \sum_a \theta(a|i) U_\alpha(i,a) \\
&\geq \min_a \{U_\alpha(i,a)\}.
\end{aligned} \tag{4.10}$$

Since θ is arbitrary it follows that $V_\alpha(i) \geq \min_a\{U_\alpha(i,a)\}$.

We now show that the reverse inequality holds. Fix $\epsilon > 0$. Define a policy θ^* as follows: For initial state i the policy selects an action in $B_i(\alpha)$. After having done this, suppose that the next state is j. By (2.14) there exists a policy $\psi(j)$ such that $V_{\psi(j),\alpha}(j) \leq V_\alpha(j) + \epsilon$. This follows since $V_\alpha(j)$ is the infimum, and hence there must be a policy achieving within ϵ of it. Let θ^* be the infinite horizon policy that chooses $b \in B_i(\alpha)$ and then follows the appropriate policy $\psi(j)$ depending on the next state. In a manner similar to (4.10) we have

$$\begin{aligned}
V_{\theta*,\alpha}(i) &= C(i,b) + \alpha \sum_j P_{ij}(b) V_{\psi(j),\alpha}(j) \\
&\le C(i,b) + \alpha \sum_j P_{ij}(b)[V_\alpha(j) + \epsilon] \\
&= C(i,b) + \alpha \sum_j P_{ij}(b) V_\alpha(j) + \alpha\epsilon \\
&= \min_a \{U_\alpha(i,a)\} + \alpha\epsilon.
\end{aligned} \tag{4.11}$$

Thus $V_\alpha(i) \le \min_a\{U_\alpha(i,a)\} + \epsilon$. Since $\epsilon > 0$ is arbitrary, we must have $V_\alpha(i) \le \min_a\{U_\alpha(i,a)\}$, and hence (4.9) holds.

Now let W be a nonnegative solution of (4.9). It follows from Corollary 4.1.3 that $V_\alpha \le W$, and hence the discounted value function is the minimum nonnegative solution of the discount optimality equation.

Now let the stationary policy f_α realize the minimum in (4.9). Then from Corollary 4.1.3 (with $W = V_\alpha$) it follows that $V_\alpha \ge V_{f_\alpha,\alpha} \ge V_\alpha$. Hence $V_\alpha(i) = V_{f_\alpha,\alpha}(i)$ for $i \in S$. Thus f_α is optimal for the infinite horizon discounted cost criterion. □

Corollary 4.1.5. If $V_\alpha(i) < \infty$ and f_α is the optimal stationary policy realizing (4.9), then

$$\lim_{n\to\infty} \alpha^n E_{f\alpha}[V_\alpha(X_n)|X_0 = i] = 0. \tag{4.12}$$

Proof: From Theorem 4.1.4 it follows that

$$V_\alpha(i) = C(i,f_\alpha) + \alpha \sum_j P_{ij}(f_\alpha) V_\alpha(j). \tag{4.13}$$

Iterating this yields (similarly to (4.3))

$$V_\alpha(i) = v_{f_\alpha,\alpha,n}(i) + \alpha^n E_{f_\alpha}[V_\alpha(X_n)|X_0 = i]. \tag{4.14}$$

From Lemma 4.1.1 and Theorem 4.1.4 it follows that the limit of the first term on the right of (4.14) exists and equals $V_\alpha(i)$. Hence the limit of the second term must exist and equal zero. (The validity of this step requires the finiteness of the discounted value function.) □

4.2 SOLUTIONS TO THE OPTIMALITY EQUATION

The following example shows that the solution to (4.9) is not unique.

Example 4.2.1. Let $S = \{0, 1, 2, \ldots\}$ with one action in each state. The transitions are $P_{ii+1} = 1$, and the costs are $C(i) \equiv 1$. The discount optimality equation is $V_\alpha(i) = 1 + \alpha V_\alpha(i+1)$. Clearly $V_\alpha(i) = 1 + \alpha + \alpha^2 + \ldots = 1/(1-\alpha)$ is a constant that satisfies the optimality equation. Unfortunately, so do many other functions.

To define a whole family of finite solutions, fix a number $z > 1/(1-\alpha)$. Then it can be shown that

$$W(i) = V_\alpha(i) + \left(\frac{z - V_\alpha(i)}{\alpha^i} \right), \qquad i \geq 0,$$

is also a solution of the optimality equation. For example, suppose that $\alpha = \frac{1}{2}$ so that $V_\alpha \equiv 2$. If $z = 3$, then $W(i) = 2 + 2^i$. Note that for every member of this family of solutions, it is the case that $\lim_{i \to \infty}(W(i) - V_\alpha(i)) = \infty$. □

It is desirable to have a condition under which a nonnegative solution to the discount optimality equation will equal V_α.

Proposition 4.2.2. Let W be a nonnegative solution of the discount optimality equation (4.9). Let f_α be an optimal stationary policy as in Theorem 4.1.4. If

$$\liminf_{n \to \infty} \alpha^n E_{f_\alpha}[W(X_n) | X_0 = i] = 0, \qquad i \in S, \tag{4.15}$$

then $W = V_\alpha$.

Proof: Since W satisifies (4.9), it follows that

$$W(i) \leq C(i, f_\alpha) + \alpha \sum_j P_{ij}(f_\alpha) W(j), \qquad i \in S. \tag{4.16}$$

Iterating (4.16) yields (similarly to (4.3))

$$W(i) \leq v_{f_\alpha, \alpha, n}(i) + \alpha^n E_{f_\alpha}[W(X_n) | X_0 = i]. \tag{4.17}$$

By Lemma 4.1.1 the limit of the first term on the right of (4.17) exists and equals $V_{f_\alpha, \alpha}(i)$. Since f_α is optimal, we have $V_{f_\alpha, \alpha} = V_\alpha$. Take the limit infimum as $n \to \infty$ in (4.17). This yields

$$\begin{aligned} W(i) &\le V_\alpha(i) + \liminf_{n \to \infty} \alpha^n E_{f_\alpha}[W(X_n)|X_0 = i] \\ &= V_\alpha(i). \end{aligned} \tag{4.18}$$

Since V_α is the minimum nonnegative solution of (4.9), it follows that $W = V_\alpha$. □

Example 4.2.3. We see that the condition in (4.15) is not satisfied for Example 4.2.1. Observe that

$$\alpha^n E[W(X_n)|X_0 = 0] = \frac{\alpha^n - 1}{1 - \alpha} + z,$$

which approaches $z - 1/(1 - \alpha) > 0$ as $n \to \infty$. □

Corollary 4.2.4. (i) Let W be a finite nonnegative solution of (4.9) that satisfies $W \le V_\alpha + B$ for some (finite) constant B. Then $W = V_\alpha$. (ii) If W is a nonnegative bounded solution of (4.9), then $W = V_\alpha$.

Proof: To prove (i), note that we have $W(X_n) \le V_\alpha(X_n) + B$. Hence

$$\alpha^n E_{f_\alpha}[W(X_n)|X_0 = i] \le \alpha^n E_{f_\alpha}[V_\alpha(X_n)|X_0 = i] + \alpha^n B. \tag{4.19}$$

Taking the limit infimum of both sides of (4.19) as $n \to \infty$ and using (4.12) yields (4.15). Hence the result follows from Proposition 4.2.2.

To prove (ii), assume that W is a nonnegative solution of (4.9) satisfying $W \le B < \infty$ for some constant B. Then the result follows from (i). □

4.3 CONVERGENCE OF FINITE HORIZON VALUE FUNCTIONS

Consider the finite horizon discounted value functions, defined in (2.10), and let the terminal cost be zero. These are denoted by $v_{\alpha,n}$.

Proposition 4.3.1. The quantity $v_{\alpha,n}$ is increasing in n and $\lim_{n\to\infty} v_{\alpha,n} = V_\alpha$. If $f_{\alpha,n}$ is a policy realizing the minimum in (3.2), then any limit point of the sequence $(f_{\alpha,n})_{n\ge 1}$ is discount optimal for the infinite horizon.

Proof: Because the costs are nonnegative, it is easy to see that $v_{\alpha,n}$ is increasing in n. Hence it forms a monotonically increasing sequence, and so $\lim_{n\to\infty} v_{\alpha,n} =: W$ exists.

Now let f_α be an optimal stationary policy as given in Theorem 4.1.4. Then from (2.10) and Lemma 4.1.1, it follows that $v_{\alpha,n} \le v_{f_{\alpha,n},\alpha,n} \le V_\alpha$. This implies that $W \le V_\alpha$.

Let f be a limit point of $(f_{\alpha,n})_{n\ge 1}$. From Definition B.1 in Appendix B, it follows that there exists a sequence n_r such that given i, we have $f_{\alpha,n_r}(i) = f(i)$ for n_r sufficiently large.

Now fix i. It follows from (3.2) that for n_r sufficiently large, we have

$$v_{\alpha,n_r}(i) = C(i,f) + \alpha \sum_j P_{ij}(f) v_{\alpha,n_r-1}(j). \tag{4.20}$$

Take the limit infimum as $r \to \infty$ of both sides of (4.20) and use the definition of W and Proposition A.1.7 to obtain

$$W(i) \ge C(i,f) + \alpha \sum_j P_{ij}(f) W(j). \tag{4.21}$$

Since this argument may be repeated for every state, it follows that (4.21) holds for all i. Then from Proposition 4.1.2 it follows that $W \ge V_{f,\alpha} \ge V_\alpha$. Since $W \le V_\alpha$, this proves that $W = V_\alpha = V_{f,\alpha}$. □

4.4 CHARACTERIZATION OF OPTIMAL POLICIES

In this section we give necessary and sufficient conditions for an arbitrary infinite horizon policy to be optimal for the expected discounted cost criterion.

Proposition 4.4.1. A policy θ for the infinite horizon is optimal for the infinite horizon expected α discounted cost criterion if and only if both of the following hold:

(i) Given initial state i, the distribution $\theta(a|i)$ is concentrated on the set $B_i(\alpha)$.

(ii) For $n \ge 1$, if h_n is a history under θ with state i_n, then the distribution $\theta(a|h_n)$ is concentrated on the set $B_{i_n}(\alpha)$.

(These conditions say that if the process finds itself in a state i at any time, then for optimality the distribution governing the choice of an action in that state must be concentrated on the set of actions realizing the minimum in (4.9).

This is not quite the same as requiring that the distribution be concentrated on $B_j(\alpha)$ for all j. The reason is that for a given initial state i, some state j may never be reached, and hence there is no necessity to restrict the choice of actions in that state. For such a state j we will never have $i_n = j$. This subtlety is illustrated in Example 4.4.2.)

Proof: Let us first prove the sufficiency of the conditions. Consider the statement:

(*) Given any infinite horizon policy θ satisfying (i–ii), we have $v_{\theta,\alpha,n} \le V_\alpha$ for $n \ge 1$.

If this can be proved, it will then follow from Lemma 4.1.1 that $V_{\theta,\alpha} \le V_\alpha$, and hence θ is optimal. Let θ satisfy (i–ii). For $n = 1$ we have

$$\begin{aligned}
v_{\theta,\alpha,1}(i) &= \sum_{a \in B_i(\alpha)} \theta(a|i)C(i,a) \\
&\le \sum_{a \in B_i(\alpha)} \theta(a|i)\left\{C(i,a) + \alpha \sum_j P_{ij}(a)V_\alpha(j)\right\} \\
&= \sum_{a \in B_i(\alpha)} \theta(a|i)U_\alpha(i,a) \\
&= \min_a\{U_\alpha(i,a)\} \\
&= V_\alpha(i).
\end{aligned} \tag{4.22}$$

The second line follows since V_α is nonnegative. The third line follows from (4.8). The fourth line follows from the definition of $B_i(\alpha)$ and Proposition A.1.1. The last line follows from (4.9).

Now assume that (*) holds for $n-1$. Assume a history $h_1 = (i, a, j)$. Let $\psi_{(i,a,j)}$ be the policy followed, under θ, from time $t = 1$ onward. If time is reindexed so that $t = 1$ becomes $t = 0$, and so on, then this policy is an infinite horizon policy with initial state j. Moreover it is the case that $\psi_{(i,a,j)}$ also satisfies (i–ii). We then have

$$\begin{aligned}
v_{\theta,\alpha,n}(i) &= \sum_{a \in B_i(\alpha)} \theta(a|i)\left\{C(i,a) + \alpha \sum_j P_{ij}(a)v_{\psi(i,a,j),\alpha,n-1}(j)\right\} \\
&\le \sum_{a \in B_i(\alpha)} \theta(a|i)\left\{C(i,a) + \alpha \sum_j P_{ij}(a)V_\alpha(j)\right\} \\
&= \min_a\{U_\alpha(i,a)\} \\
&= V_\alpha(i).
\end{aligned} \tag{4.23}$$

The second line follows from the induction hypothesis, and the other lines follow as before. This completes the induction, and hence (*) holds.

To prove the necessity, let θ be an optimal policy. We must show that it satisfies (i–ii). Look at (4.10). Since the last term equals $V_\alpha(i)$, for θ to be optimal it must be the case that both inequalities are equalities. By Proposition A.1.1 the last inequality is an equality if and only if $\theta(a|i)$ is concentrated on $B_i(\alpha)$. Hence condition (i) holds.

For the first inequality to be an equality, it must be the case that $V_{\psi(i,a,j)}(j) = V_\alpha(j)$ for each history $h_1 = (i, a, j)$. This means that $\psi_{(i,a,j)}$ must itself be an optimal policy for initial state j. But by the argument just given, this means that $\psi_{(i,a,j)}(a|j)$ must be concentrated on $B_j(\alpha)$. This proves that condition (ii) holds for $n = 1$. A repetition of this argument shows that (ii) holds for $n \geq 1$. We omit the formal argument. □

Example 4.4.2. The state space is $S = \{-1, 0, 1, 2, \ldots\}$. There is one action in states $i \geq 1$ with $P_{ii+1} = 1$ and $C(i) = 1$. We have $A_{-1} = \{a, a^*\}$ with $P_{-1-1}(a) = P_{-10}(a^*) = 1$, and costs identically equal to $C > 1$. We have $A_0 = \{b, b^*\}$ with $P_{01}(b) = P_{0-1}(b^*) = 1$ and costs identically equal to 1. See Fig. 4.1. (You are asked to verify the calculations for this example in Problem 4.7.)

There are four stationary policies that may be specified by giving the action chosen in -1 followed by the action chosen in 0. They are $f_1 = (a, b)$, $f_2 = (a^*, b)$, $f_3 = (a, b^*)$, and $f_4 = (a^*, b^*)$. We find that

$$V_{f_1,\alpha}(-1) = \frac{C}{1-\alpha}, \quad V_{f_1,\alpha}(0) = \frac{1}{1-\alpha},$$

$$V_{f_2,\alpha}(-1) = C + \frac{\alpha}{1-\alpha}, \quad V_{f_2,\alpha}(0) = \frac{1}{1-\alpha},$$

$$V_{f_3,\alpha}(-1) = \frac{C}{1-\alpha}, \quad V_{f_3,\alpha}(0) = 1 + \frac{\alpha C}{1-\alpha},$$

$$V_{f_4,\alpha}(-1) = \frac{C+\alpha}{1-\alpha^2}, \quad V_{f_4,\alpha}(0) = \frac{1+\alpha C}{1-\alpha^2}.$$

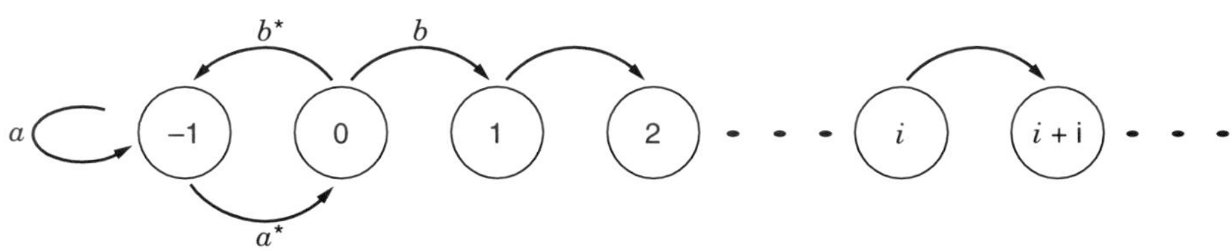

Figure 4.1 Example 4.4.2.

It is easy to see that $V_\alpha(0) = 1/(1-\alpha)$ and $V_\alpha(-1) = C + \alpha/(1-\alpha)$. The optimality equation yields

$$V_\alpha(-1) = C + \alpha \min\{V_\alpha(-1), V_\alpha(0)\},$$
$$V_\alpha(0) = 1 + \alpha \min\{V_\alpha(0), V_\alpha(-1)\}.$$

The second equation follows since $V_\alpha(1) = V_\alpha(0)$. Since $V_\alpha(0) < V_\alpha(-1)$, we obtain $B_{-1}(\alpha) = \{a^*\}$ and $B_0(\alpha) = \{b\}$. We see that both f_1 and f_2 are optimal policies for initial state 0. However, f_1 is not concentrated on $B_{-1}(\alpha)$. This does not affect its optimality, since the process never enters -1 from initial state 0 under f_1. □

4.5 ANALYTIC PROPERTIES OF THE VALUE FUNCTION

In this section we look at $V_\alpha(i)$ as a function of the discount factor α with the initial state i held fixed. Under this condition $V_\alpha(i)$: $(0,1) \rightarrow [0, \infty]$ is an extended real-valued function of a real variable. It is then possible to examine the analytic properties of this function. These include limits, continuity, and differentiability. We also examine these properties for $V_{\theta,\alpha}(i)$.

Some of the material in this section is starred, and the reader need not be overly concerned with the details. However, one result, Proposition 4.5.3, is very important to the subsequent development.

Let θ be a policy for the infinite horizon, and fix the initial state i (which we suppress in this argument). Even though $V_{\theta,\alpha}$ has not been defined for $\alpha = 0$, it is clear from (2.13) that we may set $V_{\theta,0} = E_\theta[C(X_0, A_0)]$. Thus $V_{\theta,\alpha}$: $[0, 1) \rightarrow [0, \infty]$. Let $u_n = E_\theta[C(X_n, A_n)]$. Then from (2.13) it follows that

$$V_{\theta,\alpha} = \sum_{n=0}^{\infty} \alpha^n u_n, \tag{4.24}$$

which is a power series in α. For completeness we allow $\alpha \in [0, \infty)$ in (4.24). Observe that $u_0 = E_\theta[C(X_0, A_0)] < \infty$, since the initial state i is given and A_i is finite.

The theory of power series is discussed briefly in Section A.3 of Appendix A, and the reader may review this material. As a corollary of the material on power series, we obtain the following result:

Proposition 4.5.1. Let θ be a policy. For each initial state i there exists a radius of convergence $R_i \in [0, \infty]$ for the power series (4.24). If $R_i > 0$, then $V_{\theta,\alpha}(i)$ is infinitely differentiable (and hence continuous) for $\alpha \in (0, R_i)$.

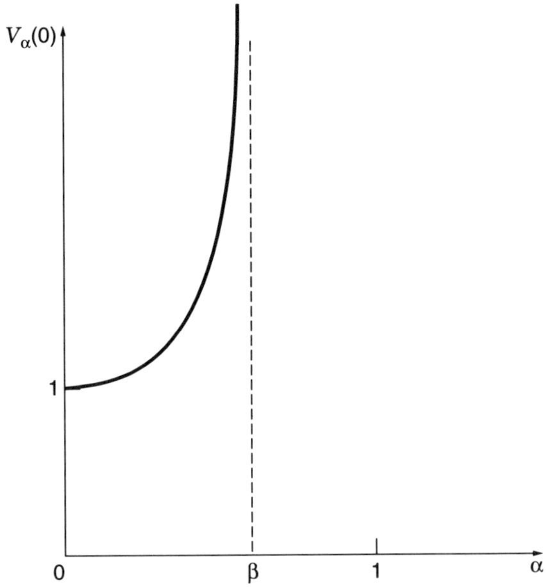

Figure 4.2 Example 4.5.2.

Example 4.5.2. Let $S = \{0, 1, 2, \ldots\}$. There is one action in each state and $P_{ii+1} = 1$ for $i \geq 0$. Fix $\beta \in (0, 1)$, and let $C(i) = \beta^{-i}$. Then $V_\alpha(0) = 1 + (\alpha/\beta) + (\alpha/\beta)^2 + \ldots$. Then $R_0 = \beta$, and the (geometric) power series converges to $1/[1 - (\alpha/\beta)]$ on $[0, \beta)$ by (A.24). We have $V_\alpha(0) = \infty$ for $\alpha \in [\beta, 1)$. See Fig. 4.2. □

The next result is needed in Chapter 6. Recall that a function $r(\alpha)$ is *rational* if there exist polynomials $p(\alpha)$ and $q(\alpha)$ such that $r = p/q$.

Proposition 4.5.3. Let S be finite, and let e be a stationary policy. Then for every initial state i, $V_{e,\alpha}(i)$ is a finite, continuous, rational function of $\alpha \in (0, 1)$.

Proof: It follows from (2.13) and (2.8) that

$$V_{e,\alpha}(i) = \sum_{n=0}^{\infty} \alpha^n \left(\sum_j C(j,e) P_{ij}^{(n)}(e) \right)$$

$$= \sum_j C(j,e) \left(\sum_{n=0}^{\infty} \alpha^n P_{ij}^{(n)}(e) \right). \tag{4.25}$$

(A stationary policy induces a Markov chain on S. For the definition of the transition matrix associated with the Markov chain, see Appendix C.) Let **P** be the (finite) transition matrix associated with e, **C** be a column vector of costs under e, and finally $\mathbf{V}_{e,\alpha}$ be a column vector of values. Then

$$\begin{aligned}\mathbf{V}_{e,\alpha} &= [\mathbf{I} + \alpha\mathbf{P} + (\alpha\mathbf{P})^2 + \ldots]\mathbf{C} \\ &= [\mathbf{I} - \alpha\mathbf{P}]^{-1}\mathbf{C}.\end{aligned} \tag{4.26}$$

The first line is (4.25) expressed in matrix notation. The second line follows from a well-known result in matrix theory. From the formula for the inverse of a matrix, it is easily seen that each entry of $[\mathbf{I} - \alpha\mathbf{P}]^{-1}$ is a rational function of α. This implies that each entry of $\mathbf{V}_{e,\alpha}$ is a rational function of α.

Since the state space is finite, the costs are bounded, say by B, and so $V_{e,\alpha} \leq B/(1-\alpha)$. Hence $V_{e,\alpha}$ is a finite function for $\alpha \in (0, 1)$. Since a rational function is continuous wherever it is defined (Apostol, 1974, p. 81), it follows that $V_{e,\alpha}$ is continuous on $\alpha \in (0, 1)$. □

Since the rest of the material in the book does not require matrix theory, it is unfortunate that a matrix theoretic proof of Proposition 4.5.3 is required. We are not aware of a simple alternative proof for this result.

We are now ready to prove some properties of the discounted value function in the general (countable state space) case.

***Proposition 4.5.4.** The discounted value function V_α is increasing and left continuous for $\alpha \in (0, 1)$.

Proof: Let θ be an arbitrary policy. Since the costs are nonnegative, it is clear from (4.24) that $V_{\theta,\alpha}$ is increasing. This means that for $0 < \alpha \leq \beta < 1$ we have $V_{\theta,\alpha} \leq V_{\theta,\beta}$. Taking the infimum over θ of both sides of this inequality yields $V_\alpha \leq V_\beta$, and hence V_α is increasing.

Now fix $0 < \beta < 1$. Since V_α is increasing, it follows that $\lim_{\alpha \to \beta -} V_\alpha =: W$ exists and is bounded above by V_β. The proof will be completed if it can be shown that $W \geq V_\beta$.

Let α_n be an increasing sequence of positive numbers converging to β. Equation (4.9) becomes

$$V_{\alpha_n}(i) = \min_a \left\{ C(i,a) + \alpha_n \sum_j P_{ij}(a) V_{\alpha_n}(j) \right\}, \qquad i \in S. \tag{4.27}$$

Take the limit infimum of both sides of (4.27). Use the definition of W and Propositions A.1.3(i) and A.1.7 to obtain

$$W(i) \geq \min_a \left\{ C(i,a) + \beta \sum_j P_{ij}(a)W(j) \right\}, \qquad i \in S. \tag{4.28}$$

Then from Corollary 4.1.3 it follows that $V_\beta \leq W$. □

Corollary 4.5.5. If there exists a (finite) constant B such that $C(i,a) \leq B$ for all state-action pairs, then V_α is finite and continuous for $\alpha \in (0, 1)$.

**Proof:* By Proposition 4.5.4 it is sufficient to prove that V_α is right continuous. Fix $0 < \beta < 1$. Since V_α is increasing, it follows that $\lim_{\alpha \to \beta+} V_\alpha =: W$ exists and $W \geq V_\beta$. The proof will be completed if it can be shown that $W \leq V_\beta$.

Let α_n be a decreasing sequence of positive numbers converging to β. We may assume that $\alpha_n \leq \alpha^* < 1$. Let f_β be an optimal stationary policy. Then from (4.9) it follows that

$$V_{\alpha_n}(i) \leq C(i,f_\beta) + \alpha_n \sum_j P_{ij}(f_\beta)V_{\alpha_n}(j), \qquad i \in S. \tag{4.29}$$

We wish to apply Corollary A.2.4 to the right side of (4.29). Since $W \leq V_{\alpha*} \leq B/(1-\alpha^*)$, it follows that for the bounding function we may take the constant $B/(1-\alpha^*)$. Taking the limit of both sides of (4.29) yields

$$W(i) \leq C(i,f_\beta) + \beta \sum_j P_{ij}(f_\beta)W(j), \qquad i \in S. \tag{4.30}$$

Iterating (4.30) yields

$$W(i) \leq v_{f_\beta,\beta,n}(i) + \beta^n E_{f_\beta}[W(X_n)|X_0 = i], \qquad i \in S. \tag{4.31}$$

Taking the limit as $n \to \infty$ and using the fact that W is bounded yields $W \leq V_\beta$. □

A much stronger result than the above is given in Problem 4.8.

4.6 ASM FOR THE INFINITE HORIZON DISCOUNTED CASE

The approximating sequence method is used to calculate both the discounted value function and an optimal stationary policy for the case when the state space is denumerably infinite.

Throughout this section let Δ be an MDC with denumerable state space, and let (Δ_N) be an approximating sequence for Δ. Then $(V_\alpha^N(i))_{i \in S_N}$ is the discounted value function in Δ_N, and f_α^N is a discount optimal stationary policy as given by Theorem 4.1.4.

The two questions of interest are as follows:

Question 1. When does $V_\alpha^N \to V_\alpha < \infty$?

Question 2. When is a limit point of $(f_\alpha^N)_{N \geq N_0}$ discount optimal for Δ?

We want to ensure both the finiteness of the discount value function in Δ and the convergence. The next example shows that the desired convergence may not hold.

Example 4.6.1. Let $S = \{0, 1, 2, \ldots\}$. There is one action in each state with $P_{i0} = P_{ii+1} = \frac{1}{2}$ and $C(i) = i^2$. Thus at each step the process is equally likely to return to 0 or to move to the next higher state. We assume that $\alpha < \frac{1}{2}$ and that the initial state is 0 (this is suppressed in the notation).

It is clear that the process eventually returns to 0. Let T be the time of a first passage back to 0. Then

$$\begin{aligned} V_\alpha(0) &= E\left[\sum_{t=0}^{T-1} \alpha^t C(X_t)\right] + E[\alpha^T]V_\alpha(0) \\ &\leq E\left[\sum_{t=0}^{T-1} C(X_t)\right] + E[\alpha^T]V_\alpha(0). \end{aligned} \tag{4.32}$$

Now $P(T = n) = 2^{-n}$ and so $E[\alpha^T] = \sum_{n=1}^{\infty}(\alpha/2)^n = (0.5\alpha)/(1 - 0.5\alpha)$. We also have

$$E\left[\sum_{t=0}^{T-1} C(X_t)\right] = \sum_{n=2}^{\infty} \frac{1^2 + \ldots + (n-1)^2}{2^n} =: D < \infty.$$

Then from (4.32) we find that

$$V_\alpha(0) \leq \frac{D(1 - \alpha/2)}{1 - \alpha} < \infty. \tag{4.33}$$

To define Δ_N, let $S_N = \{0, 1, \ldots, N\}$ for $N \geq 3$. Define the approximating distributions as follows: For $0 \leq i \leq N-2$ let $P_{i0}(N) = 0.5$, $P_{ii+1}(N) = 0.5 - 1/N$, and $P_{iN}(N) = 1/N$. Let $P_{N-10}(N) = 1 - 1/N$ and $P_{N-1N}(N) = 1/N$. Let $P_{NN}(N) = 1$.

If the process starts in state 0, it will eventually reach N. Let U be the time of first reaching N. Observe that from any state $0 \le i \le N-1$, there is a probability $1/N$ of next transitioning to N. Hence $P(U = n) = (1 - 1/N)^{n-1}/N$ and

$$E[\alpha^U] = \frac{1}{N(1 - 1/N)} \sum_{n=1}^{\infty} \left[\alpha\left(1 - \frac{1}{N}\right)\right]^n$$

$$= \frac{1}{N-1}\left(\frac{\alpha(1 - 1/N)}{1 - \alpha(1 - 1/N)}\right)$$

$$= \frac{\alpha}{N(1-\alpha) + \alpha}. \tag{4.34}$$

The last line follows from some algebra.

Since state N is absorbing, we have $V_\alpha^N(N) = N^2/(1 - \alpha)$. Then

$$V_\alpha^N(0) = E\left[\sum_{t=0}^{U-1} \alpha^t C(X_t)\right] + E[\alpha^U]V_\alpha^N(N)$$

$$\ge E[\alpha^U]V_\alpha^N(N)$$

$$= \frac{\alpha N^2}{(1-\alpha)[N(1-\alpha) + \alpha]}. \tag{4.35}$$

The second line follows since the costs are nonnegative. The third line follows from (4.34). Since $\lim_{N\to\infty} V_\alpha^N(0) = \infty$, the convergence fails. □

Example 4.6.1 shows that some assumption is necessary to have an affirmative answer to Questions 1 and 2. The theoretical development of the ASM for the infinite horizon discounted case completely parallels that for the finite horizon case given in Sections 3.2 and 3.3. The next result is the analog of Lemma 3.2.2.

Lemma 4.6.2. We have $\liminf_{N\to\infty} V_\alpha^N \ge V_\alpha$.

Proof: The discount optimality equation for Δ_N is

$$V_\alpha^N(i) = \min_a \left\{ C(i,a) + \alpha \sum_{j \in S_N} P_{ij}(a; N) V_\alpha^N(j) \right\}, \qquad i \in S_N. \tag{4.36}$$

Taking the limit infimum of both sides of (4.36) yields

$$\liminf_N V_\alpha^N(i) = \min_a \left\{ C(i,a) + \alpha \liminf_N \sum_{j \in S_N} P_{ij}(a;N) V_\alpha^N(j) \right\}$$

$$\geq \min_a \left\{ C(i,a) + \alpha \sum_j P_{ij}(a)(\liminf_N V_\alpha^N(j)) \right\}, \qquad i \in S. \tag{4.37}$$

The first line follows from Proposition A.1.3(i) and the second line from Proposition A.1.8. The result then follows from Corollary 4.1.3. □

The following *infinite horizon discounted cost* assumption for fixed α enables us to answer Questions 1 and 2.

Assumption DC(α). For $i \in S$ we have $\limsup_{N \to \infty} V_\alpha^N(i) =: W_\alpha(i) < \infty$ and $W_\alpha(i) \leq V_\alpha(i)$. □

The next result is the analog of Theorem 3.2.3.

Theorem 4.6.3. The following are equivalent:

(i) $\text{Lim}_{N \to \infty} V_\alpha^N = V_\alpha < \infty$.
(ii) Assumption DC(α) holds.

Assume that either (then both) of these holds, and let f_α^N be an optimal stationary policy for Δ_N determined by (4.36). Then any limit point of the sequence $(f_\alpha^N)_{N \geq N_0}$ is optimal for Δ.

Proof: If (i) holds, then $\limsup_N V_\alpha^N = \lim_N V_\alpha^N = V_\alpha < \infty$, and then clearly (ii) holds. If (ii) holds, then $\limsup_N V_\alpha^N \leq V_\alpha \leq \liminf_N V_\alpha^N$, where the last inequality follows from Lemma 4.6.2. Moreover the first term is finite. But this implies that all the terms are equal and finite, and thus (i) holds. This proves the equivalence of (i) and (ii).

Now assume that (i) holds. By Proposition B.5 there exists a limit point f of the sequence $(f_\alpha^N)_{N \geq N_0}$. Hence there exists a subsequence N_r such that given $i \in S$, we have $f_\alpha^{N_r}(i) = f(i)$ for N_r sufficiently large.

Now fix a state i. For N_r sufficiently large, (4.36) may be written as

$$V_\alpha^{N_r}(i) = C(i,f) + \alpha \sum_{j \in S_N} P_{ij}(f;N_r) V_\alpha^{N_r}(j). \tag{4.38}$$

This follows since $f_\alpha^{N_r}$ realizes the minimum in (4.36). Now take the limit infimum as $r \to \infty$ of both sides of (4.38). Employing (i) and Proposition A.1.8 yields

$$V_\alpha(i) \geq C(i,f) + \alpha \sum_j P_{ij}(f) V_\alpha(j). \tag{4.39}$$

Since this argument may be carried out for every state, it follows that (4.39) holds for all i. Proposition 4.1.2 implies that $V_\alpha \geq V_{f,\alpha}$, and hence $V_\alpha = V_{f,\alpha}$ and f is optimal. □

4.7 WHEN DOES DC(α) HOLD?

In this section we give various sufficient conditions for DC(α) to hold. The development parallels the results for the finite horizon in Section 3.3.

Proposition 4.7.1. Assume that there exists a (finite) constant B such that $C(i,a) \leq B$, for all state action pairs. Then DC(α) holds for $\alpha \in (0, 1)$.

Proof: We verify that Theorem 4.6.3(i) holds. Observe that $V_\alpha^N \leq B/(1-\alpha)$. Fix a subsequence N_r. It then follows from Proposition B.6 that there exist a subsubsequence N_s and a nonnegative function U, bounded above by $B/(1-\alpha)$, such that $\lim_{s\to\infty} V_\alpha^{N_s}(i) = U(i)$ for $i \in S$.

Take the limit of both sides of the optimality equation (4.36) through values of the subsequence N_s. We apply Corollary A.2.7 (with bounding function $B/(1-\alpha)$) and Proposition A.1.3(ii) to obtain

$$U(i) = \min_a \left\{ C(i,a) + \alpha \sum_j P_{ij}(a) U(j) \right\}, \qquad i \in S. \tag{4.40}$$

It then follows from Corollary 4.2.4(ii) that $U = V_\alpha$. Because every subsequence of V_α^N has a subsequence converging to V_α, it follows that Theorem 4.6.3(i) holds. □

Proposition 4.7.1. provides a complete answer to Questions 1 and 2 in the case of bounded costs. The remainder of this section is of interest only when the costs in Δ are unbounded. We develop two situations for which convergence holds. The first shows that if (Δ_N) is an ATAS that sends excess probability to a finite set, then DC(α) holds. This development is starred. If the reader wishes, the statements of the preliminary lemmas and the proof of Proposition 4.7.4 may be omitted.

To set up some notation, let e be a stationary policy for Δ, and fix states i and j in S. The notation ${}_{N*}P_{ij}^{(t)}(e)$ denotes the *taboo probability* of transitioning from i to j in t steps, while avoiding the set $S - S_N$ during the intermediate steps, that is, while remaining within S_N (except possibly at the beginning and end).

***Lemma 4.7.2.** Let e be a stationary policy for Δ. Then

$$\lim_{N \to \infty} {}_{N*}P_{ij}^{(t)}(e) = P_{ij}^{(t)}(e), \qquad i,j \in S, t \geq 1. \tag{4.41}$$

Proof: We prove (4.41) by induction on t. Reference to the policy e is suppressed in the proof. We have ${}_{N*}P_{ij} = P_{ij}$, and hence the result holds for $t = 1$.

Now assume that (4.41) holds for t. We prove that it holds for $t + 1$. Note that ${}_{N*}P_{ij}^{(t+1)} \leq P_{ij}^{(t+1)}$, and hence $\limsup_N {}_{N*}P_{ij}^{(t+1)} \leq P_{ij}^{(t+1)}$. It is thus sufficient to show that $\liminf_N {}_{N*}P_{ij}^{(t+1)} \geq P_{ij}^{(t+1)}$.

Now

$${}_{N*}P_{ij}^{(t+1)} = \sum_{k \in S_N} P_{ik}\, {}_{N*}P_{kj}^{(t)}. \tag{4.42}$$

Taking the limit infimum of both sides of (4.42) yields

$$\begin{aligned} \liminf_N {}_{N*}P_{ij}^{(t+1)} &\geq \sum_k P_{ik} P_{kj}^{(t)} \\ &= P_{ij}^{(t+1)}. \end{aligned} \tag{4.43}$$

The first line follows from Proposition A.1.8 and the induction hypothesis. The second line follows from Section C.1 of Appendix C. □

***Lemma 4.7.3.** Let $i \in S$. Assume that N is so large that $i \in S_N$, and operate under the stationary policy e until the set $S - S_N$ is reached. Let $T_i(N)$ be the number of steps in this first passage. Then

$$\lim_{N \to \infty} E_e[\alpha^{T_i(N)}] = 0. \tag{4.44}$$

Proof: Reference to e is suppressed in the proof. Observe that

$$E[\alpha^{T_i(N)}] = \sum_{n=1}^{\infty} \alpha^n P(T_i(N) = n) + 0P(T_i(N) = \infty). \tag{4.45}$$

Let $\epsilon > 0$, and choose K so large that $\alpha^K/(1 - \alpha) \leq \epsilon$. Since probabilities are bounded above by 1, we have

$$\begin{aligned} E[\alpha^{T_i(N)}] &\leq \sum_{n=1}^{K-1} \alpha^n P(T_i(N) = n) + \sum_{n=K}^{\infty} \alpha^n \\ &\leq \sum_{n=1}^{K-1} \alpha^n P(T_i(N) = n) + \epsilon. \end{aligned} \tag{4.46}$$

Suppose that

$$\lim_{N \to \infty} P(T_i(N) = n) = 0. \tag{4.47}$$

Then taking the limit supremum of both sides of (4.46) yields $\limsup_N E[\alpha^{T_i(N)}] \leq \epsilon$. Since the expectation is nonnegative and $\epsilon > 0$ is arbitrary, this proves (4.44).

So it remains to prove (4.47). Since $P(T_i(N) = n) \leq P(T_i(N) \leq n) = 1 - P(T_i(N) > n)$, it is sufficient to prove that

$$\lim_{N \to \infty} P(T_i(N) > n) = 1. \tag{4.48}$$

Now

$$P(T_i(N) > n) = \sum_{j \in S_N} {}_{N*}P_{ij}^{(n)}. \tag{4.49}$$

Take the limit infimum of both sides of (4.49) to obtain

$$\liminf_{N \to \infty} P(T_i(N) > n) \geq \sum_j P_{ij}^{(n)} = 1. \tag{4.50}$$

This follows from Proposition A.1.8 and Lemma 4.7.2. Since probabilities are bounded above by 1, this implies that (4.48) holds. □

Proposition 4.7.4. Assume that $V_\alpha < \infty$, and let (Δ_N) be an ATAS that sends excess probability to a finite set. Then DC(α) holds.

**Proof:* Let G be the finite set to which the excess probability is sent. Let $i \in S_N$ be the initial state. Consider a policy for Δ_N which operates under f_α until $S - S_N$ is reached, and then it operates under f_α^N. It follows that

$$V_\alpha^N(i) \le E_{f_\alpha}\left[\sum_{t=0}^{T_i(N)-1} \alpha^t C(X_t, A_t)\right] + E_{f_\alpha}[\alpha^{T_i(N)}] \cdot \left(\sum_{j \in G} V_\alpha^N(j)\right)$$

$$\le V_\alpha(i) + Z(N) E_{f_\alpha}[\alpha^{T_i(N)}], \tag{4.51}$$

where we have defined $Z(N) =: \sum_{j \in G} V_\alpha^N(j)$. Recall that S_N is finite and hence $Z(N) < \infty$. Equation (4.51) embodies some important observations. As long as the process has not reached $S - S_N$, then Δ and Δ_N operate exactly the same way under f_α. The first term on the right of the first line is the expected discounted cost of a first passage to $S - S_N$, and this is bounded above by the total expected discounted cost $V_\alpha(i)$. Once the process reaches $S - S_N$, it goes back to G according to some distribution, and $Z(N)$ is an upper bound for the remaining terms.

Let $\sum_{j \in G} V_\alpha(j) =: Z < \infty$. Let us add the equations in (4.51), for initial states in G, and then solve for $Z(N)$. This yields

$$Z(N) \le \frac{Z}{1 - \sum_{j \in G} E_{f_\alpha}[\alpha^{T_j(N)}]}. \tag{4.52}$$

By Lemma 4.7.3 we have $\limsup_N Z(N) \le Z$, and hence $Z(N)$ is bounded. Taking the limit supremum of both sides of (4.51) and again using Lemma 4.7.3 yields $\limsup_N V_\alpha^N(i) \le V_\alpha(i)$, and hence DC($\alpha$) holds. □

In the case that all the excess probability is sent to a distinguished state z, the optimality equation (4.36) has a simple and suggestive form.

Corollary 4.7.5. Assume that $V_\alpha < \infty$, and let (Δ_N) be an ATAS that sends the excess probability to a distinguished state z. Then DC(α) holds. If $R_\alpha^N = V_\alpha^N - V_\alpha^N(z)$ is the relative value function, then the discount optimality equation for Δ_N is

$$V_\alpha^N(i) = \alpha V_\alpha^N(z) + \min_a \left\{ C(i,a) + \alpha \sum_{j \in S_N} P_{ij}(a) R_\alpha^N(j) \right\},$$

$$i \in S_N. \tag{4.53}$$

Proof: You are asked to prove this in Problem 4.9. □

This result is utilized in the inventory model presented in Chapter 5. The final result involves Proposition 3.3.4.

Proposition 4.7.6. Assume that $V_\alpha < \infty$, and let (Δ_N) be an ATAS such that the augmentation distributions satisfy (3.20). Then $V_\alpha^N(i) \le V_\alpha(i)$ for $i \in S_N$, and hence DC(α) holds.

Proof: From Proposition 4.3.1 it follows that $v_{\alpha,n} < \infty$ for $n \ge 1$. Proposition 3.3.4 implies that $v_{\alpha,n}^N \le v_{\alpha,n}$. Taking the limit of both sides as $n \to \infty$, it follows from Proposition 4.3.1 and the hypothesis that $V_\alpha^N \le V_\alpha < \infty$. Hence DC($\alpha$) holds. □

BIBLIOGRAPHIC NOTES

Sections 4.1 through 4.4 contain mostly classical results. Most of the results in Section 4.5 can be found in the literature but are not presented in the form given here.

Many of the important works have already been referenced. Here we add Stidham (1981) and Lippman (1975).

The material on the approximating sequence method in Sections 4.6 and 4.7 is new. Recall that Langen (1991) mentioned earlier is a related treatment.

Other methods for calculation have been proposed. Fox (1971) proposed a truncation scheme that is generalized by White (1980a, b, 1982) and Hernandez-Lerma (1986). This scheme requires the rewards to be bounded. It is generalized by Cavazos-Cadena (1986) and Whitt (1978, 1979a, b). Puterman (1994) presents an approach based on these works.

The philosophy behind these methods differs from the ASM approach, and no direct comparison appears possible. In general terms, the ASM creates a sequence of finite state MDCs and these can be studied in their own right. The other schemes pass directly to a method of calculation.

PROBLEMS

4.1. Give an induction proof of (4.3).

4.2. Consider an MDC with $S = \{0, 1, 2, \ldots\}$. There is one action in each state

$i \geq 1$ such that $P_{ii-1} = 1$ and $C(i) = 1$. We have $A_0 = \{a, b\}$. Action a is associated with distribution $(p_j)_{j \geq 1}$ such that $P_{0j}(a) = p_j$. Under action b we have a similar distribution $(q_j)_{j \geq 1}$. Finally $C(0, a) = C(0, b) = 0$.

(a) Let f be the stationary policy that chooses a. Find a formula for $V_{f,\alpha}(0)$. This will involve the *generating function* $G_p(\alpha) = \sum p_j \alpha^j$. Do a similar calculation for the stationary policy e that chooses b.

(b) Determine a condition under which f is discount optimal.

(c) Assuming that f is optimal, determine $V_\alpha(i)$ for $i \geq 0$.

(d) What is the discount optimality equation (4.9)? Verify that the values you found in (c) satisfy (4.9).

4.3. Develop the discount optimality equation (4.9) for Example 2.1.2.

4.4. Develop the discount optimality equation (4.9) for Example 2.1.3.

4.5. Let Δ be an MDC with $S = \{0, 1, 2, \ldots\}$. Assume that there exists a nonnegative integer k such that for all i and a we have $P_{ij}(a) = 0$, $j > i + k$. That is, the process cannot move up more than k units in any transition. Let W be a nonnegative solution of (4.9) satisfying $W(i) \leq Di^r$ for some finite constant D and positive integer r. Use Proposition 4.2.2 to show that $W = V_\alpha$.

4.6. Show that the conclusions of Proposition 4.3.1 hold if the value functions $v_{\alpha,n}$ are defined for a nonnegative bounded terminal function.

4.7. Verify the calculations in Example 4.4.2.

***4.8.** Fix the initial state i (and suppress it). Assume that $V_{e,\alpha} < \infty$ for all stationary policies e and $\alpha \in (0, 1)$. Prove that V_α is a continuous function of $\alpha \in (0, 1)$. Note that Corollary 4.5.5 follows from this more general result.

4.9. Prove Corollary 4.7.5.

4.10. Let Δ be as in Problem 4.5. Define (Δ_N) by $S_N = \{0, 1, \ldots, N\}$, and assume that $P_{ij}(a; N) = 0$ for $j > i + k$. That is, the approximating distributions satisfy the same condition as the original distributions. Assume that there exist a finite constant D and a positive integer r such that $C(i, a) \leq Di^r$ for all state-action pairs. Show that DC(α) holds.

Hint: Show that $V_\alpha^N(i) \leq Fi^r$ for some constant F. Use this to obtain an appropriate solution W of (4.9). Then apply the result in Problem 4.5.

CHAPTER 5

An Inventory Model

In this chapter an inventory model is treated. In Section 5.1 the setup is discussed, and the model is formulated as an MDC. In Section 5.2 the discounted finite horizon and infinite horizon optimality equations for the model are obtained. In Section 5.3 an approximating sequence for the MDC is formed, and computational issues for the infinite horizon discounted cost criterion are discussed. In Section 5.4 some numerical results for a specific case of the model are presented. These utilize ProgramTwo. The chapter problems contain suggestions for additional exploration.

5.1 FORMULATION OF THE MDC

An inventory model was introduced in Example 1.1.2 and is further developed in this chapter. Our model takes into account both holding/penalty costs and actual earned revenues. Let us now discuss the particulars of the model. At the end of this section, a summary list of the operating assumptions is given for the convenience of the reader.

The time slots are referred to as periods. They may be thought of as weeks, months, quarters, or some other convenient unit. Consider the operation of the system during a single time period. At the beginning of the period there is a known inventory level x. Since unfilled orders are allowed (known as backlogging), we have $x \in \mathbf{Z}$, where $\mathbf{Z}$ is the set of integers $\{\ldots -2, -1, 0, 1, 2, \ldots\}$. In actuality, of course, the inventory level cannot be unbounded. However, it is a useful modeling device to place no a priori restrictions on the level. For example, it can be assumed that additional warehouses may be built to contain increasing inventory. Similarly it can be assumed that no orders are turned away, and thus the level of backlogging has no a priori bound.

At the beginning of the period an order for k items is placed by the inventory manager. This is the chosen action, and since action sets must be finite, we assume that k is an element of a finite nonempty set A of nonnegative integers. The order may be thought of as being filled by an outside agency or as a production level at a plant. We will speak of the action throughout as the production level, with the understanding that this should be broadly interpreted. Since there is an upper bound on the number of items produced in one period, this is called a *capacitated system.* It is assumed that the items in the order are produced (or arrive from outside) during the period. (When during the period they are assumed to actually arrive can affect the cost structure, as is discussed below.) Once the production level is set, it cannot be changed during the period. To avoid trivialities, let us assume that the largest element in A is a positive number K so that it is possible to produce some items. For example, we might have $A = \{100, 200, 300, 400\}$, so items may only be produced in lots of 100 with $K = 400$. It may or may not be an option to produce no items.

In each period there is a demand for the items that is stochastic in nature. The number y of items demanded in one period lies in a finite nonempty set D of nonnegative integers, and $d_y > 0$ is the probability that y items are demanded, where $\sum_{y \in D} d_y = 1$. It will be clear later why the demand is assumed to be bounded. The demand is revealed over the course of the period so that at the end of the period the value of y for that period is known. At the end of the period, just before the beginning of the next period, the demand is filled as much as possible. The total demand is given by y together with the backlogged inventory if $x < 0$. To avoid trivialities, let us make the following assumption: The largest element in D is a positive number Y and $D \neq \{Y\}$. This means that there is a positive probability that some items are demanded and that the demand is not constant.

In any realistic situation the demand distribution will change over time. Our assumption that the demand distribution is unchanging may be viewed as an approximation. The manager may use sales data to estimate the distribution governing demand. The optimal policy may then be computed under the assumption that demand remains constant. This gives a benchmark for setting production levels until conditions change.

We now discuss how to form an MDC for this model. The state of the system at time t is the triple (x^*, k^*, y^*), where x^* is the inventory level at the beginning of period $t-1$, k^* is the production level set at that time, and y^* is the demand occurring over period $t-1$. *The state of the system at a given period is what transpired during the previous period.* All of these quantities are known to the manager. The state space $S = \mathbf{Z} \times A \times D$.

It is helpful to fix firmly in mind that the current state is the triple of conditions that prevailed during the *previous* period. Figure 5.1 shows the process.

Here is how this formulation is initialized. The state at period 0 may be any triple (x^*, k^*, y^*) with the interpretation that x^* was the inventory level at time $t = -1$, k^* the production level at that time, and y^* the demand revealed during that period.

For the purposes of determining the transition probabilities and costs, let us assume that the state at time t is (x^*, k^*, y^*). Then the *current* inventory level is $x^* + k^* - y^*$. A couple of examples will clarify this. If the state is (4, 5, 6), then at time $t-1$ there were 4 items on hand and a decision to produce 5 items was

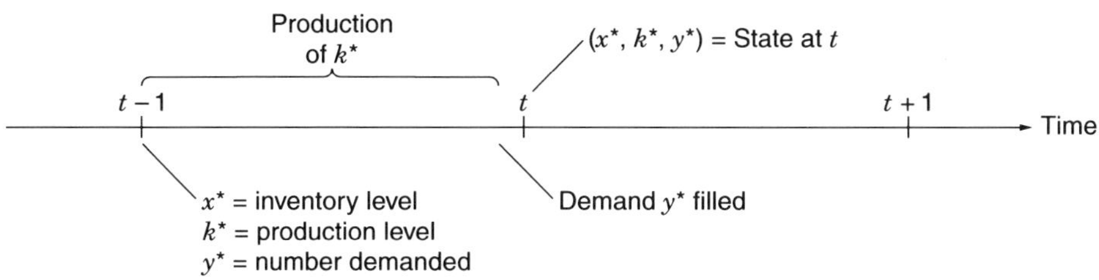

Figure 5.1 Inventory model.

made. This gave 9 items at the end of $t - 1$ and 6 were demanded, leaving 3 as the current inventory level. If the state is $(-4, 5, 6)$, then at time $t - 1$ there were 4 items backordered, and a decision to produce 5 items was made. This gave 1 item left over after the backorders were filled and 6 were demanded, leaving -5 as the current inventory level. For a given state (x^*, k^*, y^*) it is to be understood that x is the current inventory level for that state, and thus

$$\text{current inventory level} \quad x = x^* + k^* - y^*. \tag{5.1}$$

The transition probabilities are now easy to determine. Assume that a production level of k is set for period t. The state at time $t + 1$ is the triple that prevailed during period t, which is (x, k, y) with probability d_y. Formally we have

$$P_{(x*,k*,y*)(x,k,y)}(k) = d_y, \qquad y \in D. \tag{5.2}$$

We now determine the number $s = s(x^*, k^*, y^*)$ of items that were sold at the end of period $t - 1$. We claim that

$$s = \begin{cases} (x^* + k^*) \wedge y^*, & x^* \geq 0, \\ k^*, & x^* \leq -k^*, \\ -x^* + [(x^* + k^*) \wedge y^*], & -k^* < x^* < 0. \end{cases} \tag{5.3}$$

To see this observe that if $x^* \geq 0$, then there are no backlogged orders, and the number sold is the minimum of the number y^* demanded and the number $x^* + k^*$ available to meet that demand. If $x^* \leq -k^*$ then the total production goes to fill backlogged orders. If $-k^* < x^* < 0$, then the backlog is eliminated, resulting in $-x^*$ items sold, and there are $x^* + k^*$ items left over to fill the demand. (The reader may construct a few numerical examples to illustrate (5.3).)

A single expression can be given for the terms in (5.3). Let $u = 0$ for $x^* \geq 0$, and let $u = -x^*$ for $x^* < 0$. Then

$$s = u + [(x^* + k^*) \wedge y^*] \tag{5.4}$$

works in all cases. (Check it out!) When reference to s is made for a state (x^*, k^*, y^*), we are referring to s as defined in (5.4).

The cost function has several components. For a current inventory level of x, there is a nonnegative inventory cost $I(x)$. For $x \geq 0$ this is interpreted as a cost of holding x items in inventory. For $x < 0$ this is interpreted as a penalty cost for having x items backordered. As an example we might have $I(x) = 0.5\,x$ for $x \geq 0$ and $I(x) = -0.1x^3$ for $x < 0$. This is a mild penalty for a small backlog but eventually becomes much more severe. In addition there is a nonnegative cost $C(k)$ of producing k items. We could also assume a cost of changing the production level, but for simplicity we will not incorporate such a cost into the model.

It is assumed that a revenue of R is earned for each item sold so that the total revenue generated from the sale of s items is Rs. Recall from Section 2.1 that rewards may be accommodated into our model as negative costs. So as a first attempt to write the cost function, we associate with the state (x^*, k^*, y^*) and decision k the cost $I(x) + C(k) - Rs$.

Note that the inventory cost is charged on the inventory level at time t. The cost $C(k)$ is charged at time t on the production level for period t. If the items are purchased, it is assumed that the cost is incurred at that time. If the items are produced, then $C(k)$ may include labor costs and may also include a cost for holding produced items in the system during period t until the demand is cleared at the end of the period. The revenue from the items sold at the end of period $t - 1$ is accrued at time t.

Recall that in the specification of an MDC the costs must be nonnegative. As discussed in Section 2.1, rewards can be accommodated in the model as negative costs, and then a constant added to the costs to make them nonnegative. We thus require $I(x) + C(k) - Rs \geq -Rs \geq -B$ for some (finite) nonnegative constant B. This holds if $Rs \leq B$, that is, if the one period revenue is bounded above. Consider the cases in (5.3). Under the first case, $s \leq y^* \leq Y$. Under the second case, $s = k^* \leq K$, and under the third case, $s \leq -x^* + x^* + k^* = k^* \leq K$. Hence we may set $B = R(Y \vee K)$. This leads to the cost function $C[(x^*, k^*, y^*), k] = I(x) + C(k) - Rs + B$. Letting $U(s) = B - Rs$, we have formally

$$C[(x^*, k^*, y^*), k] = I(x) + C(k) + U(s), \tag{5.5}$$

where $x = x^* + k^* - y^*$ and s is given in (5.4). Note that each of the constituent functions is nonnegative.

The specification of the cost function illustrates why it is necessary to assume that the demand distribution is over a finite set. If arbitrarily large numbers of items could be demanded in one period, then the potential revenue would be unbounded, and the MDC formulation in Chapter 2 cannot handle this situation. In the Bibliographic Notes is discussed an approach for treating this case.

This completes the specification of an MDC Δ for the model. Here is a summary of the conditions that have been assumed:

1. The production level is $k \in A$, a finite set of nonnegative integers. The maximum number that may be produced is $K > 0$.
2. The demand is $y \in D$ (a finite set of nonnegative integers) with probability

$d_y > 0$. The maximum number that may be demanded is $Y > 0$ and $D \neq \{Y\}$.

3. The state at period t is (x^*, k^*, y^*), where x^* is the inventory level at $t - 1$, k^* is the production level for period $t - 1$, and y^* is the number of items demanded during that period.
4. The transition probabilities for Δ are given by (5.2).
5. $I(x)$ is the inventory function, $C(k)$ the production cost, and $U(s)$ a function incorporating the revenue earned from the sale of s items. These are nonnegative functions (finite by Remark 2.4.2).
6. The cost function for Δ is given by (5.5).

5.2 OPTIMALITY EQUATIONS

In this section we develop the finite and infinite horizon expected discounted cost optimality equations for Δ. The discount factor $\alpha \in (0, 1)$ is assumed fixed throughout the chapter.

To develop the finite horizon optimality equation, assume that the terminal cost is zero. Then $v_{\alpha,0} \equiv 0$ and for $n \geq 1$ (3.2) becomes

$$v_{\alpha,n}(x^*, k^*, y^*) = I(x) + U(s) + \min_k \left\{ C(k) + \alpha \sum_{y \in D} d_y v_{\alpha,n-1}(x, k, y) \right\}. \quad (5.6)$$

Similarly the infinite horizon optimality equation (4.9) becomes

$$V_\alpha(x^*, k^*, y^*) = I(x) + U(s) + \min_k \left\{ C(k) + \alpha \sum_{y \in D} d_y V_\alpha(x, k, y) \right\}. \quad (5.7)$$

The next result gives two situations in which V_α is finite. These require some mild additional assumptions on the model.

Proposition 5.2.1. Let

$$F(x) = \sum_{n=0}^{\infty} \alpha^n I(x + nK), \qquad x > 0,$$

$$F^*(x) = \sum_{n=0}^{\infty} \alpha^n I(x - nY), \qquad x < 0. \quad (5.8)$$

Then V_α is finite under either of the following conditions:

(i) $Y \leq K$ and for $x > 0$, I is increasing and F is finite.

(ii) $0 \in A$ and for $x < 0$, I is decreasing and $F*$ is finite.

Proof: There exists a finite upper bound W for the term $U(s) + C(k)$. For any policy θ let W_θ be the infinite horizon expected discounted inventory cost under θ. Then clearly $V_\alpha \leq W_\theta + W/(1-\alpha)$, and it is sufficient to find a policy for which $W_\theta < \infty$.

Let us assume that (i) holds, and let f be the stationary policy that always orders K. Assume that the process starts in a state (x^*, k^*, y^*) such that the current inventory level $x \leq 0$. The maximum number Y that can be demanded is always less than or equal to the number ordered. Moreover by condition 2 (Section 5.1) sometimes less than Y will be demanded. It is clear that in a finite expected amount of time, the process will reach a state with positive current inventory level. During this first passage the penalty cost will not exceed the maximum of the numbers $\{I(x), I(x+1), \ldots, I(0)\}$.

By this reasoning it is sufficient to assume that the process starts in (x^*, k^*, y^*) such that $x > 0$. There is an initial holding cost of $I(x)$, and during each subsequent period the process will either stay in the same state or will move to a greater inventory level. Because I is increasing on positive inventory, it is clear that an upper bound is obtained by assuming that there is never any demand. Thus

$$W_f(x^*, k^*, y^*) \leq I(x) + \sum_{n=1}^{\infty} \alpha^n I(x + nK). \tag{5.9}$$

The right side of (5.9) is $F(x)$ which is finite by assumption.

Now assume that (ii) holds, and let e be the stationary policy that never orders. The argument is the mirror of that above and is given as Problem 5.1. □

Problems 5.2–3 show that if I is composed of appropriate polynomials (so that it is nonnegative), then the functions in (5.8) are finite. Problem 5.4 looks at the possibility that I may have an exponential form. If $Y > K$ and (ii) fails, then the situation is more complicated. We do not treat this case.

5.3 AN APPROXIMATING SEQUENCE

In this section we develop an approximating sequence for Δ and discuss issues surrounding the computation of V_α. In developing the AS, we make some further simplifying assumptions. In addition to conditions 1 through 6 given in Section 5.1, assume that

7. We have $0 \in D$, and thus it is possible to have no items demanded.
8. Proposition 5.2.1(ii) holds.
9. $C(k)$ is increasing in k and $C(0) = 0$.

It then follows from Proposition 5.2.1 that $V_\alpha < \infty$.

Here is how we define the ATAS for this model. Let $G_N = \{-N, \ldots, -1, 0, 1, \ldots, N\}$ and $S_N = G_N \times A \times D$. Let the distinguished state $z = (0, 0, 0)$. Define the ATAS by sending the excess probability to z.

There are two ways to look at the calculation of V_α. The first way involves approximating V_α by $v_{\alpha,n}$. Then $v_{\alpha,n}$ may be approximated by $v^N_{\alpha,n}$. Under this method the discounted value function in Δ is approximated by the finite horizon value function in Δ. This in turn is approximated by the finite horizon value function in Δ_N.

The second way involves approximating V_α by V^N_α. Then V^N_α may be approximated by $v^N_{\alpha,n}$. Under this method the discounted value function in Δ is approximated by the discounted value function in Δ_N. This in turn is approximated by the finite horizon value function in Δ_N. Both ways end up in precisely the same place.

We elaborate on the first way. For a fixed finite set of states in S, and for n and N sufficiently large, we have

$$V_\alpha \approx v_{\alpha,n} \approx v^N_{\alpha,n} \quad \text{and} \quad f_\alpha \approx f_{\alpha,n} \approx f^N_{\alpha,n}. \tag{5.10}$$

The first approximations (for the functions and policies) follow from Proposition 4.3.1. The second approximations follow from Corollary 3.3.3 and Theorem 3.2.3. Note that we need $v_{\alpha,n} < \infty$, which follows from Propositions 4.3.1 and 5.2.1. (If we followed the second way, we would first appeal to Corollary 4.7.5 and Theorem 4.6.3, and then to Proposition 4.3.1 applied to Δ_N.)

In summary, our method of calculation is to compute in Δ_N the finite horizon discounted value function and corresponding optimal policy for large N and horizon. We now develop the optimality equation (3.19). We have $v^N_{\alpha,0} \equiv 0$.

Let (x^*, k^*, y^*) be a state in S_N such that $x \notin G_N$. (Recall that x^* is the inventory level at the beginning of the previous period, and x from (5.1) is the current inventory level, which figures in the state at the beginning of the subsequent period.) In this case it follows from (5.2) that all transitions from state (x^*, k^*, y^*) end up outside of S_N. This means that the summation on the right side of (3.19) vanishes. By condition 9 the minimization reduces to $\min_k\{C(k)\} = C(0) = 0$. This yields

$$v^N_{\alpha,n}(x^*, k^*, y^*) = \alpha v^N_{\alpha,n-1}(z) + I(x) + U(s), \qquad x \notin G_N, \tag{5.11}$$

and the optimal decision is to not produce.

Now consider states (x^*, k^*, y^*) for which $x \in G_N$. For these states all transitions remain within S_N, and hence (3.19) becomes

$$v_{\alpha,n}^{N}(x^{*},k^{*},y^{*}) = \alpha v_{\alpha,n-1}^{N}(z) + I(x) + U(s)$$

$$+ \min_{k} \left\{ C(k) + \alpha \sum_{y \in D} d_y r_{\alpha,n-1}^{N}(x,k,y) \right\},$$

$$x \in G_N. \tag{5.12}$$

Equations (5.11–12) are the equations used to compute the desired approximations.

Our work is not quite done because our interest is not in V_α but in a related quantity. Recall that the cost structure given in (5.5) involves the addition of the constant B to all costs to make them nonnegative. To obtain the true minimum expected discounted cost, we must subtract $B/(1 - \alpha)$ from V_α. The resulting quantity then involves the incurred costs minus the earned revenues. The negative of this quantity will involve the earned revenues minus the incurred costs and thus is the *maximum expected discounted profit* P_α. So what we wish to approximate is the quantity

$$P_\alpha = \frac{B}{1-\alpha} - V_\alpha. \tag{5.13}$$

Now $V_\alpha \approx v_{\alpha,n}^{N} = r_{\alpha,n}^{N} + v_{\alpha,n}^{N}(z)$, and hence

$$P_\alpha \approx \frac{B}{1-\alpha} - r_{\alpha,n}^{N} - v_{\alpha,n}^{N}(z). \tag{5.14}$$

When the calculations in (5.11–12) have been carried out up to the specified horizon length, then the optimal policy and the quantity in (5.14) are printed out. These tell the manager approximately how much profit may be expected and what an optimal production policy is.

Keep in mind that the quantity in (5.14) is the maximum expected discounted profit, namely the maximum expected discounted revenue minus inventory/production cost over the infinite horizon. This is under the assumption, of course, that demand and monetary conditions (which might affect the value of the discount factor) remain constant.

5.4 NUMERICAL RESULTS

In this section we discuss ProgramTwo. This program carries out the calculation discussed in Section 5.3 under the special case that $A = \{0, 1, \ldots, K\}$ and $D = \{0, 1, \ldots, Y\}$, where $Y \le K$. The inventory costs are given by

$$I(x) = \begin{cases} Hx, & x \geq 0, \\ -Px, & x < 0, \end{cases} \tag{5.15}$$

where H and P are positive constants. The cost of production is given by

$$C(k) = C_1 + C_2 k, \qquad 1 \leq k \leq K, \tag{5.16}$$

where C_1 is the setup cost (the fixed cost that is incurred whenever some items are produced) and C_2 is the marginal cost of producing one item. These are nonnegative numbers.

The user is prompted for the values of R, H, P, C_1, C_2, α, and the values of d_y. The values of N, K, Y, and the horizon length are constants that may be changed in subsequent runs of the program. The program operates much as ProgramOne, and we will not repeat that discussion here.

Remark 5.4.1. Suppose that H, P, R, C_1, and C_2 are each multiplied by a positive constant Q. The effect is to multiply $v_{\alpha,n}^N$ by Q. (You are asked to show this in Problem 5.5.) Since $B = RK$, the effect is to multiply P_α by Q. This means that the values of P, R, C_1, and C_2 may be scaled relative to the value of H. In all scenarios (other than those specifically for checking the operation of the program) we assume, without loss of generality, that $H = 0.1$. Thus the assumption is of a holding cost of 10 cents per item per period. This value keeps the other numbers within a smaller range. In all scenarios we set $\alpha = 0.95$. □

It is important to find some special cases in which the value of P_α may be determined. These cases are useful in confirming that the program is operating correctly.

Checking Scenario 5.4.2. Assume that $H = P = C_1 = C_2 = 0$ and $R > 0$. In this situation there are no inventory or ordering costs. Since $Y \leq K$, it is clear that the demand can always be taken care of in one period. Let $\lambda = \sum y d_y$ be the average demand.

Assume that the process starts in state (x^*, k^*, y^*), where $x \geq 0$. In this case it is clear that all future demands can be met without cost, and we have $P_\alpha(x^*, k^*, y^*) = R[s+(\alpha\lambda)/(1-\alpha)]$. This can be reasoned as follows: The revenue earned immediately is Rs, where s from (5.4) is the amount sold at the end of the previous period. In each future period the average revenue is $R\lambda$, and this amount is discounted by α because the revenue from the number sold in the current period is not earned until the following period.

ProgramTwo was run for $Y = 3$, $K = 4$, $R = 10$, $N = 25$, and $n = 100$. The demand distribution is given by $d_0 = 0.2$, $d_1 = 0.3$, $d_2 = 0.1$, and $d_3 = 0.4$, which gives $\lambda = 1.7$. Note that $P_\alpha(x^*, k^*, y^*) = 10s + 323$. For example, $P_\alpha(7, ., 0) = 323$, while the program gives 325.72. Moreover $P_\alpha(7, ., 1) = 333$, while the program gives 335.72.

Now assume that the process starts in state (x^*, k^*, y^*) with $x < 0$. In this case it would pay, at least for a while, to reduce the backlog as fast as possible. For example, we have $P_\alpha(-12, 0, 0) = (\alpha + \alpha^2 + \alpha^3 + \alpha^4 + \alpha^5)(4R) + \alpha^6(2.2R) + \alpha^7(\lambda R)/(1 - \alpha) = 425.53$, while the program gives 427.65. The reasoning for this is as follows: In the first period, 4 will be ordered and sold, and the new inventory level will be, on average, $-8 - 1.7 = -9.7$. The process is repeated, giving average inventory levels of -7.4, -5.1, -2.8, and -0.5. In this last state, on average, $0.5 + 1.7 = 2.2$ will be sold. From then on, the reasoning is as in the case of a nonnegative inventory level. □

Checking Scenario 5.4.3. We let $H = 1$, $P = R = 0$, $C_1 = C_2 = 5$, and $d_1 = 1$. In this case there are no revenues to be earned, and it is optimal to never order. Assume that the process starts in state (x^*, k^*, y^*) such that $x > 0$. Then exactly one item is demanded every period, and we have $P_\alpha(x^*, k^*, y^*) = -[x + \alpha(x - 1) + \alpha^2(x - 2) + \ldots + \alpha^{x-1}]$. This program was run with $K = 4$, $Y = 1$, $N = 25$, and $n = 50$. For example, we have $P_\alpha(4, 3, 0) = -25.36817$, which agrees with the program output. We also have $P_\alpha(5, 0, 1) = -9.51238$, which agrees with the program output. □

The checking scenarios give us confidence that the program is working properly. Let us now discuss some more typical scenarios.

Scenarios 5.4.4. Recall that for each scenario we have $H = 0.1$ and $\alpha = 0.95$. The results are summarized in Table 5.1.

Consider Scenario 1. The first box gives the values of the parameters, and we see that $Y = 3$. The second box gives the demand probabilities in increasing order of y, with $d_0 = 0.1$, and so on. The fourth box gives the values of N and the horizon length n.

The first run is for $N = 25$ and $n = 50$. A cursory look at the output does not yield the form of the optimal policy. However, a closer examination reveals two interesting facts. First and perhaps not too surprisingly, the optimal policy depends only on the current inventory level x. Second, the optimal policy is *bang-bang*. That is, if the current inventory level is no more than 1, then the manager should produce the maximum number 4 of items during that period. However, if the current inventory level is at least 2, then the manager should not produce at all. This is indicated in the third box by the notation B-B and by giving the inventory level for full production. (Once this observation was made, the program output instructions were modified to also print out the current inventory level.)

Another run was made for $n = 60$ (not shown), and it is confirmed that the policy is unchanging. It is suspected that the optimal policy is actually determined for substantially smaller values of n, but this was not tested.

However, the convergence of the value function is considerably slower. The value for the distinguished state $z = (0, 0, 0)$ is given. We see that the convergence is pulling in by $n = 125$. Here we have $P_\alpha(z) = 109.1$, which implies that

Table 5.1 Results for Scenarios 5.4.4

Scenario	1	2	3	4	5	6	7
Parameters	$P = 1.0$	$P = 5.0$	$R = 12.0$	$K = 6$	$P = 1.0$	$P = 1.0$	$P = 2.0$
	$C_1 = 5.0$	—	—	—	$C_1 = 5.0$	$C_1 = 0.0$	$C_1 = 5.0$
	$C_2 = 1.0$				$C_2 = 1.0$	$C_2 = 0.5$	$C_2 = 0.5$
	$R = 6.0$				$R = 10.0$	$R = 1.0$	$R = 10.0$
	$K = 4$				$K = 8$	$K = 7$	$K = 4$
	$Y = 3$				$Y = 2$	$Y = 2$	$Y = 4$
Probabilities	0.1	—	—	—	0.1	0.4	0.1
	0.3				0.5	0.4	0.1
	0.4				0.4	0.2	0.1
	0.2						0.4
							0.3
Optimal policy	B-B	B-B	B-B	B-B	B-B	s-S	B-B
	$x \leq 1$	$x \leq 2$	$x \leq 1$	$x \leq 0$	$x \leq 0$	$x \leq 1$	$x \leq 3$
						order to 2	
N	25	20	20	15	15	25	20
	50	80	80	80	50	100	80
	136.4	111.8	312.6	129.9	298.6	5.12	414.7
n	20	20	20	15	15	15	20
	100	125	125	125	80	125	125
	110.6	106.4	303.1	122.1	214.3	4.54	409.3
$P_\alpha(z)$	20				15	15	
	125				125	135	
	109.1				193.6	4.45	

the maximum expected discounted profit that can be made under these conditions is about \$109.

Scenarios 2, 3, and 4 explore changing one parameter in Scenario 1. Scenario 2 is as in Scenario 1 with the exception that the penalty cost is 5 times as much. This is an important calculation, since all the parameters except P can be accurately estimated. The penalty for backlogged orders is, in essence, a guess, and it is important to see how sensitive the optimal policy is to a change in this guess. Here the optimal policy is slightly more aggressive, producing the maximum number when the current inventory is no greater than 2. Note that $P_\alpha(z)$ is just slightly less than in Scenario 1. This indicates that by adopting a more aggressive production policy, the expected profit does not decrease by much.

Scenario 3 is as in Scenario 1 with the exception that the revenue per item is doubled. The optimal policy remains the same. Here we have $P_\alpha(z) = 303.1$. The reader may feel that this number should be roughly double the value in Scenario 1. But this is not the case. Problem 5.6 asks you to explore this.

Scenario 4 is as in Scenario 1 with the exception that the maximum production level is raised to 6. The optimal policy is still bang-bang, with maximum production called for when the inventory level is no more than 0. The maximum expected discounted profit in z is greater than in Scenario 1. This indicates that an increased profit can be obtained by increasing the production capacity. The manager could continue to explore this option, choosing greater production capacities to determine the one that gives the maximum value of $P_\alpha(z)$.

Scenario 5 explores a situation in which K is substantially greater than Y. The optimal policy is still bang-bang. However, this does not always hold, as we see in Scenario 6. Here the optimal policy has the so-called s-S form. *This is a standard name for this type of policy and should not be confused with our notation.* In this type of policy, if the inventory level is no more than s, then the optimal decision is to produce enough to bring the level up to (but no greater than) S, or as close as possible to this goal. If the inventory level exceeds s, then no items should be produced. In Scenario 6 the optimal policy is 1-2.

In Scenario 7 we have $K = Y$, and the demand is concentrated on $y = 3$ or 4. The optimal policy goes into full production whenever the inventory level is no more than 3. Note the aggressive policy, which holds since the demand is, on average, almost 3 items every period. Again the manager can explore increasing the value of K to see if an increased profit results. □

Remark 5.4.5. Because of results in the literature (see the Bibliographic Notes), one suspects that as $K \rightarrow \infty$ the optimal policy is of s-S form. This is dependent on the parameter values in (5.15–16) and might not hold for other choices. As discussed earlier, we have chosen to focus on computational issues rather than attempting to prove that optimal policies have certain structures. □

Remark 5.4.6. The array sizes required in ProgramTwo are of obvious concern. Besides simply increasing memory, there are two simple strategies that may help in smaller dimensional problems.

Suppose that a computation for a relatively small value of N indicates that the optimal policy is bang-bang, and suppose that one desires a great deal of accuracy in the expected profits. The program can be rewritten to eliminate all production choices except $\{0, K\}$.

Another possibility is to allow the maximum inventory level and maximum backlog to be different. Suppose that $M(N)$ is a sequence in N such that $\lim_{N \to \infty} M(N) = \infty$. For example, N might be a multiple of 10 and $M(N) = N/10$. Then $G_N = \{-M(N), \ldots, -1, 0, 1, \ldots, N\}$ will also work in the AS. In this way a more computationally efficient choice for S_N can be made. □

BIBLIOGRAPHIC NOTES

Various versions of this model have been extensively treated, usually from a theoretical framework. Seminal work is contained in Scarf (1960), Veinott (1966), and Schal (1976).

Bertsekas (1987, 1995a) treats a version in which the revenue to be gained from the sale of items is ignored. Models in Denardo (1982) and Puterman (1994) incorporate revenue. However, it is incorporated as an expectation of revenue to be gained in a single period rather than as actual revenue gained. So when the optimization is performed, it involves finding the expected discounted value function of an expected revenue. Puterman (1994) allows a time-dependent demand process. If one wishes to give a treatment in which rewards may be unbounded, then Puterman (1994, Sec. 6.10) may be applied. Federgruen and Zipkin (1986a, b) are further references.

The major emphasis of the literature has been on theoretically deriving the form of an optimal ordering policy. In contrast, our focus is on showing how an optimal policy may be computed.

PROBLEMS

5.1. Prove Proposition 5.2.1(ii).

5.2. Assume that $Y \le K$ and $I(x) = Hx^r$ for $x > 0$, where H is a positive constant and r is a positive integer. Show that the condition in Proposition 5.2.1(i) holds.

5.3. Assume that $0 \in A$ and $I(x) = P|x^r|$ for $x < 0$, where P is a positive constant and r is a positive integer. Show that the condition in Proposition 5.2.1(ii) holds.

5.4. Assume that $Y \le K$. Let $\beta > 1$ and $I(x) = \beta^x$ for $x > 0$. Determine when the conditions in Proposition 5.2.1(i) hold.

5.5. Verify the claim made in Remark 5.4.1.

5.6. It is desired to explain the difference between the value of P_α for Scenario 3 and its value for Scenario 1. Note that the optimal policy is the same. Let P_α denote the value for Scenario 1 and P_α^* the value for Scenario 3. Using the interpretation of the quantity in (5.14), develop an expression for $P_\alpha^* - P_\alpha$. *Note:* An interesting inference concerning the average number of items sold each period can be drawn from this. Do you see what it is?

5.7. Develop the appropriate counterparts to the equations in this chapter when a cost for changing production levels is present.

5.8. Run ProgramTwo under the following scenarios and discuss the output:

(a) $\alpha = 0.9$, $H = 0.1$, $P = 1.0$, $C_1 = 5.0$, $C_2 = 0.5$, $R = 3.5$, $Y = 2$, and $K = 4$. Assume that $d_0 = 0.5$, $d_1 = 0.3$, and $d_2 = 0.2$.

(b) As in (a) but with $K = 10$.

(c) As in (a) but with $K = 15$.

5.9. Run ProgramTwo under some interesting scenarios of your own construction.

5.10. Let us develop an inventory model similar to the one presented in this chapter but for which backlogging is not allowed. Let the state space be (x^*, k^*, y^*) such that $x^* \geq 0$. If the demand during period $t - 1$ is not met, then there is a fixed penalty cost of $P > 0$ assessed at the beginning of period t.

(a) What is the current inventory level x? Assume a nonnegative holding cost $H(x)$ assessed on current inventory. The other costs and revenues are as in Section 5.1.

(b) Develop an MDC for this model.

(c) Give the optimality equations, as in Section 5.2, for this model.

(d) Prove that if $0 \in A$, then $V_\alpha < \infty$.

(e) Assume that $0 \in D$. Develop an appropriate ATAS for this model by sending the excess probability to z. Give the optimality equations for the AS under condition 9.

***5.11.** Suppose that we wish to treat the model in Section 5.4 with the exception that $I(x) = Px^2$ for $x < 0$. Make a copy of ProgramTwo and rename the copy. Examine the code to see what should be modified to handle a quadratic penalty cost. Make the appropriate modifications and run the program for some scenarios of your construction. What conclusions can be drawn?

CHAPTER 6

Average Cost Optimization for Finite State Spaces

This chapter treats the average cost criterion when the MDC has a finite state space S. In contrast to the finite horizon and infinite horizon discounted cost optimization criteria, we will not treat the finite and denumerably infinite state space cases together. Very special results hold when S is finite, and these results are developed in this chapter.

Section 6.1 presents a fundamental relationship linking the discounted cost and average cost under a fixed policy. This relationship holds for arbitrary countable state spaces. In the remainder of the chapter, it is assumed that the state space is finite. In Section 6.2 we prove that there always exists an optimal stationary policy f for the average cost criterion. In Section 6.3 an average cost optimality equation (ACOE) satisfied by f is developed.

In Section 6.4 we obtain a strengthening of the ACOE under the assumption that the minimum average cost is a constant J. We also give various conditions for the minimum average cost $J(i)$ to be constant. In Section 6.5 we examine what can be proved if one can find *some* solution to the ACOE. Can we then be assured that the minimum average cost and an optimal policy have been found?

In Section 6.6 we develop a computational method, based on finite horizon value iteration, for finding a solution to the average cost optimality equation and an average cost optimal stationary policy. This development applies to any MDC with a *finite state space and constant minimum average cost.* Section 6.7 illustrates the value iteration method using a simple example for which the calculations may be carried out by hand.

6.1 A FUNDAMENTAL RELATIONSHIP FOR S COUNTABLE

In this section we have an MDC Δ with a countable state space S. The policy θ denotes an arbitrary infinite horizon policy. Let i be the initial state. At this time the reader may wish to reiew the definitions of the average cost $J_\theta(i)$, the minimum average cost $J(i)$, an average cost optimal policy, and $J_\theta^*(i)$ and $J^*(i)$. These definitions are in Section 2.4.

We now present a fundamental relationship that will be seen shortly to provide a crucial link between the infinite horizon discounted cost and the average cost optimization criteria.

Proposition 6.1.1. For any policy θ and initial state i, we have

$$J_\theta^*(i) \le \liminf_{\alpha \to 1^-} (1-\alpha)V_{\theta,\alpha}(i) \le \limsup_{\alpha \to 1^-} (1-\alpha)V_{\theta,\alpha}(i) \le J_\theta(i). \tag{6.1}$$

The following are equivalent:

(i) All the terms in (6.1) are equal and finite.

(ii) $J_\theta^*(i) = J_\theta(i) < \infty$, and hence the quantity in (2.15) is obtained as a limit.

(iii) $\mathrm{Lim}_{\alpha \to 1^-}(1-\alpha)V_{\theta,\alpha}(i)$ exists and is finite.

Proof: We have discussed in Section 4.5 how the value function $V_{\theta,\alpha}(i)$ may be expressed as a power series. Recall from (4.24) that if $u_n = E_\theta[C(X_n, A_n)|X_0 = i]$, then $V_{\theta,\alpha}(i)$ becomes the power series $U(\alpha)$ given in (A.20).

Similarly $J_\theta(i)$ (respectively, $J_\theta^*(i)$) is the rightmost (respectively, leftmost) term in (A.28). Then the statements in Proposition 6.1.1 follow immediately from Theorem A.4.2. □

6.2 AN OPTIMAL STATIONARY POLICY EXISTS

Throughout the rest of Chapter 6, we assume that Δ is an MDC with a *finite* state space S. For complete clarity this assumption is repeated in the hypotheses of each result.

Here is an example showing that we may have $J_\theta^* \ne J_\theta$.

Example 6.2.1. The MDC has $S = \{0, 1\}$, with $A_0 = \{a, a^*\}$ and $A_1 = \{b, b^*\}$. All transitions are deterministic and are shown in Fig. 6.1. The costs depend only on the states and satisfy $C(0) = 1$ and $C(1) = 0$.

Assume that the initial state is 0. The policy θ is constructed to realize the sequence of values in Example A.5.1. This may be done as follows: Choose a $q_1 - 1$ times, then choose a^* (giving a transition to state 1), then choose b $q_1 - 1$ times, then choose b^* (giving a transition to state 0), and so on. Then the deterministic sequence of generated costs is precisely the sequence in Example A.5.1. Under Choice One in that example, we have $J_\theta^*(0) = \frac{1}{2}$ and $J_\theta(0) = \frac{2}{3}$. □

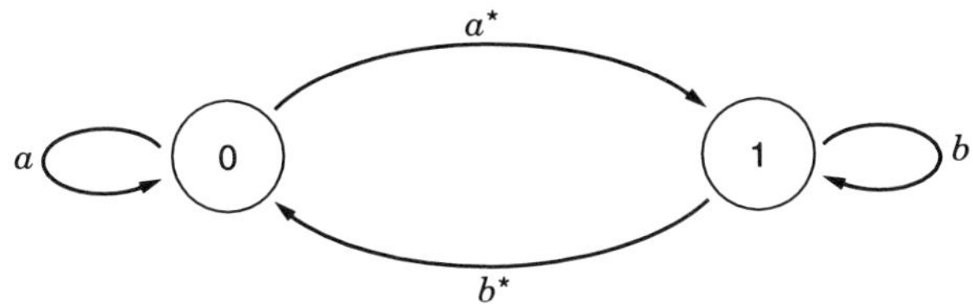

C(0) = 1 C(1) = 0

Figure 6.1 Example 6.2.1.

The next result shows that this behavior cannot occur when the policy is stationary.

Proposition 6.2.2. Let e be a stationary policy in an MDC with a finite state space S. Then

$$\begin{aligned} J_e(i) &= \lim_{\alpha \to 1^-} (1-\alpha)V_{e,\alpha}(i) \\ &= \lim_{n \to \infty} \frac{v_{e,n}(i)}{n}, \qquad i \in S. \end{aligned} \tag{6.2}$$

Proof: Since the state space is finite, there exists a (finite) upper bound B on all the costs. This readily implies that $(1-\alpha)V_{e,\alpha}(i) \le B$ and $J_e(i) \le B$ (show it!).

Now fix the initial state i and suppress it in the rest of the proof. From Proposition 4.5.3 it follows that $V_{e,\alpha}$ is a finite continuous rational function of $\alpha \in (0,1)$. Hence $(1-\alpha)V_{e,\alpha}$ has the same properties. A rational function can have at most a finite number of critical points and inflection points. (You are asked to show this in Problem 6.1). This means that a rational function cannot oscillate an infinite number of times, and hence that left (or right) limits must exist, although they may be infinite. Because $(1-\alpha)V_{e,\alpha}$ is bounded, the limit as $\alpha \to 1^-$ exists and is finite. Then (6.2) follows immediately from Proposition 6.1.1. □

Here is the major result of this section, showing the existence of an average cost optimal stationary policy of a very special type.

Proposition 6.2.3. Let Δ be an MDC with a finite state space S. Then the following hold:

(i) There exist $\alpha_0 \in (0,1)$ and a stationary policy f such that f is α discount optimal for $\alpha \in (\alpha_0, 1)$.

(ii) The policy f is average cost optimal.

(iii) We have

$$J(i) = \lim_{\alpha \to 1^-} (1 - \alpha)V_\alpha(i)$$

$$= \lim_{n \to \infty} \frac{v_{f,n}(i)}{n}, \qquad i \in S. \tag{6.3}$$

Proof: Since S and each action set are finite, it is the case that there are only a finite number of stationary policies for Δ. (Problem 6.2 asks you to given an expression for the number of stationary policies.) Associated with each $\alpha \in (0, 1)$ is an α discount optimal stationary policy. Because the number of stationary policies is finite, there must exist a sequence $\alpha_n \to 1^-$ and a stationary policy f such that f is α_n optimal for all n.

We claim that (i) holds for f. This is proved by contradiction. Suppose that it fails. Then there exists a sequence $\beta_n \to 1^-$ such that f is not β_n optimal. Because the state space is finite, this implies that there exist $i_0 \in S$ and a subsequence of β_n (call it δ_n for convenience) such that $V_{\delta_n}(i_0) < V_{f,\delta_n}(i_0)$, for all n.

By the same argument used to obtain f, we obtain a subsequence of δ_n (call if γ_n for convenience) and a stationary policy e such that e is γ_n optimal for all n.

We have the following situation. There are sequences of discount factors α_n and γ_n converging to 1, and stationary policies f and e such that

$$V_{f,\alpha_n}(i_0) \le V_{e,\alpha_n}(i_0),$$
$$V_{e,\gamma_n}(i_0) < V_{f,\gamma_n}(i_0). \tag{6.4}$$

This requires the function $V_{e,\alpha}(i_0)$ to dip below the function $V_{f,\alpha}(i_0)$ for infinitely many values but to be equal to or greater than it for infinitely many values. It follows from Proposition 4.5.3 that both $V_{f,\alpha}(i_0)$ and $V_{e,\alpha}(i_0)$ are (finite) continuous rational functions of $\alpha \in (0, 1)$. Such functions cannot exhibit the behavior in (6.4). (To see this, it helps to draw a picture of this behavior.) Thus (i) must hold for the policy f.

We now show that f is average cost optimal. Let i be an arbitrary initial state. It follows from (i) that $(1 - \alpha)V_{f,\alpha}(i) = (1 - \alpha)V_\alpha(i)$ for $\alpha \in (\alpha_0, 1)$. By Proposition 6.2.2 the limit of the quantity on the left exists and equals $J_f(i)$, and hence so does the limit of the quantity on the right.

New let θ be an arbitrary policy. From Proposition 6.1.1 and this argument, it follows that

$$J_f(i) = \lim_{\alpha \to 1^-} (1 - \alpha)V_\alpha(i) \le \limsup_{\alpha \to 1^-} (1 - \alpha)V_{\theta,\alpha}(i) \le J_\theta(i). \tag{6.5}$$

This proves that $J(i) = J_f(i)$. Equation (6.3) follows from (6.2). □

The conclusion of this section is the existence of a stationary policy that is discount optimal on an interval $(\alpha_0, 1)$ and also average cost optimal. (Such a policy is said to be *Blackwell optimal*.) In Example 6.2.1 the stationary policy f with $f(0) = a^*$ and $f(1) = b$ is average cost optimal with $J_f \equiv 0$.

6.3 AN AVERAGE COST OPTIMALITY EQUATION

In this section we construct an average cost optimality equation (ACOE) for the optimal policy found in Section 6.2.

Let f and α_0 be as in Proposition 6.2.3. The stationary policy f induces a Markov chain with costs. The cost at i is $C(i,f) = C(i,f(i))$ and the probability of transitioning from i to j is $P_{ij}(f) = P_{ij}(f(i))$.

The structure of the Markov chain induced by f is discussed in Section C.3 of Appendix C. In the general case, with S finite, this Markov chain may have multiple positive recurrent classes $R_1, R_2, \ldots, R_K$ as well as a set U of transient states. From each $i \in U$ a positive recurrent class is reached in finite expected time and with finite expected cost. Let $p_k(i)$ be the probability that class R_k is reached (first), where $\sum_k p_k(i) = 1$.

It is the case that the average cost under f is constant on R_k, and we denote it by J_k. Moreover we have $J(i) = \sum_k p_k(i)J_k$ for $i \in S$.

For $1 \le k \le K$ select a distinguished state $z_k \in R_k$, and let $Z = \cup\{z_k\}$ be the set of distinguished states. If the process starts in transient state i and reaches class R_k, then it will reach z_k in finite expected time (denoted $m_{i|k}(f)$) and with finite expected cost (denoted $c_{i|k}(f)$). Keep in mind that these quantities are conditioned on the class reached. The quantity $\sum_k p_k(i)[c_{i|k}(f) - J_k m_{i|k}(f)]$ $= c_{iZ}(f) - \sum_k J_k p_k(i) m_{i|k}(f)$ is fundamental to the development. Notice that if $i \in R_k$, then it equals $c_{iz_k}(f) - J_k m_{iz_k}(f)$ (why?).

We emphasize that all of these concepts relate to the Markov chain induced by the average cost optimal stationary policy f from Section 6.2. For details on these ideas, the reader may review Appendix C.

Now let $W_\alpha(i) =: \sum_k p_k(i) V_\alpha(z_k)$ for $i \in S$. Note that if $i \in R_k$, then $W_\alpha(i) = V_\alpha(z_k)$. Lastly define $w_\alpha(i) =: V_\alpha(i) - W_\alpha(i)$. This *relative value function* is central to the development of the ACOE.

Here is what may be proved concerning the ACOE in the general case in which the Markov chain induced by f may have multiple positive recurrent classes, with unequal values of J_k.

Theorem 6.3.1. Let the state space S be finite, and let W_α and w_α be as defined above. Then for all $i \in S$, the following hold:

(i) $J(i) = \lim_{\alpha \to 1^-} (1 - \alpha) W_\alpha(i)$.

(ii) $\operatorname{Lim}_{\alpha \to 1^-} w_\alpha(i) =: w(i) = \sum_k p_k(i)[c_{i|k}(f) - J_k m_{i|k}(f)]$.

(iii) $\operatorname{Lim}_{n \to \infty} E_f[w(X_n)|X_0 = i]/n = 0$.

(iv) The average cost optimality equation is

$$J(i) + w(i) = C(i,f) + \sum_j P_{ij}(f)w(j)$$

$$\geq \min_a \left\{ C(i,a) + \sum_j P_{ij}(a)w(j) \right\}, \qquad i \in S. \tag{6.6}$$

(v) If e is a stationary policy realizing the minimum in (6.6) and the Markov chain induced by e is positive recurrent at i, then (6.6) is an equality at i and $J_e(i) = J(i)$.

Proof: (We have not starred this proof because some of the techniques are used later. However, the interested reader need not be overly concened with all the details.)

The proof of (i) is given as Problem 6.3. We now prove (ii). First assume that $i = z_k$. Then $w_\alpha(i) \equiv 0$, and hence $w(i) = 0$. The expression in (ii) becomes $c_{z_k z_k}(f) - J_k m_{z_k z_k}(f)$, and this equals 0 by Proposition C.2.1(ii). This proves (ii) for the states in Z.

Now assume that $i \notin Z$. For $\alpha > \alpha_0$ we know that f is discount optimal. Moreover the system operating under f reaches Z in finite expected time and with finite expected cost. Let T be the time to reach Z. Suppressing the initial state $X_0 = i$, we have

$$V_\alpha(i) = E_f\left[\sum_{t=0}^{T-1} \alpha^t C(X_t,f)\right] + E_f[\alpha^T V_\alpha(X_T)]. \tag{6.7}$$

Subtract $W_\alpha(i)$ from both sides, and observe that the result may be expressed as

$$w_\alpha(i) = E_f\left[\sum_{t=0}^{T-1} \alpha^t C(X_t,f)\right] - \sum_k [(1-\alpha)V_\alpha(z_k)] \cdot \left(\frac{p_k(i) - \sum_{t=1}^{\infty} \alpha^t P_f(T = t, X_T = z_k)}{1-\alpha} \right). \tag{6.8}$$

Now let $\alpha \to 1^-$. The limit of the first term on the right of (6.8) exists and equals $c_{iZ}(f)$. In the second term the limit of the summand in square brackets equals J_k. Hence the proof will be completed if it can be shown that the summand in round brackets approaches $p_k(i)m_{i|k}(f)$.

We have

$$\frac{p_k(i) - \sum_{t=1}^{\infty} \alpha^t P_f(T = t, X_T = z_k)}{1 - \alpha}$$

$$= \sum_{t=1}^{\infty} \left(\frac{1 - \alpha^t}{1 - \alpha} \right) P_f(T = t, X_T = z_k)$$

$$= \sum_{t=1}^{\infty} (1 + \alpha + \ldots + \alpha^{t-1}) P_f(T = t, X_T = z_k)$$

$$\to \sum_{t=1}^{\infty} t P_f(T = t, X_T = z_k) \qquad \text{as } \alpha \to 1^-$$

$$= p_k(i) \sum_{t=1}^{\infty} t P_f(T = t | X_T = z_k)$$

$$= p_k(i) m_{i|k}(f), \tag{6.9}$$

where the convergence follows from Corollary A.2.4. This completes the proof of (ii).

We now prove (iv). By conditioning on the first state visited, we see that $p_k(i) = \sum_j P_{ij}(f) p_k(j)$. This holds even if i is in a positive recurrent class. Multiplying both sides of this by $V_\alpha(z_k)$ and summing yields

$$W_\alpha(i) = \sum_j P_{ij}(f) W_\alpha(j). \tag{6.10}$$

For $\alpha > \alpha_0$ the discount optimality equation (4.9) may be written as

$$(1 - \alpha) W_\alpha(i) + w_\alpha(i) = C(i, f) + \alpha \sum_j P_{ij}(f) w_\alpha(j), \qquad i \in S. \tag{6.11}$$

This follows since f is discount optimal. It is obtained from (4.9) by adding and subtracting $W_\alpha(i)$ from the left side, and by subtracting $\alpha W_\alpha(i)$ from both sides and using (6.10).

Now take the limit of both sides of (6.11) as $\alpha \to 1^-$. Since S is finite, the limit may be passed through the summation on the right. Using (i–ii) yields (6.6), and this proves (iv).

To prove (iii), we let the process operate under f and first show that

$$E_f[J(X_t) | X_0 = i] = J(i), \qquad i \in S, t \geq 0. \tag{6.12}$$

This is clearly true for $t = 0$. Now assume that $t \ge 1$. Recall that $p_k(i) = \sum_j P_{ij}(f)p_k(j)$. Iterating this, we see that $p_k(i) = \sum_j P_{ij}^{(t)}(f)p_k(j)$. Using this and the fact that $J(j) = \sum_k p_k(j)J_k$, it follows that

$$\begin{aligned} E_f[J(X_t)|X_0 = i] &= \sum_j P_{ij}^{(t)}(f)J(j) \\ &= \sum_k J_k\left(\sum_j P_{ij}^{(t)}(f)p_k(j)\right) \\ &= \sum_k J_k p_k(i) \\ &= J(i). \end{aligned} \tag{6.13}$$

Suppressing the initial state $X_0 = i$, it follows from (6.6) that

$$J(X_t) + w(X_t) = C(X_t, f) + E_f[w(X_{t+1})|X_t], \qquad t \ge 0. \tag{6.14}$$

Taking the expectation of both sides of (6.14), then using a property of expectation (namely $E[E[Y|X]] = E[Y]$) together with (6.12), yields

$$J(i) + E_f[w(X_t)] = E_f[C(X_t, f)] + E_f[w(X_{t+1})], \qquad t \ge 0. \tag{6.15}$$

These expectations are all finite (why?). Move the last term of (6.15) to the left of the equality, add the terms, for $t = 0$ to $t = n - 1$, and divide by n to obtain

$$J(i) + \frac{w(i) - E_f[w(X_n)]}{n} = \frac{v_{f,n}(i)}{n}. \tag{6.16}$$

Then (iii) follows from (6.3).

To prove (v), let e be a stationary policy realizing the minimum in (6.6). By assumption we may write

$$J(i) + w(i) = C(i, e) + \Phi(i) + \sum_j P_{ij}(e)w(j), \qquad i \in S, \tag{6.17}$$

where Φ is a nonnegative *discrepancy* function; that is, its value is what must be added to the minimum in (6.6) to obtain equality.

Let R be a positive recurrent class in the Markov chain induced by e, and let $\pi_i(e)$ be the steady state probability associated with $i \in R$. Multiply both sides of (6.17) by $\pi_i(e)$, and sum over $i \in R$. From Proposition C.2.1(i) it follows that

$$\sum \pi_i(e)J(i) + \sum \pi_i(e)w(i) \\ = J_R(e) + \sum \pi_i(e)\Phi(i) + \sum_i \pi_i(e) \sum_j P_{ij}(e)w(j), \qquad (6.18)$$

where $J_R(e)$ is the constant average cost on R. Using Proposition C.1.2(i), we see that the terms involving w on each side of (6.18) are equal (and finite), and hence they cancel. Moreover we know that $J(i) \le J_e(i) = J_R(e)$. This together with the nonnegativity of the discrepancy function yields

$$J_R(e) \ge \sum \pi_i(e)J(i) = J_R(e) + \sum \pi_i(e)\Phi(i) \ge J_R(e). \qquad (6.19)$$

Hence these terms are all equal and the discrepancy function is 0. This proves that (6.6) is an equality on R.

To complete the proof, note that $\sum \pi_i(e)\,(J_R(e) - J(i)) = 0$. Since each summand is nonnegative, it follows that $J(i) \equiv J_R(e)$ and e is average cost optimal on R. □

The following example shows why it is difficult to strengthen the statement of Theorem 6.3.1 in the general case.

Example 6.3.2. Consider the MDC Δ whose transition structure is shown in Fig. 6.2. There are single actions in the states 1, 2, and 3, with $C(1) = 0$, $C(2) = 2$, and $C(3) = 1$. In state 0 there are three actions with

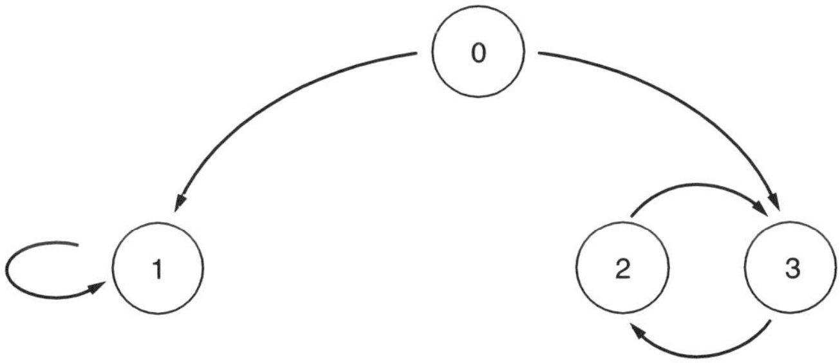

Figure 6.2 Example 6.3.2.

$$\begin{array}{lll} P_{01}(a) = P_{03}(a) = \frac{1}{2} & C(0,a) = 0 & \textit{Policy } f \\ P_{01}(b) = P_{03}(b) = \frac{1}{2} & C(0,b) = 1 & \textit{Policy } g \\ P_{01}(c) = \frac{1}{8}\ P_{03}(c) = \frac{7}{8} & C(0,c) = \frac{1}{8} & \textit{Policy } e \end{array} \tag{6.20}$$

Problem 6.4 asks you to confirm the result we give here. It is clear that $V_\alpha(1) = 0$, and we see that $V_\alpha(2) = (2+\alpha)/(1-\alpha^2)$ and $V_\alpha(3) = (2\alpha+1)/(1-\alpha^2)$.

It can be shown that f is discount optimal and that $V_\alpha(0) = \alpha(2\alpha+1)/2(1-\alpha^2)$. The chain induced by f has two positive recurrent classes. Let us choose $z_1 = 1$ and $z_2 = 2$. Then it can be shown that $w(1) = w(2) = 0$, $w(3) = -\frac{1}{2}$, and $w(0) = -1$.

It is the case that $J(1) = 0$, $J(2) = J(3) = 3/2$, and $J(0) = 3/4$. For state 0, f and g are average cost optimal but e is not. Moreover (6.6) becomes

$$\begin{aligned} \frac{3}{4} - 1 &= 0 - \frac{1}{4} \\ &\geq \min\left\{-\frac{1}{4}, \frac{3}{4}, -\frac{5}{16}\right\}, \end{aligned}$$

where the numbers in the minimum are associated with actions a, b, and c respectively. The minimum is $-5/16$, associated with nonoptimal policy e. The optimal policy g is associated with $3/4$.

This example shows that the inequality in (6.6) may be strict. Moreover a stationary policy realizing the minimum may not be optimal, and an optimal stationary policy may not realize the minimum. □

The following important result is crucial to the development in the next section:

Proposition 6.3.3. Assume that the hypotheses of Theorem 6.3.1 hold. Let $w^*(i) = w(i) - \sum_k p_k(i)\,(\sum_{s\in R_k} \pi_s(f)w(s))$. Then for $i \in S$ and $\alpha \in (0, 1)$, it follows that

$$V_\alpha(i) = \frac{J(i)}{1-\alpha} + w^*(i) + \epsilon_\alpha(i), \tag{6.21}$$

where $\epsilon_\alpha(i)$ is a function that approaches 0 as $\alpha \to 1^-$.

**Proof:* Observe that $\sum_{s\in R_k} \pi_s(f)w(s)$ is the average value of w on R_k. Hence we may regard w^* as a *normalized* version of w.

Subtracting $\sum_k p_k(i)\,(\sum_{s\in R_k} \pi_s(f)w(s))$ from both sides of (6.6) and using the fact that $p_k(i) = \sum_j P_{ij}(f)p_k(j)$, we see that

$$J(i) + w^*(i) = C(i,f) + \sum_j P_{ij}(f)w^*(j), \qquad i \in S. \tag{6.22}$$

The same argument used to derive (6.15) may be applied to (6.22), yielding

$$E_f[C(X_t,f)] = J(i) + E_f[w^*(X_t)] - E_f[w^*(X_{t+1})], \qquad t \geq 0. \tag{6.23}$$

Now multiply both sides of (6.23) by α^t and sum over t. Since f is discount optimal for $\alpha \in (\alpha_0, 1)$, this yields

$$V_\alpha(i) = \frac{J(i)}{1-\alpha} + w^*(i) - (1-\alpha)\sum_{t=0}^{\infty} \alpha^t E_f[w^*(X_{t+1})],$$
$$\alpha \in (\alpha_0, 1). \tag{6.24}$$

For $\alpha \in (0, \alpha_0]$ the function $\epsilon_\alpha(i)$ may be defined to realize an equality in (6.21). For $\alpha \in (\alpha_0, 1)$ it is defined as the last term in (6.24). In this case it may be expressed as

$$\epsilon_\alpha(i) = \frac{(1-\alpha)w^*(i)}{\alpha} - \frac{1-\alpha}{\alpha}\sum_{t=0}^{\infty} \alpha^t E_f[w^*(X_t)]. \tag{6.25}$$

The first term of (6.25) approaches 0 as $\alpha \to 1^-$. Focus on the second term, and ignore the α in the denominator. What remains is $(1-\alpha)$ times the expected discounted w^* *cost* over the infinite horizon, for initial state i and under policy f. Consider for a moment the average w^* *cost*. Using results from Appendix C, this is obtained as a limit and equals $\sum_k p_k(i)\,(\sum_{s\in R_k} \pi_s(f)w^*(s))$. Using the definition of w^*, this is easily seen to be 0 (check it out!).

The desired result then follows from (A.28) in Appendix A. *Note:* This result was proved for nonnegative terms and the function w^* may be negative. However, it is bounded below, say by $-L$, and we may apply the theorem to $w^* + L \geq 0$, yielding the desired result. □

6.4 ACOE FOR CONSTANT MINIMUM AVERAGE COST

The situation illustrated in Example 6.3.2 is rather abnormal. The more typical and important situation is that in which the minimum average cost is constant. When $J(i) \equiv J$, it follows that the average cost is independent of the initial state, a property that holds in many models. The next result presents conditions for this to hold. Recall from Section C.3 of Appendix C that a Markov chain with a finite state space is unichain if it has a single positive recurrent class.

Proposition 6.4.1. Let the hypotheses be as in Theorem 6.3.1. Consider the following statements:

(i) Every stationary policy induces a unichain Markov chain.

(ii) The optimal stationary policy f induces a unichain Markov chain.

(iii) There exist $z \in S$ and a (finite) constant L such that $|V_\alpha(i) - V_\alpha(z)| \le L$ for $i \in S$ and $\alpha \in (0, 1)$.

(iv) Given $x \in S$, there exists a (finite) constant L such that $|V_\alpha(i) - V_\alpha(x)| \le L$ for $i \in S$ and $\alpha \in (0, 1)$.

(v) Given states $i \neq j$, there exists a stationary policy $e(i,j)$ such that i leads to j in the Markov chain induced by $e(i,j)$.

(*) We have $J(i) \equiv J$ for $i \in S$.

Then

$$\begin{array}{ccc} \text{(i)} & & \\ \Downarrow & & \\ \text{(ii)} & & \text{(v)} \\ \downarrow & & \Downarrow \\ \text{(iii)} & \Leftrightarrow \text{(iv)} \Leftrightarrow & (^*) \end{array} \tag{6.26}$$

Proof: We will first show that (i) $\Rightarrow$ (ii) $\Rightarrow$ (*) $\Rightarrow$ (iv) $\Rightarrow$ (iii) $\Rightarrow$ (*). It is clear that (i) implies (ii). If f induces a chain with a single positive recurrent class R then the minimum average cost is constant, and hence (*) holds.

We now show that (*) implies (iv). Fix $x \in S$. It follows from (6.21) and (*) that $|V_\alpha(i) - V_\alpha(x)| \le |w^*(i)| + |w^*(x)| + |\epsilon_\alpha(i)| + |\epsilon_\alpha(x)|$. Since S is finite there exists a bound Q on the absolute values of w^*.

Fix a state i. For $\alpha \in (0, \alpha_0]$ it is easy to see from (6.21) that $\epsilon_\alpha(i)$ is bounded. For $\alpha \in (\alpha_0, 1)$ it may be seen from (6.25) that it is bounded. Since S is finite, there exists a (finite) constant E such that $|\epsilon_\alpha(i)| \le E$ for $i \in S$ and $\alpha \in (0, 1)$. It follows that $2\,(Q + E)$ will serve as the desired bound, and hence (iv) holds.

Clearly (iv) implies (iii). We now show that (iii) implies (*). Observe that $(1-\alpha)\,V_\alpha(i) = (1-\alpha)\,(V_\alpha(i) - V_\alpha(z)) + (1-\alpha)V_\alpha(z)$. From Proposition 6.2.3(iii) it follows that the left side approaches $J(i)$, and the last term approaches $J(z)$. From (iii) it follows that the second term approaches 0. Hence $J(i) = J(z)$ for every i, and hence (*) holds.

It remains to prove that (v) implies (*). Recall that J_k is the average cost on the positive recurrent class R_k under the average cost optimal stationary policy f, where $1 \le k \le K$. Let $J_{\min}$ be the smallest value and $J_{\max}$ the largest value. We know that any $J(i)$ is a convex combination of the values of J_k. If it can be shown that $J_{\min} = J_{\max}$, then (*) will hold.

Choose and fix an element i^* of the positive recurrent class associated with $J_{\max}$ and an element j^* of the positive recurrent class associated with $J_{\min}$. Let

e denote the stationary policy $e(i^*, j^*)$ given in (v). By assumption, there exists $n \geq 1$ such that $P^{(n)}_{i*j*}(e) > 0$.

It follows from (4.9) that

$$
\begin{aligned}
V_\alpha(i) &\leq C(i,e) + \alpha \sum_j P_{ij}(e) V_\alpha(j) \\
&\leq C(i,e) + \sum_j P_{ij}(e) V_\alpha(j), \qquad i \in S. \qquad (6.27)
\end{aligned}
$$

Iterating (6.27) $n - 1$ times yields

$$V_\alpha(i) \leq v_{e,n}(i) + \sum_j P^{(n)}_{ij}(e) V_\alpha(j). \qquad (6.28)$$

We now let $i = i^*$ and multiply both sides of (6.28) by $1 - \alpha$. This yields

$$(1-\alpha)V_\alpha(i*) \leq (1-\alpha) v_{e,n}(i^*) + \sum_j P^{(n)}_{i*j}(e)[(1-\alpha)V_\alpha(j)]. \qquad (6.29)$$

Now let $\alpha \to 1^-$. The term on the left of (6.29) approaches $J_{\max}$, and the first term on the right approaches 0. It follows from (6.3) that $\lim_{\alpha \to 1^-}(1-\alpha)V_\alpha(.) = J(.) \leq J_{\max}$. This yields

$$J_{\max} \leq P^{(n)}_{i*j*}(e) J_{\min} + (1 - P^{(n)}_{i*j*}(e)) J_{\max}. \qquad (6.30)$$

Since the coefficient of $J_{\min}$ is positive, this leads to a contradiction unless $J_{\min} = J_{\max}$. This proves that (*) holds. □

Observe that (v) is a particularly useful condition that holds in many models and is easily checked.

Proposition 6.4.1 has given us conditions under which $J(i) \equiv J$. Under the single assumption that $J(i) \equiv J$, the following result derives a strengthened form of the average cost optimality equation (ACOE) for Δ.

Theorem 6.4.2. Assume that $J(i) \equiv J$. Fix a distinguished state z, let $h_\alpha(i) =: V_\alpha(i) - V_\alpha(z)$, and let L be a bound for $|h_\alpha|$ as in Proposition 6.4.1.

(i) For $i \in S$ we have $\lim_{\alpha \to 1^-} h_\alpha(i) =: h(i) = w^*(i) - w^*(z)$.

(ii) The average cost optimality equation is

$$J + h(i) = \min_a \left\{ C(i,a) + \sum_j P_{ij}(a)h(j) \right\}. \qquad i \in S, \tag{6.31}$$

and the optimal stationary policy f realizes the minimum.

(iii) If e is a stationary policy realizing the minimum in (6.31), then e is average cost optimal. Moreover $\lim_{n\to\infty} E_e[h(X_n)|X_0 = i]/n = 0$.

(iv) Define $d_n(i) =: h(i) + nJ - v_n(i)$ for $i \in S$ and $n \geq 0$. (Recall that v_n is the minimum n horizon expected cost for a terminal cost of 0.) Then $|d_n| \leq L$.

(v) For $i \in S$ we have $\lim_{n\to\infty} v_n(i)/n = J$.

(vi) If f is unichain and z is the distinguished state in R_f from Section 6.3, then $h(i) = w(i) = c_{iz}(f) - Jm_{iz}(f)$ for $i \in S$.

Proof: It follows from the assumption of constant minimum average cost and (6.21) that $h_\alpha(i) = w^*(i) + \epsilon_\alpha(i) - w^*(z) - \epsilon_\alpha(z)$. Then (i) follows from Proposition 6.3.3. Note that $|h| \leq L$.

To prove (ii), observe that the discount optimality equation (4.9) may be written as

$$(1-\alpha)V_\alpha(z) + h_\alpha(i) = \min_a \left\{ C(i,a) + \alpha \sum_j P_{ij}(a)h_\alpha(j) \right\},$$
$$i \in S, \tag{6.32}$$

where f realizes the minimum for $\alpha > \alpha_0$. Equation (6.32) follows from (4.9) by adding and subtracting $V_\alpha(z)$ from the left side and subtracting $\alpha V_\alpha(z)$ from both sides.

We may then let $\alpha \to 1^-$ in (6.32). Using (i), (6.3), and the assumption of constant minimum average cost, and finally Proposition A.1.3(ii) and the fact that the summation is over a finite set, we obtain (6.31), and this proves (ii).

Now let e realize the minimum in (6.31). In a manner similar to the derivation of (6.16), we obtain

$$\frac{v_{e,n}(i)}{n} = J + \frac{h(i) - E_e[h(X_n)|X_0 = i]}{n}$$
$$\leq J + \frac{h(i) + L}{n}. \tag{6.33}$$

Using (6.2), we may take the limit of both sides of the inequality in (6.33) to obtain $J_e(i) \leq J$. But this implies that $J_e(i) \equiv J$, and hence e is average cost optimal. Taking the limit of the equality in (6.33) and using the optimality of e yields the second claim in (iii).

Let d_n be as in (iv), and observe that the bound holds for $n = 0$. Now assume that $n \geq 1$. We have $v_n \leq v_{e,n}$ where e is as in (6.33). Using this and multiplying the inequality in (6.33) through by n proves that $d_n \geq -L$.

We now obtain the upper bound. Let θ_n be the n horizon optimal policy. We may operate the process under θ_n for n steps and then switch to a discount optimal policy. Suppressing the initial state $X_0 = i$, this yields

$$\begin{aligned} V_\alpha(i) &\leq v_{\theta_n,\alpha}(i) + \alpha^n E_{\theta_n}[V_\alpha(X_n)] \\ &\leq v_n(i) + \alpha^n E_{\theta_n}[V_\alpha(X_n)]. \end{aligned} \tag{6.34}$$

The second line follows since an expected discounted n horizon cost under a policy is bounded above by the expected n horizon cost under the same policy, and this policy is n horizon optimal. We now subtract $V_\alpha(z)$ from both sides of (6.34) and add and subtract $\alpha^n V_\alpha(z)$ to the right side to obtain

$$\begin{aligned} h_\alpha(i) &\leq v_n(i) + \alpha^n E_{\theta_n}[h_\alpha(X_n)] - \left(\frac{1-\alpha^n}{1-\alpha}\right)[(1-\alpha)V_\alpha(z)] \\ &\leq v_n(i) + L - \left(\frac{1-\alpha^n}{1-\alpha}\right)[(1-\alpha)V_\alpha(z)]. \end{aligned} \tag{6.35}$$

Now let $\alpha \to 1^-$. It follows from (i), the assumption of constant minimum average cost and (6.3), that $h(i) \leq v_n(i) + L - nJ$. This yields $d_n \leq L$ and proves (iv).

Now write $-L \leq d_n \leq L$, then divide through by n, and pass to the limit to obtain $d_n/n \to 0$. Using this and the definition of d_n yields (v).

It follows from the assumptions in (vi) and Theorem 6.3.1(ii) that $w(i) = c_{iz}(f) - Jm_{iz}(f)$. It was also shown in the proof of that result that $w(z) = 0$. From the definition in Proposition 6.3.3, it follows that $w^*(i) = w(i) - u$, where $u = \sum_{s \in R_f} \pi_s(f) w(s)$. It then follows from (i) that $h(i) = w(i) - u - (w(z) - u) = w(i)$, and this completes the proof of (vi). □

6.5 SOLUTIONS TO THE ACOE

In this section we start with an MDC Δ with a finite state space but with no additional restrictions. We address the following question: Suppose that we have been able to find *some* constant and *some* function satisfying the ACOE, then are we assured of having found the minimum average cost and an optimal stationary policy? The answer is in the next result.

Proposition 6.5.1. Let Δ be an MDC with a finite state space S.

(i) Assume that we have a (finite) constant F and a (finite) function r such that

$$F + r(i) \geq \min_a \left\{ C(i, a) + \sum_j P_{ij}(a) r(j) \right\}, \qquad i \in S. \tag{6.36}$$

If e is a stationary policy realizing the minimum in (6.36), then $J_e(i) \leq F$ for $i \in S$. Hence, if F is a lower bound on the average costs, then the minimum average cost equals the constant F, and e is average cost optimal.

(ii) Assume that we have a (finite) constant F and a (finite) function r such that

$$F + r(i) = \min_a \left\{ C(i, a) + \sum_j P_{ij}(a) r(j) \right\}, \qquad i \in S. \tag{6.37}$$

Then the minimum average cost equals the constant F, and any stationary policy realizing the minimum in (6.37) is average cost optimal. If h is as in Theorem 6.4.2, then $r(i) = h(i) + r(z_k) - h(z_k)$ for $i \in R_k$. (Recall that this is a positive recurrent class under f with distinguished state z_k.) For i transient under f we have $r(i) \leq h(i) + \sum_k p_k(i)[r(z_k) - h(z_k)]$.

(iii) Assume that (6.37) holds, and let e be a stationary policy realizing the minimum. Assume that both e and f are unichain with common positive recurrent state x. If x is the distinguished state in Theorem 6.4.2 and in Section 6.3 and if $r(x) = 0$, then $r(i) = h(i) = c_{ix}(e) - Jm_{ix}(e)$ for all i.

Proof: To prove (i), assume that (6.36) holds, and let e be a stationary policy realizing the minimum. Using reasoning similar to that in (6.14–16), we obtain for initial state i that

$$\begin{aligned} \frac{v_{e,n}(i)}{n} &\leq F + \frac{r(i) - E_e[r(X_n)]}{n} \\ &\leq F + \frac{r(i) + M}{n}, \end{aligned} \tag{6.38}$$

where $-M$ is a lower bound for r. Taking the limit of both sides yields that $J_e(i) \leq F$. If F is a lower bound on the average costs, then $J(i) \leq J_e(i) \leq F \leq J(i)$. Hence we have $J(i) = J_e(i) \equiv F$. This completes the proof of (i).

If (6.37) holds, then it follows that

$$F + r(i) \le C(i,f) + \sum_j P_{ij}(f)r(j), \qquad i \in S. \tag{6.39}$$

Using similar reasoning as above, we obtain

$$\begin{aligned}\frac{v_{f,n}(i)}{n} &\ge F + \frac{r(i) - E_f[r(X_n)]}{n}\\ &\ge F + \frac{r(i) - M}{n},\end{aligned} \tag{6.40}$$

where M is an upper bound for r. Taking the limit of both sides and using the optimality of f yields $F \le J(i)$. Hence the first statement in (ii) follows from the second statement in (i).

We now have for all i that

$$\begin{aligned}J + r(i) &\le C(i,f) + \sum_j P_{ij}(f)r(j),\\ J + h(i) &= C(i,f) + \sum_j P_{ij}(f)h(j),\end{aligned} \tag{6.41}$$

where the second equation follows from Theorem 6.4.2(ii), and $F = J$ is the minimum average cost. Let $b = r - h$. Subtracting the second equation from the first in (6.41) yields

$$b(i) \le \sum_j P_{ij}(f)b(j), \qquad i \in S. \tag{6.42}$$

Iterating (6.42), then adding the terms and dividing by n, we see that

$$b(i) \le \sum_j b(j)\left(\frac{1}{n}\sum_{t=1}^{n} P_{ij}^{(t)}(f)\right), \qquad i \in S. \tag{6.43}$$

We know from Section C.1 of Appendix C that the limit of the quantity in round brackets exists. If j is transient, then the limit is 0, whereas if $j \in R_k$, then the limit is $p_k(i)\pi_j(f)$.

First assume that $i \in R_k$, and let $n \to \infty$ in (6.43) to obtain

$$b(i) \leq \sum_{j \in R_k} b(j)\pi_j(f), \qquad i \in R_k. \tag{6.44}$$

This says that every value of the function is less than or equal to a convex combination of the values. This is possible if and only if the function b is a constant B_k. This implies that $B_k = b(z_k) = r(z_k) - h(z_k)$. Hence $r(i) = h(i) + B_k = h(i) + r(z_k) - h(z_k)$.

Now assume that i is transient under f. Let $n \to \infty$ in (6.43), and use the fact that $b \equiv B_k$ on R_k to obtain

$$b(i) \leq \sum_k p_k(i)B_k. \tag{6.45}$$

From (6.45) it is easy to see that the last statement holds. This completes the proof of (ii).

Assume that the hypotheses in (iii) hold. Note that $h(x) = r(x) = 0$ by assumption, so we may assume that $i \neq x$. It follows from (ii) that $r(i) \leq h(i) + r(x) - h(x) = h(i)$.

It follows from (6.31) that $J + h(i) \leq C(i, e) + \sum_j P_{ij}(e)h(j)$ for all i. Using reasoning similar to that in (6.14–16), we obtain for $X_0 = i$,

$$E_e\left[\sum_{t=0}^{k-1} C(X_t, e)\right] - Jk \geq h(i) - E_e[h(X_k)]. \tag{6.46}$$

If T is the first passage time from i to x under e, then $m_{ix}(e) = \sum kP_e(T = k) < \infty$ by assumption. Let us multiply each term of (6.46) by $P_e(T = k)$ and sum over $k \geq 1$. This yields

$$c_{ix}(e) - Jm_{ix}(e) \geq h(i) - h(x) = h(i). \tag{6.47}$$

From (6.37) it follows that $J + r(i) = C(i, e) + \sum_j P_{ij}(e)r(j)$ for all i. Repeating the above argument on this equation yields

$$c_{ix}(e) - Jm_{ix}(e) = r(i) - r(x) = r(i). \tag{6.48}$$

The result now follows from (6.47–48) and the fact that $r \leq h$. This completes the proof of (iii). □

6.6 METHOD OF CALCULATION

Sections 6.2 through 6.5 give us a good theoretical understanding of the nature of solutions to the ACOE. In this section we discuss how a solution may be computed.

To compute a solution to (6.37), it is easier to work with finite horizon value iteration than with the infinite horizon discounted value function. For this reason we seek to construct a solution based on the finite horizon value function v_n rather than on the function h_α from Theorem 6.4.2.

Before proving any results, we discuss the plan of action. Let x be a distinguished state of S (to be specified later), and let us define the finite horizon relative value function $r_n(i) =: v_n(i) - v_n(x)$, where the terminal cost of the finite horizon value iteration is 0. The finite horizon optimality equation (3.2) may be written.

$$[v_n(x) - v_{n-1}(x)] + r_n(i) = \min_a \left\{ C(i,a) + \sum_j P_{ij}(a) r_{n-1}(j) \right\}, \quad n \geq 1, i \in S. \tag{6.49}$$

This is obtained by subtracting $v_{n-1}(x)$ from both sides of (3.2) and adding and subtracting $v_n(x)$ from the left side.

Suppose that it could be shown that the term in brackets approached a number F. Suppose in addition that we knew that r_n converged to some function r. Then taking the limit of both sides of (6.49) would yield a solution to (6.37) and hence J and an average optimal policy.

Carrying out this plan will require an additional assumption. To see why, consider the following example:

Example 6.6.1. Let $S = \{0, 1\}$ with a single action in each state. Let $C(0) = 0$, $C(1) = 1$, and $P_{01} = P_{10} = 1$. Let $x = 1$.

One can easily show that $v_n(1)$ equals $0.5n$ for n even and equals $0.5\ (n + 1)$ for n odd. Then $v_n(1) - v_{n-1}(1)$ equals 0 for n even, and equals 1 for n odd. Hence the first term in (6.49) has no limit, and the program cannot be carried out. □

Example 6.6.1 is a positive recurrent Markov chain with *period* 2. If the chain begins in state 0, then it can only return to state 0 at $t = 2, 4, 6, \ldots$. The concept of an aperiodic positive recurrent class, introduced in Section C.1 of Appendix C, rules out this behavior. Here is a lemma related to this notion.

Lemma 6.6.2. Assume that the minimum average cost is a constant J, and let d_n be as in Theorem 6.4.2(iv). Assume that e is an optimal stationary policy inducing an MC with an aperiodic positive recurrent class R. Then there exists a (finite) constant D such that $\lim_{n\to\infty} d_n(i) = D$ for $i \in R$.

**Proof:* We first show that

$$\sum_j P_{ij}(e)d_{n-1}(j) \le d_n(i), \qquad n \ge 1, i \in R. \tag{6.50}$$

From (3.2) it follows that $v_n(i) \le C(i,e) + \sum_j P_{ij}(e)v_{n-1}(j)$. It follows from (6.31) that $J + h(i) \le C(i,e) + \sum_j P_{ij}(e)h(j)$. Since e is optimal, it is easy to see that we must have equality at $i \in R$ (a proof similar to that in Theorem 6.3.1(v) will work). Hence we have $J + h(i) = C(i,e) + \sum_j P_{ij}(e)h(j)$ for $i \in R$. Subtracting the above inequality from this yields

$$J + h(i) - v_n(i) \ge \sum_j P_{ij}(e)[h(j) - v_{n-1}(j)], \qquad i \in R. \tag{6.51}$$

Then adding $(n-1)J$ to both sides of (6.51) yields (6.50).

Let us omit notational reference to the policy e in the remainder of the proof. It follows from Theorem 6.4.2(iv) that d_n is bounded. From Proposition B.6 it follows that there exist a subsequence n_k and a function d, with $-L \le d \le L$, such that $d_{n_k} \to d$.

We fix n and iterate (6.50) m times to obtain

$$\sum_j P_{ij}^{(m)} d_n(j) \le d_{n+m}(i), \qquad n \ge 0, m \ge 1, i \in R. \tag{6.52}$$

Now hold n fixed, and let $m \to \infty$ through values such that $n+m$ are members of the sequence n_k. Using the aperiodicity of R yields

$$\sum_{j\in R} \pi_j d_n(j) \le d(i), \qquad n \ge 0, i \in R. \tag{6.53}$$

Now let $n = n_k \to \infty$ in (6.53). This yields

$$\sum_{j\in R} \pi_j d(j) \le d(i), \qquad i \in R. \tag{6.54}$$

By the same argument given regarding (6.44), it follows that $d(i) \equiv D$ for $i \in R$.

Suppose that we had another subsequence n_u giving rise to a function g, equal to a constant G on R. Letting $n = n_u \to \infty$ in (6.53) yields

$$G = \sum_{j \in R} \pi_j g(j) \le D. \tag{6.55}$$

Because this argument could have been reversed, we must have $G = D$. This proves that $\lim_{n \to \infty} d_n(i) = D$ for $i \in R$. □

Here is the assumption that will allow the plan to be carried out.

Assumption OPA. Let e be an optimal stationary policy. Then every positive recurrent class in the MC induced by e is aperiodic. □

Note that OPA stands for *Optimal Policies are Aperiodic*, and it is used here to remind us that under this assumption every optimal stationary policy can have only aperiodic positive recurrent classes. The next result shows that under Assumption OPA the program may be carried out. Proposition 6.6.6 shows how to carry out the program when Assumption OPA fails.

Proposition 6.6.3. Assume that the minimum average cost is a constant J and that Assumption OPA holds. Let x be any distinguished state in S. Then $\lim_{n \to \infty}[v_n(x) - v_{n-1}(x)] = J$ and $\lim_{n \to \infty} r_n(i) =: r(i)$ exists. Hence (6.49) may be used to compute a solution to the ACOE (6.37). Any limit point of the finite horizon optimal stationary policies f_n realizes the minimum in (6.37) and hence is average cost optimal.

Proof: Let f be the optimal stationary policy from Section 6.2. It must have at least one positive recurrent class. Let R be one of the classes, and assume first that $x \in R$. At the end of the proof, we will argue that x may be chosen arbitrarily.

The relationships

$$\begin{aligned} r_n(i) &= h(i) - h(x) - d_n(i) + d_n(x), \\ v_n(x) - v_{n-1}(x) &= J - d_n(x) + d_{n-1}(x), \end{aligned} \tag{6.56}$$

enable us to translate results about d_n into results about the quantities in (6.49).

It follows from Lemma 6.6.2 and the fact that $x \in R$ that there exists a constant D such that $d_n(x) \to D$. The second equation in (6.56) then yields $v_n(x) - v_{n-1}(x) \to J$.

Recall from Theorem 6.4.2(iv) that d_n is bounded. It then follows from (6.56) that r_n is bounded. Let r_* (respectively, r^*) be the limit infimum (respectively, limit supremum) of r_n.

Take the limit infimum of both sides of (6.49) to obtain

$$J + r_*(i) \geq \min_a \left\{ C(i,a) + \sum_j P_{ij}(a) r_*(j) \right\}, \qquad i \in S. \tag{6.57}$$

Let e be a stationary policy realizing the minimum in (6.57). It follows from Proposition 6.5.1(i) that e is optimal.

Returning to (6.49), we see that

$$[v_n(x) - v_{n-1}(x)] + r_n(i) \leq C(i,e) + \sum_j P_{ij}(e) r_{n-1}(j), \qquad i \in S. \tag{6.58}$$

Taking the limit supremum of both sides of (6.58) yields

$$J + r^*(i) \leq C(i,e) + \sum_j P_{ij}(e) r^*(j), \qquad i \in S. \tag{6.59}$$

Using (6.59) and the fact that e realizes the minimum in (6.57) yields

$$C(i,e) - J + \sum_j P_{ij}(e) r_*(j) \leq r_*(i) \leq r^*(i) \leq C(i,e) - J + \sum_j P_{ij}(e) r^*(j). \tag{6.60}$$

Let $s = r^* - r_*$, and note that $s \geq 0$. It follows from (6.60) that

$$s(i) \leq \sum_j P_{ij}(e) s(j), \qquad i \in S. \tag{6.61}$$

Let $U_1, U_2, \ldots, U_M$ be the positive recurrent classes under e, and let $q_m(i)$ be the probability of reaching U_m from state i under e. As in (6.42–44) we obtain

$$s(i) \leq \sum_m q_m(i) \left(\sum_{j \in U_m} \pi_j(e) s(j) \right), \qquad i \in S. \tag{6.62}$$

We claim that the right side of (6.62) is 0. Recall that $d_n(x) \to D$. Fix attention on some U_m. It follows from Lemma 6.6.2 and Assumption OPA that there exists a constant E such that $d_n(j) \to E$ for $j \in U_m$. Then from (6.56) it follows that $r_n(j) \to h(j) - h(x) - E + D$. Since the limit exists, it follows that $s(j) = 0$. This proves that the right side of (6.62) is 0. Since $s \geq 0$, it follows that $s \equiv 0$. This means that $r(i) =: \lim_{n \to \infty} r_n(i)$ exists for $i \in S$. This completes the proof of the first statement when the distinguished state is chosen in a positive recurrent class under f.

Now suppose that the distinguished state is chosen to be some arbitrary state y. In this case $v_n(y) - v_{n-1}(y) = r_n(y) + [v_n(x) - v_{n-1}(x)] - r_{n-1}(y)$. Using what has been proved above shows that $\lim_{n\to\infty}[v_n(y) - v_{n-1}(y)] = r(y) + J - r(y) = J$. Similarly $v_n(i) - v_n(y) = r_n(i) - r_n(y)$, and hence $\lim_{n\to\infty}(v_n(i) - v_n(y)) = r(i) - r(y)$. It remains to prove the last statement. Let e be a limit point of f_n. Then there exists a subsequence n_k such that $f_{n_k} \to e$. For a fixed i and n_k sufficiently large, we have

$$[v_{n_k}(x) - v_{n_k-1}(x)] + r_{n_k}(i) = C(i,e) + \sum_j P_{ij}(e) r_{n_k-1}(j), \tag{6.63}$$

and e realizes the minimum in (6.49) for i. Passing to the limit and using what has been proved, we see that e realizes the minimum in (6.37) for the fixed i. Since this argument may be carried out for each i, it follows that e realizes the minimum in (6.37). By Proposition 6.5.1(ii) it follows that e is average cost optimal. □

One may wonder how Assumption OPA can be verified without already knowing the optimal stationary policies. In practice it will be shown that Assumption OPA holds for all stationary policies. In certain cases it might be possible to argue that the optimal stationary policies fall within a class of stationary policies each member of which satisfies Assumption OPA. If there is any doubt that Assumption OPA holds, then the method to be presented in Proposition 6.6.6 should be used.

The above development allows us to give the following *value iteration algorithm* (VIA) for obtaining a solution to (6.37).

Value Iteration Algorithm 6.6.4. Let Δ be an MDC with a finite state space. Assume that the minimum average cost is constant and that Assumption OPA holds. Let x be a distinguished state and ϵ a small positive number.

VIA Version 1:

1. Set $n = 0$ and $u_0 \equiv 0$.
2. Set $w_n(i) = \min_a\{C(i,a) + \sum_j P_{ij}(a)u_n(j)\}$.
3. If $n = 0$, set $\delta = 1$. If $n \geq 1$, then set $\delta = |w_n(x) - w_{n-1}(x)|$. If $\delta < \epsilon$, go to 6.
4. Set $u_{n+1}(i) = w_n(i) - w_n(x)$.
5. Go to 2, and replace n by $n + 1$.
6. Print $w_n(x)$ and a stationary policy realizing $\min_a\{C(i,a) + \sum_j P_{ij}(a)u_n(j)\}$.

VIA Version 2:

Follow Version 1, but set $\delta = \max_{i \in S} |w_n(i) - w_{n-1}(i)|$ in Step 3.

Justification: One can show, by induction on n, that $u_n = r_n$ and $w_n = r_{n+1} + v_{n+1}(x) - v_n(x)$ for $n \geq 0$. Hence $w_n(x) = v_{n+1}(x) - v_n(x)$. (You are asked to verify these claims in Problem 6.5.) The validity of the VIA then follows from Proposition 6.6.3. □

Since we know that $w_n(x) \to J$, it will be a good approximation to J for n large. We know that u_n converges to some function r. A stationary policy realizing the minimum in Step 6 is optimal for the $n + 1$ horizon problem at time 0. We have seen that any limit point of this sequence of policies is average cost optimal. Hence this policy will be close to optimal for large n.

Note that there are two versions of the VIA. The more stringent Version 2 is suitable when the state space of the model is naturally finite and one is applying the VIA a single time to compute an optimal policy. In Chapter 8 an approximating sequence method is developed for the computation of an optimal stationary policy when the state space is denumerable. This method uses the VIA for a sequence of finite state MDCs with increasing state spaces. In this case it is our opinion that Version 1 works well. The reason is that the VIA will be executed several times for increasing state spaces in order to be confident that a good approximation to an optimal policy for the original MDC with a denumerably infinite state space has been obtained. For a particular finite state space approximation, it is therefore less important to be sure that we are very close to the minimum average cost for that single approximation. Of course Version 2 could also be applied in this case; however, it would increase the computation time.

We now discuss an approach to take when Assumption OPA fails (or we suspect it may fail). This involves a transformation of the MDC which causes Assumption OPA to be satisfied for the transformed MDC. (In fact in the transformed MDC every stationary policy has only aperiodic positive recurrent classes.)

Quantities in the transformed MDC will be superscripted with an asterisk. First fix a number τ with $0 < \tau < 1$. The state space and action sets of Δ^* are the same as those of Δ. The costs are given by $C^*(i, a) = \tau C(i, a)$. The transition probabilities are given by

$$\begin{aligned} P^*_{ij}(a) &= \tau P_{ij}(a), \qquad j \neq i, \\ P^*_{ii}(a) &= \tau P_{ii}(a) + (1 - \tau). \end{aligned} \tag{6.64}$$

A stationary policy induces an MC in both Δ and Δ^*, and the next result relates properties of these two Markov chains.

Lemma 6.6.5. For a fixed stationary policy, the following properties hold for the Markov chains induced by the policy:

(i) The communicating classes are identical in Δ and Δ^*, and hence the positive recurrent classes are identical.

(ii) Every positive recurrent class in Δ^* is aperiodic.

(iii) For a positive recurrent class R we have $\pi_j^* = \pi_j$ for $j \in R$, and $J_R^* = \tau J_R$.

Proof: Let MC denote the Markov chain (with costs) induced in Δ, and let MC* be the one induced in Δ^*.

It is clear that i leads to j in MC if and only if i leads to j in MC*, and thus the communicating classes are identical. If a class R is positive recurrent in MC and transient in MC*, then this readily leads to a contradiction (why?). Hence this (or the reverse situation) cannot happen. This proves (i).

Now let R be a positive recurrent class. From (6.64) it follows that $P_{ii}^* > 0$ for $i \in R$, and hence R is aperiodic in MC*. This proves (ii).

For positive recurrent class R it is easy to see that $(\pi_j)_{j \in R}$ satisfies the steady state equation in Proposition C.1.2(i) for MC*. Hence $\pi_j^* = \pi_j$ for $j \in R$. Moreover the second statement is clear, and this proves (iii). □

Here is how the transformed MDC may be used to compute a solution to (6.37).

Proposition 6.6.6. Assume that the minimum average cost in Δ is a constant and that Δ has been transformed to yield Δ^*. Then the minimum average cost in Δ^* is a constant, and Assumption OPA holds. Hence the results of Proposition 6.6.3 are valid for Δ^* and yield J^* and a function r^*. The pair $(J^*/\tau, r^*)$ satisfies (6.37) and hence produces an optimal stationary policy for Δ.

(The essence of this result is the following: The VIA may fail for Δ if Assumption OPA fails to hold. By means of the transformation we are still able to construct a solution to (6.37) based on value iteration. However, the value iteration takes place not in Δ but in the transformed Δ^*.)

Proof: From Lemma 6.6.5 and previous results, it follows that the minimum average cost in Δ^* equals τJ and hence is constant. (You are asked to prove this in Problem 6.8.) Since Lemma 6.6.5 yields that Assumption OPA holds, we may apply Proposition 6.6.3 to produce $J^* = \tau J$ and a function r^* satisfying the ACOE for Δ^*. This yields

$$
\begin{aligned}
J^* + r^*(i) &= \min_a \left\{ C^*(i,a) + \sum_j P^*_{ij}(a) r^*(j) \right\} \\
&= \min_a \left\{ \tau C(i,a) + \tau \sum_j P_{ij}(a) r^*(j) \right\} + (1-\tau) r^*(i), \\
& \qquad i \in S. \qquad (6.65)
\end{aligned}
$$

From (6.65) we immediately obtain that $(J^*/\tau, r^*)$ is a solution of the ACOE (6.37). □

6.7 AN EXAMPLE

In this section we present a simple example to illustrate the VIA.

Example 6.7.1. Single-Server Queue with Finite Waiting Room. There is a single server and the actions are the (geometric) service rates a_1 and a_2, where $0 < a_1 < a_2 < 1$. There is a probability p of a new customer arriving in any slot, where $0 < p < 1$. The waiting room can hold at most two customers, one in service and one waiting for service. If a customer arrives to find a full waiting room, it is turned away. If a customer arrives to an empty system, then it may enter service immediately. Note that this is different from our typical assumption.

The state space is $S = \{0, 1, 2\}$, where $i \in S$ denotes the number in the system. There is a holding cost of Hi, where H is a positive constant. The service costs are $C(a)$, for $a = a_1, a_2$. In this model the action set $A = \{a_1, a_2\}$ is available in every state. In state 0 the server chooses a service rate in anticipation of an arriving customer (if any). In each slot there is the opportunity to choose anew a service rate.

It is clear that every stationary policy induces an irreducible aperiodic MC on S, and hence the assumptions of VIA 6.6.4 hold. The VIA equations are seen to be

$$
\begin{aligned}
w_n(0) &= \min_a \{C(a) + [1 - p + ap]u_n(0) + (1-a)pu_n(1)\}, \\
w_n(1) &= H + \min_a \{C(a) + a(1-p)u_n(0) \\
&\qquad + [ap + (1-a)(1-p)]u_n(1) + (1-a)pu_n(2)\}, \\
w_n(2) &= 2H + \min_a \{C(a) + a(1-p)u_n(1) + [1 - a(1-p)]u_n(2)\}. \qquad (6.66)
\end{aligned}
$$

For example, we have $P_{00}(a) = 1 - p + ap$, since the system will remain in 0 if there is no arrival, or if there is an arrival (which goes into service immediately) and a service completion. The other transition probabilities are obtained similarly.

The specific calculation given here has $H = 1$, $p = 0.5$, $a_1 = 0.4$, $a_2 = 0.7$, $C(a_1) = 1$, and $C(a_2) = 2$. Note that for distinguished state $x = 0$, it follows from Step 4 of the VIA that $u_n(0) \equiv 0$. Using this and the specific values, it can be seen that (6.66) becomes

$$\begin{aligned} w_n(0) &= 1 + \min\{0.3u_n(1), 1 + 0.15u_n(1)\}, \\ w_n(1) &= 2 + 0.5u_n(1) + \min\{0.3u_n(2), 1 + 0.15u_n(2)\}, \\ w_n(2) &= 3 + \min\{0.2u_n(1) + 0.8u_n(2), 1 + 0.35u_n(1) + 0.65u_n(2)\}. \end{aligned} \tag{6.67}$$

Performing the VIA by hand is typically not feasible. However, for this simple example it is possible to perform a number of iterations. The calculations given in Table 6.1 were done by hand and confirmed on an Excel spreadsheet.

We employ Version 2. The entries under u_n and w_n are the values for states 0, 1, and 2 with $u_n(0)$ omitted. The fourth column keeps track of the policy realizing the minimum in Step 2. Only two policies arise. Policy e_1 always chooses a_1, whereas policy e_2 chooses a_1 in states $\{0, 2\}$ and a_2 in state 1. It is seen that $J \cong 2.15$ and e_2 is the optimal average cost policy. Continuing to 25 iterations yields $J = 2.20$ accurate to two decimal places, with e_2 as the optimal policy. Note that in state 1 it is optimal to serve at the faster rate in order to attempt to prevent the system from transitioning to state 2 and incurring a larger holding cost. However, in state 2 the capacity constraint on the waiting room limits new customers from entering, and in this case it is optimal to drop to the slower rate. This type of behavior will not generally arise when the buffer has infinite capacity. □

BIBLIOGRAPHIC NOTES

The treatment given in this chapter owes most to Derman (1970), Bertsekas (1987, 1995, vol. 2), and Puterman (1994).

Proposition 6.3.3 is from Derman (1970). The argument for the optimality of f in Proposition 6.2.3 comes from Bertsekas (1987). The latter approach has been expanded in Bertsekas (1995, vol. 2). A large part of this development utilizes matrix methods. We have not favored this approach because it does not generalize to the denumerable state space case. Rather we have favored probabilistic methods, as in Derman (1970), because they do suggest the proper generalizations, which will be seen to be extremely fruitful in Chapter 7.

Puterman (1994) has an extensive treatment. It begins with results for the average cost determined by an MC and continues to various classes of finite state MDCs. An MDC is *unichain* if every stationary policy induces an MC with a single positive recurrent class. An ACOE is developed under the unichain

Table 6.1 Results for Example 6.7.1

n	u_n	w_n	f_n	δ
0	0	1	e_1	1
	0	2		
		3		
1	1	1.3	e_1	1.8
	2	3.1		
		4.8		
2	1.8	1.54	e_1	1.36
	3.5	3.95		
		6.16		
3	2.41	1.723	e_1	1.018
	4.62	4.591		
		7.178		
4	2.868	1.8604	e_1	0.760
	5.455	5.0705		
		7.9376		
5	3.2101	1.96303	e_1	0.566
	6.0772	5.42821		
		8.50378		
6	3.46518	2.039554	e_1	0.422
	6.54075	5.694815		
		8.925636		
7	3.655261	2.096578	e_2	0.314
	6.886082	5.860543		
		9.239918		
8	3.763965	2.129189	e_2	0.228
	7.14334	5.953483		
		9.467465		
9	3.824294	2.147288	e_2	0.168
	7.338275	6.012888		
		9.635479		
10	3.8656	2.15968	e_2	0.128
	7.488191	6.056029		
		9.763673		

assumption. The Laurent series expansion treated in Puterman (1994) generalizes Proposition 6.3.3.

Extensive further results for the multichain case are given in Puterman (1994). The crucial Proposition 6.6.3 was suggested by Theorem 9.4.4 of Puterman (1994). This result (in a much more general form) is originally due to Schweitzer and Federgruen (1978).

The value iteration algorithm developed in Section 6.6 is also developed in Puterman (1994) and Bertsekas (1995, vol. 2) under the unichain assumption. Other important references for Section 6.6 include D. White (1963), Odoni (1969), and Federgruen and Schweitzer (1980).

The aperiodicity transformation is taken from Puterman (1994, p. 371) and is due to Schweitzer (1971).

PROBLEMS

6.1. Show that a rational function has at most a finite number of critical points and inflection points.

6.2. Let Δ be an MDC with a finite state space. Give an expression for the total number of stationary policies for Δ and show that there are a finite number of them. (*Hint:* A stationary policy may be considered as a point in what product space?)

6.3. Prove Theorem 6.3.1(i).

6.4. Confirm the results given in Example 6.3.2.

6.5. Verify the statements made in the justification of VIA 6.6.4.

6.6. Consider a service system with a total of K servers. At the beginning of each time slot, there is a probability p that a new customer arrives to the system, where $0 < p < 1$. At this time, if there is a free server, then the new customer is assigned to one of the free servers. Its service may not start until the beginning of the following slot. If all of the servers are busy when a new customer arrives, then that customer is turned away.

In this system there is no queueing. The state space $S = \{0, 1, \ldots, K\}$, where $i \in S$ denotes the number of busy servers. In state $1 \le i \le K - 1$ the decision maker has actions $A_i = \{0, a_1, \ldots, a_M\}$, where $0 < a_1 < a_2 < \ldots < a_M < 1$. Action 0 means that the servers are idle during that slot. If action a is chosen, then all of the busy servers will serve at geometric rate a during the next slot. This means that the probability that any server finishes service during that slot equals a. We assume that $A_K = \{a_1, \ldots, a_M\}$ so that service must be rendered when the system is full. (This assumption is not necessary to make the theory work, but it is evident that one would want to make it.) The services are independent and independent of the arrival process. At the beginning of the next slot, a new action may be chosen.

There is a cost $C(a)$ of choosing to serve at rate a and a cost $H(i)$ of having i customers in the system. It is reasonable to assume that $C(a)$ is increasing in the service rate with $C(0) = 0$ and that $H(i)$ is increasing in i with $H(0) = 0$.

(a) Set this model up as an MDC. Develop the transition probabilities.

(b) Prove that any stationary policy induces an irreducible aperiodic MC on S.

(c) Give the ACOE (6.31) with $z = 0$.

6.7. This is a service system with a single server and buffers for high-priority (HP) customers and low-priority (LP) customers. The state space S consists of ordered pairs (x, y) where x is the number of HP customers and y is the number of LP customers present at the beginning of a slot. If one of the buffers is empty but the other is not, then the server serves a customer from the nonempty buffer in one slot (perfect service). If both buffers contain customers, then the server may make decision a to (perfectly) serve a HP customer or decision b to (perfectly) serve a LP customer. There is a holding (delay) cost of $H(.)$ imposed on the HP customers and of $H^*(.)$ imposed on the LP customers. Reasonable assumptions on these costs are that they are increasing in the number of customers, are 0 when no customers are present, and that $H(x) > H^*(x)$; that is to say, it costs more to delay HP customers.

The arrival process of new HP customers is Bernoulli (p) and the arrival process of new LP customers is Bernoulli (q), where we have $0 < p, q < 1$. The arrival processes are independent and independent of the services provided. There is a capacity K on the buffer of HP customers and the same capacity K on the buffer of LP customers. If, say, there are K HP customers present at the beginning of a slot, then a new arrival is turned away. This happens before service (if any) is provided to that buffer.

(a) Set this up as an MDC and develop the transition probabilities. There are a number of cases to consider.

(b) Verify that Proposition 6.4.1(v) holds and hence that the minimum average cost is a constant J.

(c) Prove that Assumption OPA holds. *Hint:* Show that given any stationary policy e and state (x, y), we have $P_{(x,y)(x,y)}(e) > 0$.

6.8. Complete the proof of Proposition 6.6.6.

6.9. Confirm the validity of (6.66-67), and verify the entries in Table 6.1.

CHAPTER 7

Average Cost Optimization Theory for Countable State Spaces

Chapter 6 dealt with the average cost optimization criterion for an MDC with a finite state space S. It was possible in the finite state space case to prove very special results. In this chapter we present the general existence theory of average cost optimization for countable state spaces. The results are primarily of interest when S is denumerably infinite, but the theory is general and also applies when S is finite.

In Chapter 8 we develop a method for the computation of optimal average cost stationary policies. This method is based on approximating sequences and relies primarily on the results proved in Chapter 6, although occasionally selected results from this chapter are called upon. The reader whose primary interest is in computation may prefer to skip this chapter entirely. When certain results from this chapter are called upon later, they can be read at that time. The reader who wishes to obtain a complete picture of both the existence and computation of optimal average cost stationary policies should read this chapter.

We saw in Chapter 6 that an average cost optimal stationary policy always exists when S is finite. The examples in Section 7.1 show that this is no longer the case when S is denumerably infinite and that, indeed, an optimal policy of any sort may not exist. These examples illustrate that some assumptions are necessary to obtain the existence of an optimal stationary policy in the countable state space case.

In addition to guaranteeing the existence of an optimal stationary policy, it is also useful to require the minimum average cost to be constant. In Section 7.2 we present a set (SEN) of assumptions under which both goals are met. The accompanying existence theorem obtains an inequality for the average cost criterion, known as the average cost optimality inequality (ACOI). The (SEN) assumptions are the centerpiece of Chapter 7, and the remainder of the chapter is an exploration of various ramifications of these assumptions.

Section 7.3 presents a technical example showing that under the (SEN) assumptions an optimal stationary policy may induce a null recurrent MC. This example may be omitted on first reading.

In Section 7.4 we present various results pertaining to the ACOI. Under quite weak assumptions it is shown that the ACOI is an equality and hence the average cost optimality equation (ACOE) holds.

One way to verify the existence of an optimal stationary policy is to show that the (SEN) assumptions hold. In Section 7.5 useful sufficient conditions are given for (SEN) to hold. These include the (BOR) and (CAV) sets of assumptions which are usually easier to verify than (SEN) itself. An important result in this section shows that under the (BOR) assumptions strong inferences concerning the behavior of the MC induced by an optimal stationary policy may be made.

In Section 7.6 we present three examples illustrating how the existence of an average cost optimal stationary policy may be efficiently verified.

Section 7.7 contains a set (H) of assumptions that is weaker than (SEN). Although normally more difficult to verify than (SEN), the (H) set is useful in certain models. An example is given for which the (H) assumptions are valid but for which one of the (SEN) assumptions fails to hold.

7.1 COUNTEREXAMPLES

In this section we have an MDC with a denumerably infinite state space S. We present examples showing that the nice results obtained in Chapter 6 for finite state spaces need no longer hold. Indeed, as we will see, quite pathological situations may be created.

We first prove a result giving a basic property of the average cost. Namely, in most cases, the costs accumulated over any fixed finite number of transitions do not affect the average cost. This result is useful to keep in mind when we present the counterexamples. In this chapter finite horizon value functions will assume a terminal cost of 0.

Proposition 7.1.1. Let K be a positive integer. Let i be an initial state and θ a policy such that $v_{\theta,K}(i) < \infty$. Then

$$J_\theta(i) = \limsup_{n \to \infty} \frac{1}{n-K} E_\theta\left[\sum_{t=K}^{n-1} C(X_t, A_t) | X_0 = i\right]. \tag{7.1}$$

Proof: For $n > K$ and initial state $X_0 = i$, we have

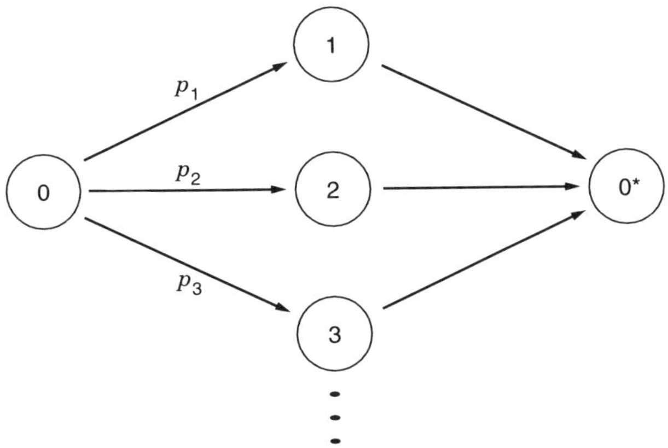

Figure 7.1 Example 7.1.2.

$$\frac{v_{\theta,n}(i)}{n} = \frac{v_{\theta,K}(i)}{n} + \frac{1}{n-K} E_\theta \left[\sum_{t=K}^{n-1} C(X_t, A_t) \right] \left(\frac{n-K}{n} \right). \tag{7.2}$$

The limit of the first term on the right of (7.2) equals 0. Taking the limit supremum of both sides of (7.2), we see that (7.1) follows. □

The right side of (7.1) is the expected average cost from time K onward. A similar result to that in Proposition 7.1.1 holds for J_θ^*. The next example shows that (7.1) may be invalid if some of the expected costs are infinite.

Example 7.1.2. The structure of Δ is shown in Fig. 7.1. There is a null action in each state. Here (p_i) is a probability distribution on $i \geq 1$. The costs are $C(0) = C(0^*) = 0$, and $C(i)$ chosen to satisfy $\sum_i C(i)p_i = \infty$ for $i \geq 1$.

There is a single policy, and it has the property that $E[C(X_1)|X_0 = 0] = \infty$. This means that $J(0) = \infty$. However, we see that $E[C(X_t)|X_0 = 0] = 0$ for $t \geq 2$. Hence, if $K \geq 2$, then the right side of (7.1) equals 0. □

We are now ready to present the counterexamples. The first example shows that an average cost optimal policy may not exist.

Example 7.1.3. The state space is shown in Fig. 7.2. For each i^* there is a null action, and we have $P_{i^*i^*} = 1$ and $C(i^*) = 1/i$. For each i there are two actions, and we have $P_{ii+1}(a) = 1$ and $P_{ii^*}(b) = 1$. Finally $C(i) = 1$, for $i \geq 1$.

Once the process reaches the lower level, it remains there. On the upper level the controller may choose at any time to enter the lower level, or to advance to one higher state on the upper level.

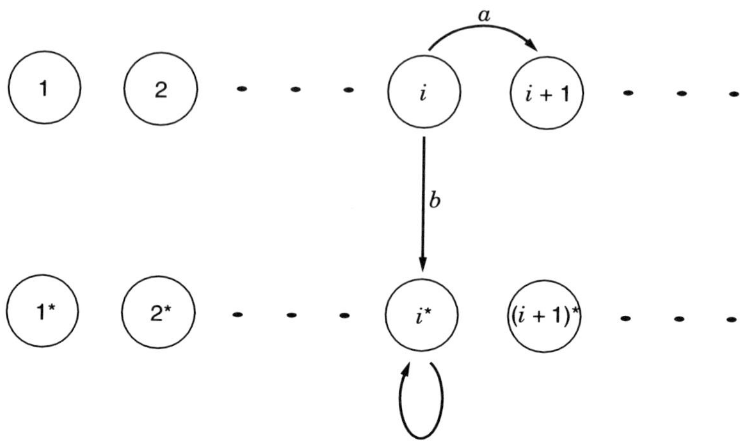

Figure 7.2 Example 7.1.3.

Let f_K, $K \geq 1$, be the stationary policy that chooses a in states $1 \leq i \leq K - 1$, then chooses b in state K. It follows from Proposition 7.1.1 that $J_{f_K}(1) = 1/K$. It is clear that $J(1) = 0$ and that no policy achieves this value. However, given $\epsilon > 0$, there exists a stationary policy f_K for which $J_{f_K}(1) < \epsilon$. □

The next example shows that even if an average cost optimal policy exists, it may be other than stationary.

Example 7.1.4. We have $S = \{1, 2, 3, \ldots\}$. There are two actions in each state with $P_{ii+1}(a) = P_{ii}(b) = 1$, $C(i, a) \equiv 1$, and $C(i, b) = 1/i$. At any time we may advance to the next state and pay 1 unit or choose to stay where we are and pay $1/i$ units.

Let f_K be the stationary policy that chooses a in states $1 \leq i \leq K - 1$ and chooses b in state K. Then it follows from Proposition 7.1.1 that $J_{f_K}(1) = 1/K$.

Let θ operate as follows: When the process enters state i, choose b i times, then choose a. For $X_0 = 1$ the sequence of costs generated under θ is

$$1, 1, \tfrac{1}{2}, \tfrac{1}{2}, 1, \tfrac{1}{3}, \tfrac{1}{3}, \tfrac{1}{3}, 1, \tfrac{1}{4}, \tfrac{1}{4}, \tfrac{1}{4}, \tfrac{1}{4}, 1, \ldots. \tag{7.3}$$

Problem 7.1 asks you to show that $J_\theta(1) = 0$. □

In both of the above examples it is the case that there exists a stationary policy that is within ϵ of $J(1)$. If this were always the case, we would probably be satisfied to know that we could produce a stationary policy with any desired degree of closeness to the optimal value. However, the next example shows that that hope is illusory.

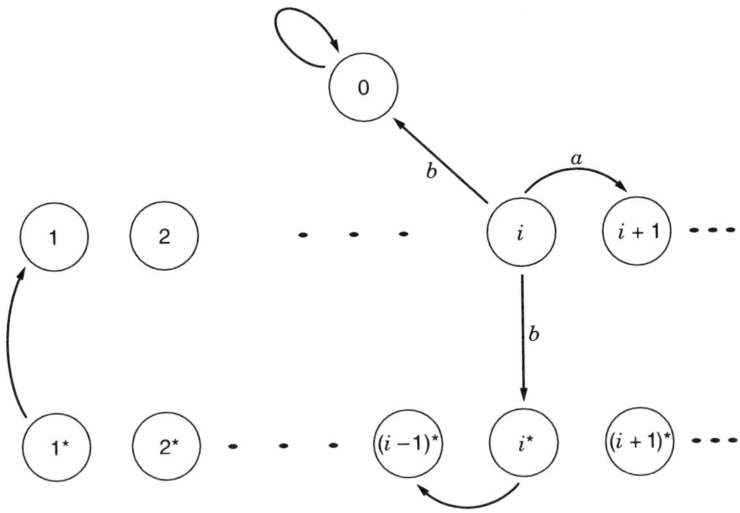

Figure 7.3 Example 7.1.5.

Example 7.1.5. The state space is given in Fig. 7.3. For states i^* on the lower level, there is a null action with transitions $P_{i^*(i-1)^*} = P_{1^*1} = 1$ and costs identically 0. State 0 satisfies $P_{00} = 1$. For states $i \geq 1$ there are two actions. For action a we have $P_{ii+1}(a) = 1$. For action b we have $P_{ii^*}(b) = p_i < 1$ and $P_{i0}(b) = 1 - p_i$. The probabilities p_i will be specified shortly. All costs in $i \geq 0$ equal 1.

Notice how this MDC operates. It is desirable for the process to be in the * states because in those states there is no cost. However, if an attempt is made to reach those states from some $i \geq 1$, then there is a probability of ending up in the absorbing state 0, and hence of incurring a cost of 1 per unit time from then on.

Let the initial state be 1, and let f be a stationary policy. If f always chooses a, then $J_f(1) = 1$. Suppose that f chooses b for the first time in state K. Then every time the process enters state K there is a positive probability $1 - p_K$ that it will end up in state 0. Because this "trial" is repeated over and over, eventually the process will end up in state 0. Then it follows from Proposition 7.1.1 that $J_f(1) = 1$.

The key to this example is that there exist a choice of p_i and a policy θ for which $J_\theta(1) < 1$. Let θ operate as follows: It first chooses b. If it succeeds in reaching 1 again, it then chooses a, moves to 2, and chooses b. If it succeeds in reaching 1 again, it then chooses a twice, moves to 3, chooses b, and so on. On every successive return to 1, the process moves to one higher state before attempting to reach the * states.

Let S_n be the proportion of time spent in * states during $[0, n-1]$. Let E_n be the event that state 0 is not entered during that time. Note that $E_n \to E$, where E is the event that 0 is never entered by the process. Now $P(E)$ is the product $p_1 p_2 p_3 \ldots$, and it is possible to choose the probabilities so that their product equals $\frac{3}{4}$. (For material on infinite products, see Apostol, 1972.)

Then, suppressing the dependence on $X_0 = 1$, we see that $v_{\theta,n}/n = 1 - E_\theta[S_n] = 1 - E_\theta[S_n|E_n]P(E_n) - E_\theta[S_n|E_n^c]P(E_n^c) \le 1 - E_\theta[S_n|E_n]P(E_n)$. We have $P(E_n) \to P(E) = \frac{3}{4}$. Moreover $E_\theta[S_n|E_n] \to \frac{1}{2}$. This is so because if the process has not entered 0, then the proportion of time it spends in the * states approaches $\frac{1}{2}$. This reasoning implies that $J_\theta(1) \le 1 - \frac{3}{8} = \frac{5}{8}$. □

7.2 THE (SEN) ASSUMPTIONS

The examples in Section 7.1 show that some assumptions are necessary to guarantee the existence of an average cost optimal stationary policy. It is also useful for these assumptions to imply that $J(i) \equiv J < \infty$ for $i \in S$. This means that the minimum average cost is a (finite) constant J, independent of the initial state of the process. The property of constant minimum average cost holds in the models of interest to us.

Thus we desire a set of assumptions under which there exist a stationary policy f and a (finite) constant J such that $J(i) = J_f(i) \equiv J$ for $i \in S$. Proposition 6.2.3 suggests that f might be obtained as a limit point of discount optimal stationary policies, as the discount factor approaches 1, and it will be shown that this is possible under suitable assumptions.

Proposition 6.4.1(iii) is a necessary and sufficient condition for the minimum average cost to be a constant when S is finite. This result suggests that we might take this as our assumption when S is countable. This is a viable approach. However, when S is infinite, it turns out that Proposition 6.4.1(iii) is far too strong an assumption and fails to hold in many models. A subtle modification of it will accomplish our goals.

(As an important reminder, we need to keep in mind throughout this chapter that quantities that were automatically finite in Chapter 6 may become infinite when S is infinite. This possibility must be taken into account in all of our proofs.)

This reasoning leads to the following set of (SEN) assumptions. Let z be a distinguished state in S.

(SEN1). The quantity $(1-\alpha)V_\alpha(z)$ is bounded, for $\alpha \in (0, 1)$. (This implies that $V_\alpha(z) < \infty$ and hence we may define the function $h_\alpha(i) =: V_\alpha(i) - V_\alpha(z)$ without fear of introducing an indeterminate form.)

(SEN2). There exists a nonnegative (finite) function M such that $h_\alpha(i) \le M(i)$ for $i \in S$ and $\alpha \in (0, 1)$.

(SEN3). There exists a nonnegative (finite) constant L such that $-L \le h_\alpha(i)$ for $i \in S$ and $\alpha \in (0, 1)$.

Note that $h_\alpha(z) = 0$, and hence we may always take $M(z) = 0$. The first assumption is related to the requirement that the minimum average cost be finite. Notice that the second and third assumptions comprise basically the condition in Proposition 6.4.1(iii), though modified to allow the upper bound for h_α to be a function rather than a constant. In Section 7.7 a set of assumptions allowing the lower bound to also be a function is developed. This approach requires additional assumptions to carry out the development. The requirement of a constant lower bound simplifies the presentation and suffices for many models.

Here is an important lemma.

Lemma 7.2.1. Let e be a stationary policy. Assume that there exist a (finite) constant J and a (finite) function h that is bounded below in i such that

$$J + h(i) \geq C(i, e) + \sum_j P_{ij}(e)h(j), \qquad i \in S. \tag{7.4}$$

Then $J_e(i) \leq J$ for $i \in S$.

Proof: By assumption, there exists a (finite) nonnegative constant L such that $h(i) \geq -L$ for $i \in S$. The proof is similar to the development in (6.14–16). However, we present all the details here.

Let $X_0 = i, X_1, X_2, \ldots$ be the sequence of values of the process operating under the policy e and suppress the initial state in what follows. Then from (7.4) it follows that

$$J + h(X_t) \geq C(X_t, e) + E_e[h(X_{t+1})|X_t], \qquad t \geq 0. \tag{7.5}$$

We claim that $E_e[h(X_t)] < \infty$, and to show this, we prove by induction on t that $E_e[h(X_t)] \leq tJ + h(i)$. This is clearly true for $t = 0$. Now assume that it is true for t. Then from (7.5) it follows that $E_e[h(X_{t+1})|X_t] \leq J + h(X_t)$. Taking the expectation of both sides and using a property of expectation (i.e., that $E(E[X|Y]) = E[X]$), we find that $E_e[h(X_{t+1})] \leq J + E_e[h(X_t)] \leq J + tJ + h(i) = (t+1)J + h(i)$. Here the second inequality follows from the induction hypothesis. This completes the induction.

Now take the expectation of both sides of (7.5) to obtain

$$E_e[C(X_t, f)] \leq J + E_e[h(X_t)] - E_e[h(X_{t+1})], \qquad t \geq 0. \tag{7.6}$$

What has just been proved assures us that we have not created the indeterminate form $\infty - \infty$. Add the terms in (7.6), for $t = 0$ to $n - 1$, and divide by n to obtain

$$\frac{v_{e,n}(i)}{n} \le J + \frac{h(i) - E_e[h(X_n)]}{n}$$
$$\le J + \frac{h(i) + L}{n}. \tag{7.7}$$

Taking the limit supremum of both sides of (7.7) yields the result. □

Before proceeding with our main result, we present a definition. It is given in general terms independent of (SEN).

Definition 7.2.2.

(i) Let z be a distinguished state, and assume that $V_\alpha(z) < \infty$ for $\alpha \in (0, 1)$. This implies that the function $h_\alpha(i) = V_\alpha(i) - V_\alpha(z)$ involves no indeterminate form. Let $\alpha_n \to 1^-$. Assume that there exist a subsequence (call it β_n for convenience) and a function h on S such that

$$\lim_{n\to\infty} h_{\beta_n}(i) = h(i), \qquad i \in S. \tag{7.8}$$

Then h is a *limit function* (of the sequence h_{α_n}).

(ii) Let f_α be a stationary policy realizing the discount optimality equation, and let $\alpha_n \to 1^-$. Assume that there exist a subsequence β_n and a stationary policy f such that $\lim_{n\to\infty} f_{\beta_n} = f$. This means that for a given i and sufficiently large n (dependent on i), we have $f_{\beta_n}(i) = f(i)$. Then f is a *limit point* (of f_{α_n}). (This is Definition B.1 repeated here for convenience.)

(iii) Let f be a limit point. The limit function h is *associated with* f if there exists a sequence β_n such that $\lim_{n\to\infty} h_{\beta_n} = h$ and $\lim_{n\to\infty} f_{\beta_n} = f$. This means that there exists a sequence such that both quantities converge with respect to this sequence. □

The following existence theorem is the major result of this chapter:

Theorem 7.2.3. Let Δ be an MDC for which the (SEN) assumptions hold.

(i) There exists a finite constant $J =: \lim_{\alpha\to 1^-}(1-\alpha)V_\alpha(i)$ for $i \in S$.

(ii) There exists a limit function. Any such function h satisfies $-L \le h \le M$ and

$$J + h(i) \geq \min_a \left\{ C(i,a) + \sum_j P_{ij}(a)h(j) \right\}, \qquad i \in S. \tag{7.9}$$

Let e be a stationary policy realizing the minimum in (7.9). Then e is average cost optimal with (constant) average cost J and

$$\lim_{n \to \infty} \frac{1}{n} E_e[h(X_n)|X_0 = i] = 0, \qquad i \in S. \tag{7.10}$$

(iii) Any limit point f is average cost optimal. There exists a limit function associated with f. Any such function h satisfies

$$J + h(i) \geq C(i,f) + \sum_j P_{ij}(f)h(j), \qquad i \in S, \tag{7.11}$$

and

$$\lim_{n \to \infty} \frac{1}{n} E_f[h(X_n)|X_0 = i] = 0, \qquad i \in S. \tag{7.12}$$

(iv) The average cost under any optimal policy is obtained as a limit.

Proof: We first prove (ii). Fix a sequence $\alpha_n \to 1^-$. It follows from (SEN2–3) and Proposition B.6 that there exists a limit function of the sequence h_{α_n}.

Now let h be any such limit function as in (7.8). It follows from (SEN2–3) that $-L \leq h \leq M$. Using (SEN1) we see that $(1-\beta_n)V_{\beta_n}(z)$ is a bounded sequence of real numbers. Any such sequence has a convergent subsequence. Hence there exist a subsequence (call it δ_n for convenience) and a (finite) number J such that

$$\lim_{n \to \infty} (1 - \delta_n)V_{\delta_n}(z) = J. \tag{7.13}$$

Note that $(1 - \alpha)V_\alpha(i) = (1 - \alpha)h_\alpha(i) + (1 - \alpha)V_\alpha(z)$. Let $\alpha = \delta_n$, and let $n \to \infty$. The last term approaches J. It follows from (7.8) and the finiteness of h that the second term approaches 0. Hence

$$\lim_{n \to \infty} (1 - \delta_n)V_{\delta_n}(i) = J, \qquad i \in S. \tag{7.14}$$

The discount optimality equation (4.9) may be written as

$$(1 - \alpha)V_\alpha(z) + h_\alpha(i) = \min_a \left\{ C(i, a) + \alpha \sum_j P_{ij}(a) h_\alpha(j) \right\}, \qquad i \in S. \tag{7.15}$$

This is obtained from (4.9) by subtracting $\alpha V_\alpha(z)$ from both sides and by adding and subtracting $V_\alpha(z)$ from the left side.

Now fix a state i, and consider the sequence $f_{\delta_n}(i)$ of discount optimal actions in i. Because the action sets are finite, it is the case that there exist an action $a(i)$ and a subsequence γ_n (dependent on i) such that $f_{\gamma_n}(i) \equiv a(i)$. For the fixed state i and $\alpha = \gamma_n$, (7.15) becomes

$$(1 - \gamma_n)V_{\gamma_n}(z) + h_{\gamma_n}(i) = C(i, a(i)) + \gamma_n \sum_j P_{ij}(a(i)) h_{\gamma_n}(j). \tag{7.16}$$

Taking the limit infimum of both sides of (7.16) as $n \to \infty$ and using (7.8), (7.13), and Proposition A.2.1 yields

$$\begin{aligned} J + h(i) &\geq C(i, a(i)) + \sum_j P_{ij}(a(i)) h(j) \\ &\geq \min_a \left\{ C(i, a) + \sum_j P_{ij}(a) h(j) \right\}. \end{aligned} \tag{7.17}$$

Because this argument may be repeated for each i, it follows that (7.9) holds.

Now let e be a stationary policy realizing the minimum in (7.9). Then (7.4) holds for e. To prove that e is optimal, let θ be an arbitrary policy, and fix an initial state i. Then

$$J_e(i) \leq J \leq \limsup_{\alpha \to 1^-} (1 - \alpha)V_\alpha(i) \leq \limsup_{\alpha \to 1^-} (1 - \alpha)V_{\theta,\alpha}(i) \leq J_\theta(i). \tag{7.18}$$

The leftmost inequality follows from Lemma 7.2.1. The next inequality follows from (7.14) and the definition of the limit supremum. The next inequality follows, since $V_\alpha \leq V_{\theta,\alpha}$, and the rightmost inequality follows from (6.1). This proves that e is average cost optimal. Moreover, by setting $\theta = e$, we see that $J_e(i) \equiv J$, and hence J is the minimum average cost.

Recall that the whole argument was carried out with respect to the sequence α_n. Given this sequence, we obtained a subsequence δ_n such that (7.14) holds for the minimum average cost J. This means that given any sequence, there exists a subsequence such that (7.14) holds for the fixed value J. This implies that the limit exists, and hence (i) holds.

We prove (iv) and then return to the proof of (7.10). To prove (iv), let ψ be an arbitrary average cost optimal policy. Note that all that is known is that $J \equiv J_\psi(i)$. We have

$$J = \lim_{\alpha \to 1^-} (1 - \alpha)V_\alpha(i) \leq \liminf_{\alpha \to 1^-} (1 - \alpha)V_{\psi,\alpha}(i) \leq \limsup_{\alpha \to 1^-} (1 - \alpha)V_{\psi,\alpha}(i) \leq J. \tag{7.19}$$

The leftmost equality follows from (i). The next inequality follows from the fact that $V_\alpha \leq V_{\psi,\alpha}$, and the rightmost inequality follows from (6.1) and the optimality of ψ. Hence all the terms in (7.19) are equal to J, and it follows that $\lim_{\alpha \to 1^-} (1 - \alpha)V_{\psi,\alpha}(i)$ exists. Then (iv) follows from Proposition 6.1.1.

Let us now prove (7.10). Using the optimality of e and (iv), it follows that we may take the limit of both sides of (7.7) to obtain (7.10).

It remains to prove (iii). We sketch the proof and leave the details to Problem 7.2. Let f be a limit point. Then there exists a sequence β_n such that $\lim_{n \to \infty} f_{\beta_n} = f$. Using (SEN2–3) and Proposition B.6 yields a subsequence ϵ_n and a limit function h of h_{ϵ_n}. Then h is associated with f (the sequence ϵ_n works in Definition 7.2.2(iii)).

Now let h be associated with f and assume that the sequence β_n works in Definition 7.2.2(iii). Fix a state i and choose n so large that $f_{\beta_n}(i) = f(i)$. Letting $\alpha = \beta_n$ in (7.15) and recalling that f_{β_n} is discount optimal yields

$$(1 - \beta_n)V_{\beta_n}(z) + h_{\beta_n}(i) = C(i,f) + \beta_n \sum_j P_{ij}(f)h_{\beta_n}(j). \tag{7.20}$$

Taking the limit infimum of both sides of (7.20) as $n \to \infty$, and using (i), (7.8), and Proposition A.2.1 yields (7.11). The optimality of f follows immediately from Lemma 7.2.1. Finally (7.12) follows as in the proof of (7.10). □

Notice that Theorem 7.2.3 encompasses two viewpoints. In (ii) we show that an arbitrary limit function may be used to construct an average cost optimal stationary policy, namely the one realizing the minimum in (7.9). In (iii) we show that any limit point of a sequence of discount optimal stationary policies is average cost optimal.

The rest of this chapter is spent elucidating the consequences of this theorem and showing how the (SEN) assumptions may be verified. Here we address the following question: Assuming that (SEN) holds for a distinguished state z, can it fail if z is replaced by another state? Proposition 6.4.1 suggests that the answer is no, and the following result confirms this.

Proposition 7.2.4. Assume that the (SEN) assumptions hold for a distinguished state z. Then (SEN) holds if z is replaced by any other state.

Proof: Assume that (SEN) holds for z, and let $x \neq z$. We wish to show that it holds with z replaced by x. Let these assumptions be denoted $(\text{SEN})_x$. By Theorem 7.2.3(i) it follows that $\lim_{\alpha \to 1^-} (1 - \alpha)V_\alpha(x)$ exists and is finite. This, together with the fact that $V_\alpha(x)$ is increasing in α (and hence $V_\alpha(x) < \infty$ for all α), implies that $(\text{SEN1})_x$ holds.

Now $V_\alpha(i) - V_\alpha(x) = h_\alpha(i) - h_\alpha(x)$. This implies that $-L - M(x) \leq V_\alpha(i) - V_\alpha(x) \leq M(i) + L$. Hence $(\text{SEN2})_x$ holds for the function $M_x(i) = M(i) + L$, and $(\text{SEN3})_x$ holds for the constant $L_x = M(x) + L$. □

7.3 AN EXAMPLE

We know that a stationary policy e for the MDC Δ induces an MC with costs. The transition probabilities of the MC are given by $P_{ij}(e(i)) = P_{ij}(e)$ and the costs by $C(i, e)$. Sections C.1 and C.2 of Appendix C give background material on Markov chains with countable state spaces.

No implication concerning the structure of the MC induced by an optimal stationary policy can be drawn from the (SEN) assumptions. This is easily seen as follows: Let Δ be an MDC with any desired transition structure whatsoever but with identically 0 costs. Then (SEN) holds, and all policies are optimal.

Here is a more interesting example. It shows that an optimal stationary policy may induce a null recurrent MC and that the inequalities in (7.9) and (7.11) may be strict. This example may be omitted by the reader whose primary interest is in applications.

****Example 7.3.1.*** The state space $S = \{0, 1, 2, \ldots\}$. In state $i \geq 1$ there is a null action with $P_{ii-1} = 1$ and $C(i) = 1$. In state 0 there are actions a and b with $C(0, a) = 0$ and $C(0, b) = 1$. Let (p_i) and (q_i) be probability distributions on $i \geq 1$ to be specified later. The transition probabilities are given by $P_{0i}(a) = p_i$ and $P_{0i}(b) = q_i$.

To summarize, when in state $i \geq 1$, the process decreases one state at a time at a cost of 1 per slot. When in state 0, there are two choices of "fanning out" to the states $i \geq 1$. One choice costs 0, and the other costs 1.

Let f (respectively, e) be the stationary policy that chooses a (respectively, b) when in state 0. The costs under e are identically 1, and hence $V_{e,\alpha}(0) = 1/(1 - \alpha)$. It is clearly the case that $V_{f,\alpha}(i) \leq 1/(1 - \alpha)$ for $i \geq 0$. Hence $V_{f,\alpha}(0) = 0 + \alpha \sum p_i V_{f,\alpha}(i) \leq \alpha/(1 - \alpha) < 1$. Hence f is discount optimal for $\alpha \in (0, 1)$.

We verify that (SEN) holds with $z = 0$. Observe that (SEN1) holds in any MDC with bounded costs. For $i \geq 1$ it is easily seen that $V_\alpha(i) = (1 - \alpha^i)/(1 - \alpha) + \alpha^i V_\alpha(0)$. After some algebraic manipulation we have

$$h_\alpha(i) = \left(\frac{1 - \alpha^i}{1 - \alpha} \right) [1 - (1 - \alpha)V_\alpha(0)]. \tag{7.21}$$

The first term on the right of (7.21) is bounded above by i. The second term lies between 0 and 1. Hence $0 \le h_\alpha(i) \le i$, and (SEN2-3) hold. This proves that (SEN) holds. It follows from (7.21) and Theorem 7.2.3(i) that $h(i) = i(1 - J)$.

Moreover from Theorem 7.2.3(iii) it follows that f is average cost optimal. Assume that $X_0 = 0$, and let S_n be the proportion of time, during $t = 0$ to $t = n - 1$, that the process is in state 0 when operating under f. Then it is easy to see that $J = 1 - \lim_{n\to\infty} S_n$. If we choose (p_i) such that $\sum ip_i = \infty$, then the MC induced by f is null recurrent and $S_n \to 0$. This yields $J = 1$ and $h \equiv 0$.

We now examine (7.9) and (7.11) for $i = 0$. The left side is $J + h(0) = 1$. The right side of (7.9) is $\min\{0 + 0, 1 + 0\} = 0$ achieved by the policy f. This shows that there is strict inequality in both equations. □

(The example will fail if $\lambda =: \sum ip_i < \infty$. In this case f induces a positive recurrent MC and $S_n \to \pi_0 = (1 + \lambda)^{-1}$. Then $J = 1 - \pi_0 = \lambda/(1+\lambda)$ and $h(i) = i/(1+\lambda)$. At $i = 0$ the left side of (7.9) is $\lambda/(1+\lambda)$, and it is easy to see that this equals the minimum on the right side, and that this minimum is realized by f. This suggests that if the optimal stationary policy f is positive recurrent at state i, then (7.9) is an equality there. This idea is proved in the next section.)

The reader may have noticed that no further mention has been made of the distribution (q_i). This may be chosen arbitrarily, and we have $J_e \equiv 1$. It implies that e is also average cost optimal. However, e does not realize the minimum in (7.9) at $i = 0$. If the distribution satisfies $\sum iq_i < \infty$, then this gives an example of an MDC satisfying (SEN) and for which there exist two average cost optimal stationary policies. The one arising from the discount optimal stationary policies is null recurrent. The other is positive recurrent yet fails to realize the minimum in (7.9).

7.4 AVERAGE COST OPTIMALITY INEQUALITY

Assume that the (SEN) assumptions hold. Equation (7.9) is known as the *average cost optimality inequality* (ACOI). Theorem 7.2.3(ii) tells us that any stationary policy realizing the minimum on the right of the ACOI is average cost optimal with constant average cost J. Example 7.3.1 shows that the inequality in the ACOI may be strict. If (7.9) is an equality, we refer to it as the *average cost optimality equation* (ACOE).

In this section we give conditions under which the ACOE holds. As part of this development, some important properties of any limit function h are derived. These properties are related to some of the results in Chapter 6. It turns out that the ACOE is "almost always" valid and will certainly hold in the models of interest to us.

We first develop some notation. Let G be a nonempty subset of S. Then $\Re(i, G)$ is the set of policies θ satisfying $P_\theta(X_n \in G \text{ for some } n \ge 1 | X_0 = i) = 1$ and the expected time $m_{iG}(\theta)$ of a first passage from i to G is finite. This is the class of policies having the property that starting from i, the set G will be entered sometime in the future and the expected number of slots before this first happens is finite.

We let $\mathfrak{R}^*(i,G)$ be the class of policies $\theta \in \mathfrak{R}(i,G)$ such that the expected cost $c_{iG}(\theta)$ of a first passage from i to G is finite. If $G = \{x\}$, then $\mathfrak{R}(i,G)$ (respectively, $\mathfrak{R}^*(i,G)$) is denoted $\mathfrak{R}(i,x)$ (respectively, $\mathfrak{R}^*(i,x)$).

The proofs of the following two lemmas are closely related, and hence we present these results together. The first result gives a sufficient condition for (SEN2) to hold. The second result gives an upper bound for h under the assumption that (SEN) holds.

Lemma 7.4.1. Assume that $V_\alpha(z) < \infty$, for a distinguished state z and $\alpha \in (0,1)$. Given $i \neq z$, assume that there exists a policy $\theta_i \in \mathfrak{R}^*(i,z)$. Then $h_\alpha(i) \leq c_{iz}(\theta_i)$, and hence (SEN2) holds for z with $M(i) = c_{iz}(\theta_i)$.

Proof: If the process begins in state $i \neq z$ and follows θ_i, it will reach state z at some time in the future. Let T be a random variable denoting this time. Let the policy ψ follow θ_i until z is reached, and then follow an α discounted optimal policy f_α.

Then

$$\begin{aligned} V_\alpha(i) &\leq V_{\psi,\alpha}(i) \\ &= E_\psi\left[\sum_{t=0}^{T-1} \alpha^t\, C(X_t,A_t)|X_0=i\right] + E_\psi[\alpha^T|X_0=i]V_\alpha(z) \\ &\leq E_\psi\left[\sum_{t=0}^{T-1} C(X_t,A_t)|X_0=i\right] + V_\alpha(z) \\ &= c_{iz}(\theta_i) + V_\alpha(z). \end{aligned} \tag{7.22}$$

The result then follows by subtracting $V_\alpha(z)$ from both sides. □

Lemma 7.4.2. Assume that the (SEN) assumptions hold. Assume that for some fixed state i and nonempty set G, there exists a policy $\theta \in \mathfrak{R}(i,G)$ such that $\sum_{j\in G} M(j)P_\theta(X_T=j) < \infty$, where T is the first passage time from i to G and M is the function from (SEN2). Then for any limit function h we have

$$h(i) \leq c_{iG}(\theta) - Jm_{iG}(\theta) + E_\theta[h(X_T)|X_0=i]. \tag{7.23}$$

Proof: (Note that if $\theta \notin \mathfrak{R}^*(i,G)$, then the right side of (7.23) is infinite.) Let us suppress the initial state i in the proof. In a derivation very similar to that in (7.22), we obtain

$$V_\alpha(i) \le c_{iG}(\theta) + E_\theta[\alpha^T V_\alpha(X_T)]. \tag{7.24}$$

(Problem 7.3 asks you to supply the details.) Then (7.24) may be written

$$h_\alpha(i) \le c_{iG}(\theta) - (1-\alpha)V_\alpha(z)\left(\frac{1 - E_\theta[\alpha^T]}{1-\alpha}\right) + E_\theta[\alpha^T\, h_\alpha(X_T)]. \tag{7.25}$$

This follows by subtracting $V_\alpha(z)$ from both sides and by adding and subtracting $E_\theta[\alpha^T]V_\alpha(z)$ from the right side.

Now

$$\begin{aligned}\frac{1 - E_\theta[\alpha^T]}{1-\alpha} &= \sum_{t=1}^{\infty}\left(\frac{1-\alpha^t}{1-\alpha}\right) P_\theta(T=t)\\ &= \sum_{t=1}^{\infty}(1+\alpha+\ldots+\alpha^{t-1})P_\theta(T=t).\end{aligned} \tag{7.26}$$

The term in parenthesis is increasing in α and converges to t as $\alpha \to 1^-$. We may apply Corollary A.2.4 with bounding function $w(t) = t$ to conclude that the limit of the left side of (7.26) exists and equals $m_{iG}(\theta)$.

Let us assume that the limit function h is defined in terms of the sequence β_n as in (7.8). Now take the limit of both sides of (7.25) as $\alpha = \beta_n \to 1^-$. Using what has just been proved and Theorem 7.2.3(i) yields

$$\begin{aligned} h(i) &\le c_{iG}(\theta) - Jm_{iG}(\theta) + \lim_{n\to\infty} E_\theta[(\beta_n)^T h_{\beta_n}(X_T)]\\ &= c_{iG}(\theta) - Jm_{iG}(\theta) + \lim_{n\to\infty}\sum_{j\in G}\sum_{t=1}^{\infty}(h_{\beta_n}(j)(\beta_n)^t P_\theta(T=t, X_T=j)).\end{aligned} \tag{7.27}$$

To justify passing the limit through the summation, we will employ Corollary A.2.4. Note that the index set of the summation is the set of pairs (j,t). The function $u_n(j,t) = h_{\beta_n}(j)(\beta_n)^t$ which converges to $h(j)$. The bounding function is $w(j) = \max\{L, M(j)\}$, where L is from (SEN3). The assumption allows us to apply Corollary A.2.4 to (7.27), which yields (7.23). □

We now give sufficient conditions for the ACOE to hold.

Theorem 7.4.3. Assume that the (SEN) assumptions hold, and let e be a stationary policy realizing the minimum in the ACOI (7.9). Define the nonnegative *discrepancy function* Φ to satisfy

$$J + h(i) = C(i, e) + \Phi(i) + \sum_{j} P_{ij}(e)h(j), \qquad i \in S. \tag{7.28}$$

Then $\Phi(i) = 0$, and hence (7.9) is an equality at the particular state i under any of the following conditions:

(i) There exists a nonempty set G such that e satisfies the assumptions in Lemma 7.4.2. This also implies that $e \in \mathfrak{R}^*(i, G)$ and $h(i) = c_{iG}(e) - Jm_{iG}(e) + E_e[h(X_T)|X_0 = i]$, where T is the time of a first passage.
(ii) We have $e \in \mathfrak{R}(i, z)$. This also implies that $e \in \mathfrak{R}^*(i, z)$ and $h(i) = c_{iz}(e) - Jm_{iz}(e)$.
(iii) The MC induced by e is positive recurrent at i.
(iv) We have $\sum_j P_{ij}(a)M(j) < \infty$ for $a \in A_i$.

Proof: To prove equality under (i), let the process operate under e, and suppress the initial state i. As in the proof of Lemma 7.2.1, we obtain

$$J + E_e[h(X_t)] = E_e[C(X_t, e)] + E_e[\Phi(X_t)] + E_e[h(X_{t+1})], \qquad t \geq 0. \tag{7.29}$$

Rearranging terms and adding for $t = 0$ to $k - 1$ yields

$$E_e\left[\sum_{t=0}^{k-1} C(X_t, e)\right] - Jk + E_e\left[\sum_{t=0}^{k-1} \Phi(X_t)\right] = h(i) - E_e[h(X_k)]. \tag{7.30}$$

If T is the first passage time from i to G, then by assumption $m_{iG}(e) = \sum kP_e(T = k) < \infty$. Let us multiply each term of (7.30) by $P_e(T = k)$ and sum over k. This yields

$$c_{iG}(e) - Jm_{iG}(e) + E_e\left[\sum_{t=0}^{T-1} \Phi(X_t)\right] + E_e[h(X_T)] = h(i). \tag{7.31}$$

It follows from (7.31) that $c_{iG}(e) < \infty$, and hence $e \in \mathfrak{R}^*(i, G)$. We may then apply Lemma 7.4.2. The other claims follow from (7.23), (7.31), and the nonnegativity of Φ. In addition to proving that $\Phi(i) = 0$, notice that this argument proves that $\Phi \equiv 0$ during a first passage from i to G.

Claim (ii) follows from (i) by choosing $G = \{z\}$ and recalling that $h(z) = 0$. Claim (iii) follows from (i) by noting that if the MC induced by e is positive recurrent at i, then $e \in \mathfrak{R}(i, i)$. From (i) it then follows that $J = c_{ii}(e)/m_{ii}(e)$, which agrees with Proposition C.2.1(ii).

To prove equality under (iv), we return to the proof of Theorem 7.2.3 and consider (7.15). Let us take the limit of both sides as $\alpha = \gamma_n \to 1^-$. Using Proposition A.1.3(ii) and Corollary A.2.4 (with bounding function M), we obtain $J + h(i) = \min_a\{C(i,a) + \sum_j P_{ij}(a)h(j)\}$, and hence $\Phi(i) = 0$. □

Theorem 7.4.3(i) is a remarkable result. A corollary of this result is that if, starting from an arbitrary initial state i, in a finite expected amount of time the MC induced by e reaches a finite set G, then the ACOE holds. Note that G may depend on i.

7.5 SUFFICIENT CONDITIONS FOR THE (SEN) ASSUMPTIONS

We now consider the verification of the (SEN) assumptions. It is often difficult to verify them directly, and some well-chosen sufficient conditions will prove extremely useful. In this section we assume that Δ is an MDC, and we seek sufficient conditions for (SEN) to hold.

The following definition gives an important type of policy.

Definition 7.5.1. Let d be a (randomized) stationary policy. Then d is a z *standard policy* if the MC induced by d is z standard (see Definition C.2.5). □

We will usually use the letter d to refer to a z standard policy, and the reader should keep in mind that d may be either a stationary policy or a randomized stationary policy. The following preliminary result is useful:

Lemma 7.5.2. If d is a z standard policy with positive recurrent class R, then

$$J_d = (1-\alpha)\sum_{i\in R}\pi_i(d)V_{d,\alpha}(i), \qquad \alpha \in (0,1). \tag{7.32}$$

Proof: The result follows by multiplying the expression in Proposition C.2.1(iii) by α^n and summing over n. Interchanging the order of the summations is justified by the fact that the terms are nonnegative. □

The next result gives our standard method for verifying (SEN1–2).

Proposition 7.5.3. Assume that there exists a z standard policy d. Then (SEN1-2) hold for z.

Proof: From (7.32) it follows that $J_d \geq (1-\alpha)\pi_z(d)V_{d,\alpha}(z) \geq (1-\alpha)\pi_z(d)V_\alpha(z)$. Hence $(1-\alpha)V_\alpha(z) \leq J_d\pi_z^{-1}(d) = c_{zz}(d)$, by results in Appendix C. Hence (SEN1) holds.

From Lemma 7.4.1 it follows that (SEN2) holds with $M(i) = c_{iz}(d)$ for $i \neq z$. □

Corollary 7.5.4. Assume that $S = \{0, 1, 2, \ldots\}$ and that V_α is increasing in i for $\alpha \in (0, 1)$. If there exists a 0 standard policy, then the (SEN) assumptions hold. Moreover every limit function is nonnegative and increasing in i.

Proof: Let the distinguished state be 0. It follows from Proposition 7.5.3 that (SEN1–2) hold. Since V_α is increasing, it follows that $h_\alpha \geq 0$, and hence (SEN3) holds with $L = 0$. The second statement is clear from (7.8). □

Suppose that the (SEN) assumptions have been verified. Then Example 7.3.1 shows that an optimal stationary policy may induce a MC without any positive recurrent states. The next result gives a sufficient condition for an optimal stationary policy to induce a MC with at least one positive recurrent state. This result is stated in a form independent of (SEN).

The term used in (7.33) below is defined in (C.1).

Proposition 7.5.5. Assume that the minimum average cost is a constant J and that e is an optimal stationary policy. Assume that there exist a state i and $\epsilon > 0$ such that the set $G = \{j | C(j, e) \leq J + \epsilon\}$ satisfies

$$\lim_{n\to\infty} \sum_{j\in G} Q_{ij}^{(n)}(e) = \sum_{j\in G} \lim_{n\to\infty} Q_{ij}^{(n)}(e). \tag{7.33}$$

Then the MC induced by e has at least one positive recurrent state $j \in G$, and i leads to j.

(Equation (7.33) says that the limit may be moved across the summation. This is always possible if G is finite and may be possible in certain situations if G is infinite.)

Proof: The set G must be nonempty (why?). Let $X_0 = i$, and suppress the initial state in what follows. Then, operating under e, we obtain

$$\begin{aligned}\frac{1}{n} E_e\left[\sum_{t=0}^{n-1} C(X_t, e)\right] &\geq (J + \epsilon) E_e\left[1 - \frac{1}{n}\sum_{t=0}^{n-1} I(X_t \in G)\right] \\ &= (J + \epsilon)\left(1 - \sum_{j\in G} Q_{ij}^{(n)}(e)\right).\end{aligned} \tag{7.34}$$

Here I is the indicator function. To obtain the first line in (7.34), the costs associated with visits to G have been set to 0, and the costs associated with visits outside of G have been replaced by their lower bound $J + \epsilon$. The second line follows from (C.1) and the definition of the expectation of an indicator function.

We now take the limit supremum as $n \to \infty$ of both sides of (7.34). Using the optimality of e, (7.33), and results in Section C.1, we obtain

$$J \geq (J + \epsilon)\left(1 - \sum_{j \in G} P_e(T_{ij} < \infty)\pi_j(e) \right), \tag{7.35}$$

where $\pi_j(e)$ is the steady state probability of being in j and T_{ij} is the first passage time from i to j. This yields a contradiction unless there exists $j \in G$ such that $P_e(T_{ij} < \infty) > 0$ (which means that i leads to j) and $\pi_j(e) > 0$ (which means that j is positive recurrent). □

Corollary 7.5.4 verifies (SEN3) by employing a structural result on the discount value function. This is an important method of verification. However, it is also useful to have a method that does not employ structural results. The following set (BOR) of assumptions implies that (SEN) holds, that the ACOE is valid, and that optimal stationary policies possess "nice" properties.

Theorem 7.5.6. Assume that the following set (BOR) of assumptions holds:

(BOR1). There exists a z standard policy d with positive recurrent class R_d.

(BOR2). There exists $\epsilon > 0$ such that $D = \{i | C(i,a) \leq J_d + \epsilon \text{ for some } a\}$ is a finite set.

(BOR3). Given $i \in D - R_d$, there exists a policy $\theta_i \in \mathfrak{R}^*(z,i)$.

Then:

(i) The (SEN) assumptions hold and the ACOE is valid.

(ii) The MC induced by an optimal stationary policy e has at least one positive recurrent state in the set $D(e) = \{i | C(i,e) \leq J + e\}$. Let $R(e)$ be the set of positive recurrent states. Then the number of positive recurrent classes making up $R(e)$ does not exceed $|D(e)|$, and there are no null recurrent classes.

(iii) If e is a stationary policy realizing the minimum in the ACOE, then $e \in \mathfrak{R}^*(i, D(e) \cap R(e))$ for all i. Hence, if $R(e)$ consists of a single class, then e is x standard for $x \in R(e)$.

**Proof:* We first verify (SEN). It follows from (BOR1) and Proposition 7.5.3 that (SEN1–2) hold. Let us now show that (SEN3) holds. Consider the statement:

(*) For each $\alpha \in (0, 1)$ the minimum value of V_α exists and is taken on in the set D.

Assume that (*) has been proved. Then we claim that

$$L =: \max_{j \in D - R_d} \{c_{zj}(\theta_j)\} \vee \max_{j \in R_d \cap D - \{z\}} \{c_{zj}(d)\} \tag{7.36}$$

will work in (SEN3). Proposition C.2.2(iv) shows that the second term on the right of (7.36) is finite, since $z \in R_d$ and thus $c_{zj}(d) < \infty$. The first term is finite by (BOR3). Hence $L < \infty$.

For each i and α, by (*) we may choose and fix $j \in D$ such that $V_\alpha(i) \geq V_\alpha(j)$. Then $h_\alpha(i) = (V_\alpha(i) - V_\alpha(j)) + h_\alpha(j) \geq h_\alpha(j)$.

We use the proof method of Lemma 7.4.1. If $j \in D - R_d$, then, following this proof, it can be shown that $V_\alpha(z) - V_\alpha(j) \leq c_{zj}(\theta_j) \leq L$. If $j \in R_d \cap D - \{z\}$, then it can be shown that $V_\alpha(z) - V_\alpha(j) \leq c_{zj}(d) \leq L$. Hence in either case we have $h_\alpha(j) \geq -L$.

So to complete the verification of (SEN), we need to prove (*). Fix α throughout this segment of the proof. The key to the proof is the following: For any stationary policy f and $i \notin D$, let T be the time of a first passage from i to D. Then we have

$$\begin{aligned} V_{f,\alpha}(i) \geq E_f \Bigg[& \left(\frac{J_d + \epsilon}{1 - \alpha}\right) I(T = \infty) \\ & + \left\{ \left(\frac{J_d + \epsilon}{1 - \alpha}\right) (1 - \alpha^T) + \alpha^T V_{f,\alpha}(X_T) \right\} I(T < \infty) \Bigg]. \end{aligned} \tag{7.37}$$

This follows since $J_d + \epsilon$ is a lower bound on the costs outside of D.

Since the expression on the right of (7.32) is a convex combination, it follows from (7.32) that there exists $i_\alpha \in R_d$ such that $J_d \geq (1 - \alpha)V_{d,\alpha}(i_\alpha)$. We claim that

$$J_d \geq (1 - \alpha)V_{d,\alpha}(j_\alpha) \qquad \text{for some } j_\alpha \in D. \tag{7.38}$$

The proof is by contradiction. Assume that (7.38) fails. The use (7.37) with $f = d$ and $i = i_\alpha$ (note that $I(T = \infty) = 0$) to obtain a contradiction.

Since D is finite, it follows that there exists $k_\alpha \in D$ such that $V_\alpha(j) \geq V_\alpha(k_\alpha)$ for all $j \in D$. Then from (7.38) it follows that

$$\frac{J_d}{1-\alpha} \geq V_\alpha(k_\alpha). \tag{7.39}$$

Let us now begin the process in state $i \notin D$ and operate under f_α. Applying (7.37) with $f = f_\alpha$ and using (7.39) easily yields $V_\alpha(i) \geq V_\alpha(k_\alpha)$. This proves (*).

The proof of the validity of the ACOE is given later in this proof. To prove (ii), let e be an optimal stationary policy and fix an initial state i. Since $D(e)$ is a subset of D, it is finite and hence (7.33) holds for $D(e)$ and i. It follows from Proposition 7.5.5 that the MC induced by e has at least one positive recurrent state $j \in D(e)$ such that i leads to j. This clearly proves (ii).

To prove (iii), assume that e realizes the minimum in (7.9).

We first show that $e \in \Re(i, D(e))$ for $i \notin D(e)$. Recall that $h \geq -L$. Let us define the nonnegative function $r = h + L$. Now add L to both sides of (7.9) and rearrange the terms to obtain

$$\sum_j P_{ij}(e)[r(j) - r(i)] \leq J - C(i,e) \qquad i \in S. \tag{7.40}$$

If $i \notin D(e)$, then $J - C(i,e) < -\epsilon$. The result now follows from Proposition C.1.5, and we have $m_{iD(e)}(e) \leq r(i)/\epsilon$.

We now prove that $e \in \Re(i, D(e))$ for $i \in D(e)$. Using reasoning as in Appendix C, we obtain

$$\begin{aligned} m_{iD(e)}(e) &= 1 + \sum_{j \notin D(e)} P_{ij}(e) m_{jD(e)}(e) \\ &\leq 1 + \sum_{j \notin D(e)} P_{ij}(e)\left(\frac{r(j)}{\epsilon}\right) \\ &\leq 1 + \sum_j P_{ij}(e)\left(\frac{r(j)}{\epsilon}\right) \\ &\leq 1 + \frac{J + r(i)}{\epsilon}. \end{aligned} \tag{7.41}$$

The second line follows from what has just been proved. The third line follows from the nonnegativity of r. The last line follows from (7.40).

This proves that $e \in \Re(i, D(e))$ for all i. The validity of the ACOE now follows from Theorem 7.4.3(i).

Let $F = D(e) \cap R(e)$. Let us now give an informal argument that $e \in \Re(i, F)$ for all i. For $i \notin D(e)$ it follows from the above that in finite expected time we will reach the finite set $D(e)$. Hence it is sufficient to argue that $e \in \Re(i, F)$ for $i \in D(e)$.

First assume that $i \in F$. Then $i \in R(e)$, and since $m_{ii}(e) < \infty$, it follows that $e \in \Re(i, F)$.

Now let $j \in D(e) - R(e)$. Then j is transient, and there is a probability $q_j > 0$ of not returning to j each time it is entered. Then $q =: \min\{q_j \mid j \in D(e) - R(e)\}$ is a positive lower bound on the probability of never returning to $D(e) - R(e)$ each time it is entered. Observe that $m =: \max\{m_{jD(e)}(e) \mid j \in D(e) - R(e)\}$ is a (finite) upper bound on the expected time to return to $D(e)$ from the set $D(e) - R(e)$.

Now assume that the process begins in $i \in D(e) - R(e)$. Each time the process returns to $D(e)$, it conducts a "trial" which results, with probability at least q, in entering F. Hence $m_{iF}(e) \le m/q$.

This proves that $e \in \Re(i, F)$ for all i. It follows from Theorem 7.4.3(i) that $e \in \Re^*(i, F)$ for all i. The second statement of (iii) is then clear. Problem *7.4 asks you to fill in the details of this proof. □

Remark 7.5.7. The reader may have noticed that the positivity of ϵ is not used in some parts of the proof. Problem *7.7 explores this issue and gives a weaker set (WS) of assumptions under which Theorem 7.5.6(i–ii) hold but for which (iii) may fail. In this problem you are asked to construct an example with an optimal policy from the ACOI that has a positive recurrent state but for which the expected time to reach this state is infinite. □

Remark 7.5.8. Assume that the (BOR) assumptions hold, and let e be an optimal stationary policy. Theorem 7.5.6(ii) shows that the MC induced by e has a nonempty set $R(e)$ of positive recurrent states and no null recurrent classes. It is shown in Sennott (1993) that the probability of going from a transient state to $R(e)$ is 1. However, an example is given there to show that the expected time of such a first passage may be infinite. Of course this cannot happen if e realizes the ACOE. □

Here are some sufficient conditions for the (BOR) assumptions to hold. These conditions are easy to verify and often hold when the costs of Δ are unbounded. The proofs are left as Problem 7.8.

Corollary 7.5.9. Assume that the following set (CAV) of assumptions hold:

(CAV1) = (BOR1).

(CAV2). Given $U > 0$, the set $D_U = \{i \mid C(i, a) \le U \text{ for some } a\}$ is finite.

(CAV3). Given $i \in S - R_d$, there exists a policy $\theta_i \in \Re^*(z, i)$.

Then the (BOR) assumptions hold.

Corollary 7.5.10. Assume that the following set (CAV^*) of assumptions hold:

(CAV^*1). There exists a standard policy d such that $R_d = S$.

(CAV^*2). Given $U > 0$, the set $D_U = \{i|C(i,a) \le U \text{ for some } a\}$ is finite.

Then the (CAV) assumptions hold, and hence (BOR) is valid.

7.6 EXAMPLES

It is time to put the theory to the test. Are these results of use in verifying the existence of optimal stationary policies in interesting models? In this section we present three examples that amply illustrate the practicality of the theory.

There is a set of *basic assumptions* (BA) for each example. Other assumptions may be added as necessary.

Example 7.6.1. This is Example 2.1.1 which is also treated in Section 3.4. Let $s = \sup\{j|p_j > 0\}$. The basic assumptions (BA) are as follows:

(BA1). The holding cost $H(i)$ is increasing in i with $H(0) = 0$.

(BA2). We have $0 < p_0 < 1$. □

In each slot there is a positive probability of no arrivals and a positive probability of a batch arriving. Hence it follows that $1 \le s \le \infty$. (Note that $s = \infty$ means that batches of arbitrarily large size may arrive in a single slot.) As notation we let $H^\infty = \lim_{i\to\infty} H(i)$ and $K(i) = \sum_{j=1}^{i} H(j)$ for $i \ge 1$ (set $K(0) = 0$). Recall that λ is the mean batch size. Nothing is assumed about these quantities at this time.

Lemma 7.6.2. Assume that (BA1) holds. For $\alpha \in (0, 1)$ and a zero terminal cost, $v_{\alpha,n}$ is increasing in i for $n \ge 0$. Hence V_α is increasing in i.

Proof: This is proved in Lemma 3.4.1 for $\alpha = 1$ and a terminal cost of $H(i)$. The same proof works here, and Problem 7.9 asks you to confirm this. The fact that $V_\alpha(i)$ is increasing in i follows from Proposition 4.3.1. □

Proposition 7.6.3 Assume that the (BA) assumptions hold.

(i) The (SEN) assumptions hold and h is nonnegative and increasing in i. Letting $H^*(i) = \sum_j p_j[h(i+j) - h(i)]$, the ACOI may be written

$$\begin{cases} J \geq \min\{R, H^*(0)\}, & i = 0, \\ J + \mu[h(i) - h(i-1)] \geq H(i) & \\ \qquad +\min\{R, \mu H^*(i-1) + (1-\mu)H^*(i)\}, & i \geq 1. \end{cases} \tag{7.42}$$

(ii) If $\lambda < \infty$ and $\sum K(i+j)p_j < \infty$ for $i \geq 0$ (note that if $H^\infty < \infty$, then $\lambda < \infty$ implies the second condition), then the ACOE holds.

(iii) Assume that the conditions in (ii) hold and that $H^\infty > R$. Then the (BOR) assumptions hold, and any optimal stationary policy is positive recurrent at 0.

(iv) Assume that the conditions in (iii) hold, and let e be a stationary policy realizing the ACOE. Then e is 0 standard. Assume that e breaks ties by rejecting. If e rejects in state i, then it rejects in higher states. If $H^\infty = \infty$, then there exists i^* such that $e(i^*) = r$.

Proof: We employ Corollary 7.5.4. First consider *any* stationary policy f. Since at most one packet can be served in any slot and $p_0 > 0$, it follows that $i \geq 1$ leads to 0 in the MC induced by f, and that the only path is $i \to (i-1) \to (i-2) \to \ldots \to 1 \to 0$.

Now let d be the policy that always rejects; we claim that d is 0 standard. Now $P_{00}(d) = 1$, and hence $R_d = \{0\}$. We must show that $d \in \mathfrak{R}^*(i, 0)$ for $i \geq 1$. Starting in state i, no new batches enter the system. The expected time to serve a packet is $1/\mu$, and hence $m_{i0}(d) = i/\mu$. The expected cost of serving the first packet is $(H(i) + R)/\mu$, and similarly for the second, and so on. Thus we see that $c_{i0}(d) = (K(i) + Ri)/\mu$. Then the first claim in (i) follows from Lemma 7.6.2 and Corollary 7.5.4.

Using (7.9), we can easily see that the ACOI is given by

$$J \geq \min\left\{R, \sum_j p_j h(j)\right\}, \qquad i = 0,$$

$$J + h(i) \geq H(i) + \min\{R + \mu h(i-1) + (1-\mu)h(i),$$

$$\mu \sum_j p_j h(i-1+j) + (1-\mu) \sum_j p_j h(i+j)\}, \qquad i \geq 1. \tag{7.43}$$

Then (7.42) is obtained by subtracting $\mu h(i-1) + (1-\mu)h(i)$ from both sides of (7.43). This proves (i).

To prove (ii), we employ Theorem 7.4.3(iv). Recall that for $i \geq 1$ we have $M(i) = c_{i0}(d) = (K(i) + Ri)/\mu$. Then $\sum p_j M(i+j) = \sum p_j\{K(i+j) + R(i+j)\}$, which is finite under the assumed conditions. Theorem 7.4.3(iv) then verifies that the ACOE holds.

Now assume the conditions in (iii). We verify the (BOR) assumptions for the policy d that always rejects. We have seen that (BOR1) holds and note that $J_d = R$.

Since H is increasing and $H(0) = 0$, there must exist a state x with the following property: For $i \in [0, x]$ we have $H(i) \leq R$ but $H(x+1) > R$. Choose $\epsilon > 0$ such that $R + \epsilon < H(x+1)$. Then the set D in (BOR2) is precisely the interval $[0, x]$, and hence (BOR2) holds.

To verify (BOR3), it is sufficient to construct a stationary policy f with positive recurrent class $R_f \supset [0, x]$ and such that f has finite average cost on R_f. (BOR3) will then follow from Propositions C.1.4(iv) and C.2.2(iv).

First assume that $s = \infty$, and let f be such that $f(0) = a$ and $f(i) = r$ for $i \geq 1$. Observe that f induces an irreducible MC on $[0, \infty)$. We claim that the chain is positive recurrent. Using what has been proved for d, we see that $m_{00}(f) = 1 + \sum_{j \neq 0} p_j m_{j0}(d) = 1 + \lambda/\mu$. Similarly $c_{00}(f) = \sum_{j \neq 0} p_j c_{j0}(d)$ is finite by assumption. Hence $J_f < \infty$.

Now assume that $s < \infty$. If $x = 0$, then the policy that always rejects will fulfill the conditions. Next assume that $x \geq 1$. Define $f(i) = a$ for $0 \leq i < x$, and $f(i) = r$ for $i \geq x$. It is easy to see that $[0, x - 1 + s]$ is a communicating class containing $[0, x]$. Observe that from this class no state outside the class can be reached. Hence this class forms a finite state MC, and by Section C.3 it is positive recurrent with finite average cost. This completes the verification of the (BOR) assumptions.

If e is an optimal stationary policy, then it follows from Theorem 7.5.6(ii) that the MC induced by e has a positive recurrent state i. If $i > 0$, then it leads to 0, and hence 0 must lead to i. This implies that i and 0 are in the same communicating class, and hence the MC is positive recurrent at 0. This proves (iii).

Now assume that e realizes the ACOE. We have shown that the MC induced by e is positive recurrent at 0. It is clear that there cannot be two positive recurrent classes. Hence it follows from Theorem 7.5.6(iii) that e is 0 standard.

To prove the next claim it is sufficient to prove that if $e(i) = r$, then $e(i+1) = r$. From (7.42) it is easy to see that this holds if $H^*(i)$ is increasing in i. Moreover $H^*(i)$ is increasing in i if the following holds: For each fixed j, $h(i+j) - h(i)$ is increasing in i. Because this is a sum of one-step increments, it is clear that this holds if the following statement is true.

(*) For $i \geq 0$, $h(i+1) - h(i)$ is increasing in i.

It remains to prove (*). Suppose that the process starts in state $i + 1$. It must pass through state i in a first passage to 0. Letting $G = \{i\}$, it follows from Theorem 7.4.3(i) that

$$h(i+1) - h(i) = c_{i+1,i}(e) - J m_{i+1,i}(e). \tag{7.44}$$

The quantity on the right of (7.44) is the J *revised cost* of a first passage from i

$+ 1$ to i. In each slot of the first passage, we can think of the "cost" of $C(i, e) - J$ being incurred.

It follows from Lemma 7.4.2 (with $G = \{i\}$) that if θ is any policy with finite J revised cost from $i + 1$ to i, then this J revised cost is bounded below by the right side of (7.44). This leads to the important idea that the optimal stationary policy e from the ACOE minimizes the J revised cost of a first passage from $i + 1$ to i.

The argument to prove (*) may be completed as follows: Fix states $k < j$. Probabilistically both situations are exactly the same except that the holding costs in states above j are uniformly at least as great as the corresponding holding costs in states above k. Therefore the minimum J revised cost from $j + 1$ to j must be at least as great as the minimum J revised cost from $k + 1$ to k. This proves that (*) holds.

It remains to prove that if $H^\infty = \infty$, then e must reject batches for a large enough buffer content. Let $W = \lim_{i \to \infty} H^*(i)$. If we can prove that $W = \infty$, then the result follows from (7.42). Since $s \geq 1$, we may fix $j^* \geq 1$ such that $p_{j^*} > 0$. Then $h(i + j^*) - h(i) \geq h(i + 1) - h(i) \geq H(i + 1) - J$. The last inequality follows by (7.44) and the observation that a revised cost of at least $H(i + 1) - J$ is incurred at every stage of the first passage. Hence $H^*(i) \geq p_{j^*}[h(i + j^*) - h(i)] \geq p_{j^*}(H(i + 1) - J)$. Since $H^\infty = \infty$, we must have $W = \infty$. This completes the proof of (iv). □

Example 7.6.4. This is Example 2.1.2. Let $\lambda^{(n)} = \sum j^n p_j$ be the nth moment of the arrival process, with $\lambda^{(1)} = \lambda$. The basic assumptions (BA) are as follows:

(BA1). The holding cost $H(i)$ is increasing in i.

(BA2). There exist a (finite) constant B and nonnegative integer n such that $H(i) \leq Bi^n$ for $i \geq 0$.

(BA3). We have $0 < \lambda < a_K$ and $\lambda^{(n+1)} < \infty$. □

Observe that $0 < \lambda < a_K$ implies that $0 < p_0 < 1$. It makes sense to assume that $C(a)$ is increasing in a, but suprisingly this is not necessary for our results.

Lemma 7.6.5. Assume that (BA1) holds. For $\alpha \in (0, 1)$ and a zero terminal cost, $v_{\alpha,n}$ is increasing in i for $n \geq 0$. Hence V_α is increasing in i.

Proof: Recall that in Problem 3.2 you were asked to develop the finite horizon optimality equation for this model. In Problem 3.11(i) you were asked to prove that the finite horizon value function is increasing in i. This holds for a terminal cost of 0. □

Lemma 7.6.6. Assume that (BA2–3) hold. Let d be the stationary policy that always serves at rate a_K and let $\epsilon =: a_K - \lambda > 0$. Then:

(i) Any stationary policy induces an irreducible MC on $[0, \infty)$.

(ii) The policy d is standard with $R_d = [0, \infty)$.

(iii) We have $m_{i0}(d) \le i/\epsilon$ for $i \ge 1$, and $m_{00}(d) \le 1 + \lambda/\epsilon$.

(iv) There exists a (finite) constant D such that $c_{i0}(d) \le Di^{n+1}$ for $i \ge 1$, and $c_{00}(d) \le D\lambda^{(n+1)}$.

Proof: Under any stationary policy there is a positive probability of no batches arriving as well as a positive probability of a batch arriving and a service not being completed. Hence all states communicate with 0, and (i) holds.

If we can prove (iii–iv), then (ii) will follow. To prove (iii), we apply Corollary C.1.6 with $z = 0$ and $y(i) = i$. Then the first inequality in (C.10) holds, since $\lambda < \infty$. The left side of the second inequality becomes

$$a_K \sum_j p_j[y(i-1+j) - y(i)] + (1-a_K) \sum_j p_j[y(i+j) - y(i)] = \lambda - a_K. \quad (7.45)$$

Since $\lambda - a_K = -\epsilon$, it follows that $m_{i0}(d) \le i/\epsilon$. Moreover we have $m_{00}(d) = 1 + \sum_{j \ne 0} p_j m_{j0}(d) \le 1 + \lambda/\epsilon$.

To prove (iv), we apply Corollary C.2.4 with $r(i) = Ki^{n+1}$, where K is a positive number to be specified later. The first inequality in (C.16) holds for $i \ge 0$, since $\lambda^{(n+1)} < \infty$. Note that

$$r(i+k) - r(i) = K \sum_{u=0}^{n} \binom{n+1}{u} i^u k^{n+1-u}. \quad (7.46)$$

After some algebraic manipulation we find that the left side of the second inequality in (C.16) may be written as

$$Q(i) =: K \sum_{u=0}^{n} \binom{n+1}{u} i^u \left\{ a_K \sum_j p_j(j-1)^{n+1-u} + (1-a_K)\lambda^{(n+1-u)} \right\}. \quad (7.47)$$

We see that $\lambda^{(n+1)}$ is the largest moment involved, and hence $Q(i)$ is finite; note that it is a polynomial in i of degree n. By letting $u = n$, we find that its leading coefficient is $-K\epsilon(n+1)$.

Consider the requirement $Q(i) \le -C(i, d)$ in (C.14). This is true if $Q(i) \le -(H(i) + C(a_K))$. It is equivalent to $Q(i) + H(i) + C(a_K) \le 0$. So it is clearly sufficient to prove that $U(i) =: Q(i) + Bi^n + C(a_K) \le 0$.

Now U is a polynomial in i of degree n with leading coefficient $-K\epsilon(n+1)+B$.

If K is chosen to satisfy $K > B[\epsilon(n+1)]^{-1}$, then the leading coefficient of U is negative.

A polynomial with negative leading coefficient is negative for sufficiently large i, say $i > i^*$. Then we may let $H^* = [0, i^*]$, and the hypotheses of Corollary C.2.4 hold.

Then from Corollary C.2.4 and (iii) it follows for $i \geq 1$ that $c_{i0}(d) \leq Ki^{n+1} + Fi/\epsilon \leq Di^{n+1}$, where $D = K + F/\epsilon$. Finally $c_{00}(d) = \sum_{j \neq 0} p_j c_{j0}(d) \leq D\lambda^{(n+1)}$. □

Here is the main result.

Proposition 7.6.7. Assume that the (BA) assumptions hold. Then:

(i) The (SEN) assumptions hold, and the ACOE is valid. Moreover h is nonnegative and increasing in i. Letting $H^*(i) = \sum_j p_j[h(i+j) - h(i+j-1)]$ for $i \geq 1$, the ACOE may be written as $J = \sum p_j h(j)$ for $i = 0$, and

$$J + h(i) = H(i) + \sum_j p_j h(i+j) + \min_a \{C(a) - aH^*(i)\}, \qquad i \geq 1. \tag{7.48}$$

(ii) If $H(i)$ is unbounded, then (CAV*) holds, and any optimal stationary policy e is standard with $R_e = [0, \infty)$. If $C(a) = Ca + C^*$ for a positive constant C and for $C^* \geq -H(1)$, then $e = d$.

(iii) Assume that $H(i)$ is unbounded, and let e be a stationary policy realizing (7.48). Assume that e breaks ties by always choosing to serve at the lowest rate satisfying (7.48). Then $e(i)$ is increasing in i and eventually chooses rate a_K.

Proof: The proof of (i) is very similar to the proof of Proposition 7.6.3(i–ii) and is left as Problem 7.10.

To prove (ii), observe that the (CAV*) assumptions clearly hold for d. From Theorem 7.5.6(ii) and Lemma 7.6.6(i), it follows that the MC induced by e is positive recurrent on $[0, \infty)$. Since $J_e = J < \infty$, it is the case that e is standard.

Now assume that $C(a) = Ca + C^*$ as in (ii). We wish to prove that $e = d$. We employ Proposition C.1.7. The drift $\gamma_i(e)$ is easily calculated to be $\gamma_0 = \lambda$ and $\gamma_i(e) = \lambda - e(i)$ for $i \geq 1$.

Now by Proposition C.1.7 it follows that the mean drift $\sum \pi_i(e)\gamma_i(e) = 0$. This implies that $\sum_{i=1}^{\infty} \pi_i(e)e(i) = \lambda$. Then

$$J_e = \sum_{i=1}^{\infty} \pi_i(e)(H(i) + C(e(i)))$$

$$= \sum_{i=1}^{\infty} \pi_i(e)(H(i) + C^*) + C\lambda. \tag{7.49}$$

The problem of minimizing the average cost then becomes equivalent to minimizing the average "holding cost" for a holding cost of 0 in state 0 and $H(i) + C^* \geq 0$ in $i \geq 1$. But it is clear that this is minimized by always serving at maximum rate; hence $e = d$. This proves (ii).

We now prove some facts about $H^*(i)$. Since h is increasing in i, it follows that $H^*(i) \geq 0$. Using the same argument as in the proof of Proposition 7.6.3, it may be shown that $h(i+j) - h(i+j-1)$ is the minimum J revised cost of a first passage from $i + j$ to $i+j-1$ and that this quantity is increasing in i for each fixed j. This implies that $H^*(i)$ is increasing in i. Moreover we have $H^*(i) \geq p_0[h(i) - h(i-1)] = p_0[c_{ii-1}(e) - Jm_{ii-1}(e)] \geq p_0(H(i) - J)$. Since H is unbounded, it follows that H^* is also unbounded.

From (7.48) it follows that $e(k)$ satisfies

$$C(e(k)) - C(a) \leq H^*(k)(e(k) - a), \qquad \text{all } a, \tag{7.50}$$

and that the inequality in (7.50) is strict for $a < e(k)$. Now assume that $e(i) > e(i+1)$. We wish to obtain a contradiction. Applying (7.50) and the convention to $k = i$ yields

$$H^*(i) > \frac{C(e(i)) - C(e(i+1))}{e(i) - e(i+1)}. \tag{7.51}$$

Applying (7.50) to $k = i + 1$ yields that $H^*(i + 1)$ is less than or equal to the quantity on the right of (7.51). This contradicts the fact that H^* is increasing and thus proves that $e(i)$ is increasing in i.

Now assume that e does not eventually serve at rate a_K. Because there are only finitely many rates, there must exist a rate $a^* < a_K$ and a sequence $i_r \to \infty$ such that $e(i_r) - a^*$. Then (7.51) yields

$$H^*(i_r) \leq \frac{C(a_K) - C(a^*)}{a_K - a^*}. \tag{7.52}$$

But this contradicts the fact that H^* is unbounded. It proves (iii). □

The result for linear service costs is a somewhat counterintuitive result. Perhaps it can be said that when $C^* \geq -H(1)$, then the balance requirement in

Proposition C.1.7 dominates and forces maximum service. Note that if $K = 2$, it is not necessarily true that the optimal policy equals d. The reason is that even though $C(a)$ is linear in a in this case, we may have $C^* < -H(1)$.

Example 7.6.8. This is Example 2.1.4. To simplify the presentation, we assume that $K = 2$ and that there is no cost for changing the routing decision. Hence $S = \{(i_1, i_2) | i_1$ and i_2 nonnegative integers$\}$. Rather than attempting to prove the most general result possible, we give assumptions under which (CAV*) holds. The basic assumptions (BA) are as follows:

(BA1). The holding cost $H_k(i_k)$ is increasing and unbounded for $k = 1, 2$.

(BA2). There exist a (finite) positive constant B and nonnegative integer n such that $H_k(i_k) \leq B i_k^n$ for $k = 1, 2$.

(BA3). We have $0 < p_0$, $0 < \lambda < \mu_1 + \mu_2$, and $\lambda^{(n+1)} < \infty$. □

Note that $0 < \lambda$ implies that $p_0 < 1$. A *fixed splitting* is a randomized stationary policy $d(w)$ defined by the probability distribution $(w, 1 - w)$. The interpretation is that an arriving batch is sent to the first server with probability w and to the second server with probability $1 - w$. This is implemented by a randomization that is performed before the batch size is observed. Recall that the arrival slot is taken up with routing and packets are available for service in the following slot.

We have discussed the fact that any randomized stationary policy induces a MC on S. In this case the costs are $C(\mathbf{i}, d(w)) = H_1(i_1) + H_2(i_2)$. For i_1 and i_2 both positive, we have, for example, $P_{\mathbf{i}(i_1+j, i_2-1)}(d(w)) = wp_j(1 - \mu_1)\mu_2$. Other transition probabilities are obtained similarly.

Lemma 7.6.9. Assume that (BA2–3) hold, and let $w = \mu_1/(\mu_1 + \mu_2)$. Then $d(w)$ is standard with $R_{d(w)} = S$.

Proof: One can easily see that any fixed splitting with $0 < w < 1$ induces an irreducible MC on S. Such a fixed splitting actually induces two independent MCs, one governing the first buffer and the second governing the second buffer. Each buffer behaves as in the previous example, with a fixed service rate.

Now let $w = \mu_1/(\mu_1 + \mu_2)$. We apply the result in Lemma 7.6.6. The first buffer has mean arrival rate $w\lambda$ and service rate μ_1. Since $w\lambda < \mu_1$, it follows from Lemma 7.6.6 that the induced MC is positive recurrent. It also follows that the average cost $J_{d(w)}(1)$ for the first buffer is finite. Similar remarks are valid for the second buffer.

Then the MC induced by $d(w)$ is positive recurrent and $\pi_{\mathbf{i}}(d(w)) = \pi_{i_1}(d(w))\pi_{i_2}(d(w))$, since the two buffers operate independently. Moreover we see that

$$J_{d(w)} = \sum_{\mathbf{i}} \pi_{i_1}(d(w))\pi_{i_2}(d(w))\{H_1(i_1) + H_2(i_2)\}$$

$$= \sum_{i_1} \pi_{i_1}(d(w))H_1(i_1) + \sum_{i_2} \pi_{i_2}(d(w))H_2(i_2)$$

$$= J_{d(w)}(1) + J_{d(w)}(2), \tag{7.53}$$

and hence the average cost under $d(w)$ is the sum of the average costs associated with each buffer. □

Proposition 7.6.10. Assume that the (BA) assumptions hold. Then the (CAV*) assumptions hold and any optimal stationary policy is positive recurrent at (0, 0).

Proof: Clearly Lemma 7.6.9 and (BA1) imply that (CAV*) holds. Let e be an optimal stationary policy. Then the MC induced by e has a positive recurrent state. Assume that it is $\mathbf{i} \neq (0, 0)$. It is easy to see that $\mathbf{i}$ leads to (0, 0) under any stationary policy. Hence (0, 0) and $\mathbf{i}$ must communicate, and thus the MC is positive recurrent at (0, 0). □

Problem 7.11 asks you to prove some additional properties associated with this example.

7.7 WEAKENING THE (SEN) ASSUMPTIONS

The (SEN) assumptions and the stronger (BOR) and (CAV) assumptions suffice for many models we wish to optimize under the average cost criterion. In certain models none of these assumption sets can be verified, and we need weaker assumptions. In addition there is merit in "picking apart" the proof of Theorem 7.2.3 to determine exactly what makes it work. We will not attempt to find the absolutely weakest conditions under which the conclusions of Theorem 7.2.3 hold. Rather, we give a useful set (H) of assumptions under which they hold. It will be clear from the proof of Proposition 7.7.2 how to further weaken (H) if necessary.

The set (H) of assumptions is weaker than (SEN). The idea is to weaken (SEN3) by allowing the constant L to be a function. In this case additional assumptions are required. The (H) assumptions are as follows:

(H1) = (SEN1).

(H2) = (SEN2).

(H3). There exists a nonnegative (finite) function L such that $-L(i) \le h_\alpha(i)$ for $i \in S$ and $\alpha \in (0, 1)$.

(H4). We have $\sum_j P_{ij}(a)L(j) < \infty$ for $i \in S$ and $a \in A_i$.

(H5). Let h be any limit function and e any stationary policy. Then for all initial states, we have:

(i) $-\infty < E_e[h(X_n)]$ for $n \ge 2$,
(ii) $\liminf_{n \to \infty} E_e[h(X_n)]/n \ge 0$.

The next result shows the relationship among (SEN), (H), and a slightly stronger version of (H), which we denote by (H*).

Proposition 7.7.1. Let (H*) be the set (H) of assumptions with (H5) replaced by:

(H*5). Given any stationary policy e and any initial state, we have the following:

(i) $E_e[L(X_n)] < \infty$ for $n \ge 2$,
(ii) $\lim_{n \to \infty} E_e[L(X_n)]/n = 0$.

Then (SEN) $\Rightarrow$ (H*) $\Rightarrow$ (H).

Proof: Assume that the (SEN) assumptions hold, and set $L(.) = L$ from (SEN3). Then clearly (H3-4) and (H*5) hold for the constant L. Hence (H*) holds.

To prove that (H*) $\Rightarrow$ (H), it is sufficient to show that (H*5) $\Rightarrow$ (H5). To show (H5), let h be a limit function and let e be a stationary policy. It follows from (H3) that $-E_e[L(X_n)] \le E_e[h(X_n)]$. This together with (H*5) easily implies (H5). □

Here is the existence result under (H).

Proposition 7.7.2. Let Δ be an MDC for which the (H) assumptions hold. Then the conclusions of Theorem 7.2.3 are valid where L is the function from (H3).

Proof: We will follow the approach in the proofs of Lemma 7.2.1 and Theorem 7.2.3 and indicate the necessary changes.

An examination of the proof of Theorem 7.2.3 shows that the steps continue to be valid under (H) up to and including (7.16). We may write (7.16) as

$$(1-\gamma_n)V_{\gamma_n}(z) + h_{\gamma_n}(i) + \gamma_n \sum_j P_{ij}(a(i))L(j)$$

$$= C(i,a(i)) + \gamma_n \sum_j P_{ij}(a(i))\{h_{\gamma_n}(j) + L(j)\}, \tag{7.54}$$

where (H4) implies that the term added to each side is finite. We then take the limit infimum of both sides as before and use Proposition A.1.7 to justify moving the limit infimum across the summation. This yields (7.17) and thus the ACOI (7.9).

Let e be a stationary policy realizing the minimum in (7.9), and observe that (7.4) holds for e. Now examine the proof of Lemma 7.2.1. To avoid introducing an indeterminate form in (7.6), it is necessary to have $-\infty < E_e[h(X_n)] < \infty$ for all n. The right inequality follows as in the proof of Lemma 7.2.1. The left inequality, for $n \geq 2$, is (H5)(i). For $n = 1$ it follows from (H3–4).

The proof then proceeds as before, where (7.7) becomes

$$\frac{v_{e,n}(i)}{n} \leq J + \frac{h(i) - E_e[h(X_n)]}{n}. \tag{7.55}$$

We then take the limit supremum of both sides and use (H5)(ii) to obtain

$$J_e(i) \leq J - \liminf_{n\to\infty} \frac{E_e[h(X_n)]}{n} \leq J. \tag{7.56}$$

This proves that $J_e(i) \leq J$.

Going back to the proof of Theorem 7.2.3, we see that (7.18) is valid. This proves that e is optimal with constant average cost J. The arguments for (i) and (iv) are as before.

To prove (7.10), use (iv) and take the limit infimum of both sides of (7.55) to obtain $J \leq J - \limsup_n E_e[h(X_n)]/n$. This together with (H5)(ii) yields $0 \leq \liminf_n E_e[h(X_n)]/n \leq \limsup_n E_e[h(X_n)]/n \leq 0$, which proves (7.10).

The proof of (iii) is similar, and we omit it. □

Here is a useful set of sufficient conditions for the (H*) assumptions to hold.

Proposition 7.7.3. Assume that there exists a distinguished state z such that the following conditions hold:

(i) There exists a (finite) function B such that $m_{iz}(e) \leq B(i)$ for all stationary policies e and $i \in S$.

(ii) For all $i \in S$ and any stationary policy e, we have $b_{iz}(e) < \infty$, where this is the B *cost* of a first passage.

(iii) There exists a stationary policy d such that $c_{iz}(d) < \infty$ for all $i \in S$.

Then the (H*) assumptions hold.

Proof: It follows from (i) and (iii) that d is z standard. It then follows from Proposition 7.5.3 that (H1–2) hold. Notice also that every stationary policy is z standard with respect to the B cost.

Assume that the process starts in $i \neq z$ and operates under the discount optimal stationary policy f_α. It must reach z in a finite expected amount of time; let T_i be the time to reach z. In a similar manner to (7.24), and suppressing the initial state, we obtain

$$\begin{aligned} V_\alpha(i) &= E_{f_\alpha}\left[\sum_{t=0}^{T_i - 1} \alpha^t\, C(X_t)\right] + E_{f_\alpha}[\alpha^{T_i}]V_\alpha(z) \\ &\geq E_{f_\alpha}[\alpha^{T_i}]V_\alpha(z). \end{aligned} \tag{7.57}$$

This implies that

$$h_\alpha(i) \geq -E_{f_\alpha}\left[\frac{1 - \alpha^{T_i}}{1 - \alpha}\right](1 - \alpha)V_\alpha(z). \tag{7.58}$$

By (H1) there exists a (finite) number U such that $(1 - \alpha)V_\alpha(z) \leq U$. We know that $(1 - \alpha^{T_i})/(1 - \alpha) \leq T_i$, and hence we see that $h_\alpha(i) \geq -m_{iz}(f_\alpha)U \geq -UB(i)$. Thus we may let $L(i) = UB(i)$ for $i \neq z$. This verifies (H3).

It is sufficient to verify (H4) and (H*5) for the function B and an arbitrary stationary policy e. Since e is standard with respect to the B cost, it follows from Proposition C.2.6 that the average B cost is finite and is obtained as a limit. Let the average B cost be denoted by K_e. The fact that $K_e < \infty$ implies that (H4) and (H*5)(i) hold.

Now let w_n be the expected n horizon B cost under the policy e. Then we have

$$\frac{w_{n+1}(i)}{n+1} = \frac{w_n(i)}{n}\left(\frac{n}{n+1}\right) + \frac{E_e[B(X_n)|X_0 = i]}{n}\left(\frac{n}{n+1}\right). \tag{7.59}$$

The limit of the term on the left exists and equals $K_e < \infty$, as does the first term on the right. Hence the limit of the second term must exist and equal 0, and this proves (H*5)(ii). □

Here is an example for which the (H*) assumptions hold but for which the (SEN) assumptions may fail.

Example 7.7.4. This is a priority queueing system. The setup is shown in

Fig. 1.4, but with two buffers. There is a single server and the probability of a successful service in any slot is μ, where $0 < \mu < 1$. Services slot to slot are independent. Buffer 1, containing priority customers, has priority over buffer 2.

The state of the system is (i, x), where i is the number of packets in buffer 1 and x the number in buffer 2. The probability of a batch of size j arriving to buffer 1 is p_j and (independently) the probability of a batch of size y arriving to buffer 2 is q_y. Let $\lambda^{(n)}$ (respectively, $\omega^{(n)}$) be the nth moment of the arrival process to buffer 1 (respectively, buffer 2).

The server always serves the priority queue if there are packets in its buffer. When its buffer is empty, then the server is free to serve packets in buffer 2. Observe that the priority buffer is not affected in any way by the second buffer and does not even "see" it. Control may be exercised on the second buffer. The action a results in a batch arriving to buffer 2 being admitted, whereas action r results in the batch being rejected and lost.

There is a nonnegative holding cost $H(x)$ on the content of buffer 2 and a nonnegative rejection cost of $R(i)$ for choosing action r. Observe that the cost of rejecting a batch arriving to buffer 2 may depend on the contents of buffer 1. We will be assuming that $R(i)$ is decreasing in i. As the priority buffer becomes fuller, it costs less to reject batches to the second buffer. This has the effect that when the first buffer does clear, the second buffer will not be overloaded. Formally we have $C[(i,x),a] = H(x)$ and $C[(i,x),r] = H(x) + R(i)$.

The basic assumptions (BA) for this model are as follows:

(BA1). There exist a (finite) constant U and integer $n \ge 1$ such that $H(x) \le Ux^n$.

(BA2). The rejection cost $R(i)$ is decreasing in i.

(BA3). We have $\lambda + \omega < \mu$.

(BA4). The moments $\lambda^{(n+1)}$ and $\omega^{(n+1)}$ are finite. □

Proposition 7.7.5. Assume that the (BA) assumptions hold. Then the (H*) assumptions hold. For at least one value of the rejection cost, the (SEN) assumptions fail to hold.

Proof: We verify that the conditions in Proposition 7.7.3 are satisfied. We will show that any stationary policy e is $z = (0, 0)$ standard. Let $\epsilon =: \mu - (\lambda + \omega) > 0$. We employ Corollary C.1.6 with $y(i,x) = i + x$. It is easy to see that (C.10) holds and yields $m_{(i,x)z}(e) \le (i + x)/\epsilon$ for $(i,x) \ne z$. Moreover we have $m_{zz}(e) \le \mu/\epsilon$. Hence (i) holds with $B(z) = \mu/\epsilon$ and $B(i,x) = (i + x)/\epsilon$ for $(i,x) \ne z$.

Now turn to the expected cost of a first passage to z. Since $R(i)$ is decreasing, it follows that $R(i) \le R(0)$, and hence the rejection cost is bounded. It is sufficient to show that the expected holding cost $g_{(i,x)z}(e)$ of a first passage is finite. This may be done in a similar way to the proof of Lemma 7.6.6 with $r(i,x) = K(i+x)^{n+1}$. We leave the details to Problem *7.12. This proves that (iii) holds for any stationary policy.

It remains to show that the expected B cost of a first passage is finite. This follows from the definition of B and from what we have claimed (and left as a problem) for the first passage holding costs. The reason is that in (BA1) we have assumed that $n \geq 1$, and hence by (BA4) the second moments are finite. Then the proof for the finiteness of the first passage holding costs also yields the finiteness of the first passage B costs. This verifies that the conditions in Proposition 7.7.3 hold, and hence that (H*) holds.

Let us now argue that for some value of the rejection cost, (SEN3) fails to hold. The key is to observe that the result in Lemma 7.4.2 remains valid under (H*) if the function L is bounded on the set G. This is easy to check, and we leave it to the reader. Now let d be the policy that always rejects arriving batches to buffer 2. Then it follows from Lemma 7.4.2 (with G = $\{z\}$) that $h(i,0) \leq c_{(i,0)z}(d) - Jm_{(i,0)z}(d)$.

Let m be the expected time to go from i to $i-1$ in buffer 1 (clearly these times all have the same distribution). Then $m_{(i,0)z}(d) = im$. Moreover $c_{(i,0)z}(d) = m(R(i) + R(i-1) + \ldots + R(1))$.

Let us choose $R(i) = 1/i$ for $i \geq 1$, with $R(0) \geq 1$ arbitrary. Then

$$h(i,0) \leq m\left(\left[\sum_{j=1}^{i} \frac{1}{j} - \ln i\right] + \ln i - Ji\right). \tag{7.60}$$

As $i \to \infty$, the expression in square brackets on the right of (7.60) approaches a constant γ, known as Euler's constant. See Apostol (1972) for details. Thus $\lim_{i\to\infty} h(i) \leq m[\gamma + \lim_{i\to\infty}(\ln i - Ji)] = -\infty$. This implies that (SEN3) cannot hold. □

Remark 7.7.6. Several weaker "sequence versions" of the (H) assumptions can be given. For example, assume that $f_{\alpha_n} \to f$ for some sequence of optimal discount policies with $\alpha_n \to 1^-$. We can modify (H) to guarantee only that the particular limit point f is average cost optimal. Problem *7.13 asks you to give such a set $(\mathrm{H})_f$ of assumptions and to determine how the statement of Theorem 7.2.3 should be modified. □

BIBLIOGRAPHIC NOTES

The subject treated in this chapter has a large and diverse literature, making it an especially difficult topic for a person not immersed in the details. One purpose of Chapter 7 is to organize results and fit them into an intelligible framework.

We now attempt, to the best of our ability, to give credit to the originators of the ideas in this chapter.

Reward versions of Examples 7.1.3 and 7.1.4 appear in Ross (1983) and Puterman (1994). Example 7.1.5 is due to Ross (1971) and also appears in Ross (1983). Fisher and Ross (1968) contains a more complex example of an MDC with no average cost optimal stationary policy but for which every stationary policy gives rise to an irreducible standard MC.

The results in Section 7.2 are largely from Sennott (1989a). Lemma 7.2.1 appears there. The version of the assumptions given in Sennott (1989a) differs slightly from the (SEN) assumptions. This current version, which appears in Sennott (1993), is clearer than the original version. For some additional coments, see Cavazos-Cadena (1991c).

Theorem 7.2.3 is a modification of the main result in Sennott (1989a). Part (iv) is new and is based on Proposition 6.1.1 (iii) $\Rightarrow$ (ii). This result is classical (see Bibliographic Notes for Appendix A), but we were unaware of it when Sennott (1989a) was written.

A somewhat similar set (SCH) of assumptions is presented in Schal (1993) for general (uncountable) state spaces. Problem 7.6 shows that (SEN) and (SCH) are equivalent.

The impetus for the (SEN) assumptions came from a realization that the assumptions in Ross (1983) could be weakened. The results in Ross (1983) were based on Ross (1968). Earlier pivotal work is Taylor (1965) and Derman (1966).

The example in Section 7.3 is a minor modification of the one in Cavazos-Cadena (1991b).

The idea for Lemma 7.4.1 comes from Theorem 2.4 of Ross (1983) and is used in Sennott (1986a), which is an early version of Sennott (1989a). The impetus for the rest of the material in Section 7.4 comes from Cavazos-Cadena (1991a) where many of these results are proved under somewhat stronger assumptions. We modified the proofs to hold under the (SEN) assumptions in Sennott (1993). The proofs presented here have been simplified from the earlier versions. Additional references are Derman and Veinott (1967) and Makowski and Shwartz (1994).

Lemma 7.5.2 appears in Cavazos-Cadena and Sennott (1992). The ideas in Proposition 7.5.3 and Corollary 7.5.4 appear in Sennott (1989a). Proposition 7.5.5 is based on a result in Cavazos-Cadena (1989).

The assumptions in Theorem 7.5.6 are a modification of an important line of development due to Borkar (1984, 1988, 1989). The Borkar (1991) monograph summarizes his convex analytic approach. The (BOR) assumptions given in Theorem 7.5.6 are weaker than the original assumptions in these papers.

The proof that (BOR) $\Rightarrow$ (SEN) appears in Cavazos-Cadena and Sennott (1992). The version of (BOR) given in Cavazos-Cadena and Sennott (1992) is stronger than this version, which appears in Sennott (1993). The idea for the proof of (ii) comes from Cavazos-Cadena (1989). Some parts of the proof of (iii) are in Sennott (1993) and some are new.

Theorem 7.5.6 shows that strong conclusions hold under (BOR), and for this reason it is an important set of assumptions. Remark 7.5.7 makes reference to another set (WS) of assumptions that appears in Stidham and Weber (1989). We have presented a slightly modified version of this set in Problem 7.7. It lies strictly between (BOR) and (SEN). Part (iii) of Problem 7.7 shows that the conclusions that can be drawn from (WS) are not quite as strong as those that can be drawn from (BOR).

Remark 7.5.8 mentions an example in Sennott (1993) that presents a limitation of the conclusions that can be drawn from (BOR).

The (CAV*) assumptions appear in Cavazos-Cadena (1989), and the (CAV) assumptions are a slight modification of those. An earlier paper by Wijngaard (1978) approaches the Cavazos-Cadena concept but with additional strong assumptions.

Example 7.6.1 is treated in Sennott (1989a) and again in Sennott (1993). The argument for the form of h and the monotonicity of the optimal policy appears (in a slightly less concise way) in Sennott (1993), and these ideas are a minor generalization of results in Stidham and Weber (1989). The seminal Stidham and Weber paper gives fresh ideas about proving structural properties for optimal average cost stationary policies. The arguments are for continuous time but are easily adapted to discrete time.

Example 7.6.4 is treated in Sennott (1989a) and again in Sennott (1993). The crucial argument in Lemma 7.6.6 appears in Sennott (1989a). The argument concerning the structural properties of an optimal stationary policy appears in Sennott (1993) and is based on results in Stidham and Weber (1989).

Example 7.6.8 is discussed in Sennott (1997b).

The (H*) assumptions appear in Sennott (1995) in a slightly different form. The (H*) assumptions are most closely related to Hu (1992) and a whole development due to Hordijk and other researchers. Hordijk (1976, 1977) initiates this line of development. The assumptions are related to those in Proposition 7.7.3 but are not identical to our assumptions. In Hordijk (1976) a Lyapunov condition is assumed, and this work is extended in Hordijk (1977). Other work is Federgruen and Tijms (1978), Federgruen, Hordijk, and Tijms (1979), and Federgruen, Schweitzer, and Tijms (1983). The conditions in these papers are closely related (but apparently not identical) to the assumptions in Proposition 7.7.3. Other development is presented in Spieksma (1990).

A large survey of many approaches to the existence question is presented in Arapostathis et al. (1993). Work on extending the (SEN) assumptions has been performed by Hernandez-Lerma and other researchers. Hernandez-Lerma and Lasserre (1990) extend the existence result to the case of Borel state spaces. An extension is also given in Ritt and Sennott (1992). Some of the above referenced papers also deal with more general state spaces than covered in this book. Hernandez-Lerma (1991) involves an extension of the Schal assumptions to the case of unbounded action sets. Other related work is Hernandez-Lerma (1993) and Montes-de-Oca and Hernandez-Lerma (1994). This line of development has culminated in the book by Hernandez-Lerma and Lasserre (1996).

For an example related to the average cost criterion, see Flynn (1974).

PROBLEMS

7.1. For the policy θ in Example 7.14, prove that $J_\theta(1) = 0$.

7.2. Complete the proof of Theorem 7.2.3(iii).

7.3. Fill in the details in the derivation of (7.24) in the proof of Lemma 7.4.2.

***7.4.** Fill in the omitted details of the proof of Theorem 7.5.6.

7.5. This problem shows that some tempting modifications of the (SEN) assumptions are actually equivalent to (SEN).

(a) Consider the statement: (*) There exists $\alpha_0 \in (0, 1)$ such that $(1 - \alpha)V_\alpha(z)$ is bounded for $\alpha \in (\alpha_0, 1)$. Show that (SEN1) $\Leftrightarrow$ (*). Hence nothing is gained by replacing (SEN1) with (*).

(b) Consider the statement: (**) There exist $\alpha_0 \in (0, 1)$, a finite nonnegative function M, and a finite nonnegative constant L such that $-L \le h_\alpha(i) \le M(i)$ for $i \in S$ and $\alpha \in (\alpha_0, 1)$. Show that (SEN2–3) $\Leftrightarrow$ (**). Hence nothing is gained by replacing (SEN2–3) with (**).

7.6. Define $w_\alpha = \inf_{i \in S} V_\alpha(i)$. Consider the following set (SCH) of assumptions:

(SCH1). The quantity $(1 - \alpha)w_\alpha$ is bounded for $\alpha \in (0, 1)$.

(SCH2). There exists a (finite) function W such that $V_\alpha(i) - w_\alpha \le W(i)$ for $i \in S$ and $\alpha \in (0, 1)$.

Prove that (SEN) holds if and only if (SCH) holds.

***7.7.** Consider the following set (WS) of assumptions:
(WS1) = (BOR1).

(WS2). The set $D^* = \{i | C(i, a) \le J_d$ for some $a\}$ is finite.

(WS3). Given $i \in D^* - R_d$, there exists a policy $\theta_i \in \mathfrak{R}^*(z, i)$.

(a) Prove that (BOR) $\Rightarrow$ (WS) $\Rightarrow$ (SEN). The proof of (SEN3) follows with minor modifications to the proof of Theorem 7.5.6(i). Check the details.

(b) Let e be an optimal stationary policy and $D(e) = \{i | C(i, e) \le J\}$. Prove that e has at least one positive recurrent state in $D(e)$.

(c) Construct an example for which (WS) holds but such that e realizing the minimum in (7.9) satisfies $e \notin \mathfrak{R}(i, D(e))$ for some i. This example shows that while (WS) is only slightly weaker than (BOR), it cannot guarantee that an optimal stationary policy realizing the ACOI induces an MC with a "nice" structure.

7.8. Prove Corollaries 7.5.9 and 7.5.10.

7.9. Prove Lemma 7.6.2.

7.10. Prove Proposition 7.6.7(i). *Hint:* This follows much as the proof of Proposition 7.6.3(i–ii).

7.11. This problem concerns Example 7.6.8. Assume that (BA) holds.

(a) Give the finite horizon discounted optimality equations.

(b) Prove that $v_{\alpha,n}(i_i, i_2)$ is increasing in one coordinate when the other coordinate is held fixed. *Hint:* Prove this by induction; it is only necessary to argue it for the first coordinate.

(c) Use the result in (b) to prove that the (SEN) assumptions hold even if the holding costs are bounded.

(d) Now assume that there exists a cost for changing routing decisions. Show that the (CAV*) assumptions still hold. *Hint:* What is the new state space, and how does it behave under $d(w)$?

***7.12.** Fill in the omitted details in the proof of Proposition 7.7.5.

***7.13.** Carry out what is requested in Remark 7.7.6.

7.14. There is a single server, and the probability of successfully serving a packet is μ, where $0 < \mu < 1$. This server serves two buffers (see Fig. 1.4). The probability of a batch of size j arriving to the first buffer is p_j, and the probability of a batch of size y arriving to the second buffer is q_y. The two arrival processes are independent, and $\lambda^{(n)}$ (respectively, $\omega^{(n)}$) is the nth moment of the arrival process to the first buffer (respectively, second buffer).

In each slot the decisions are a = serve (or be in front of, if it is empty) buffer 1, and b = serve (or be in front of, if it is empty) buffer 2. Let (i, x) denote the buffer status, where i is the number of packets in the first buffer and x the number in the second buffer. There are nonnegative holding costs $H(i)$ and $K(x)$ and a cost for changing buffers.

The (BA) are:

(BA1). The holding cost $H(i)$ is increasing and unbounded in i and similarly for $K(x)$.

(BA2). There exist a (finite) constant U and nonnegative integers n and m such that $H(i) \le Ui^n$ and $K(x) \le Ux^m$ for $i, x \ge 0$.

(BA3). We have $0 < p_0 < 1$, $0 < q_0 < 1$, and $\lambda + \omega < \mu$.

(BA4). The moments $\lambda^{(n+1)}$ and $\omega^{(m+1)}$ are finite.

(a) Set this model up as an MDC.

(b) Prove that the (CAV*) assumptions hold.

7.15. This problem summarizes the relationships among all the assumption sets that have been introduced in this chapter. Observe that

$$\begin{array}{ccccc}(\text{CAV}^*) \Rightarrow (\text{CAV}) \Rightarrow (\text{BOR}) \Rightarrow (\text{WS}) \Rightarrow & (\text{SEN}) & \Rightarrow (\text{H}^*) \Rightarrow (\text{H}) \\ & \Updownarrow & \\ & (\text{SCH}) & \end{array} \tag{7.61}$$

The first implication is in Corollary 7.5.10, and the second is in Corollary 7.5.9. The third and fourth implications are in Problem *7.7. The equivalence is in Problem 7.6, and the last two implications follow from Proposition 7.7.1.

Equation (7.61) provides a road map of possible assumption sets to use in verifying the existence of an average cost optimal stationary policy.

The claim is that each of the implications on the top row of (7.61) is nonreversible (but we do not concern ourselves with the last implication). The example in Problem *7.7(c) shows that (WS) does not imply (BOR). Example 7.7.4 shows that (H*) does not imply (SEN).

(a) Construct an example for which (CAV) holds but for which (CAV*) fails.

(b) Construct an example for which (BOR) holds but for which (CAV) fails.

(c) Provide an example for which (SEN) holds but for which (WS) fails.

CHAPTER 8

Computation of Average Cost Optimal Policies for Infinite State Spaces

In Chapter 7 the existence theory was developed for the case of a countable state space. By means of this theory we are able to prove that average cost optimal stationary policies exist in a wide variety of models. However, the existence theory does not yield a method for the computation of an optimal policy.

In this chapter we develop the approximating sequence method for the computation of an average cost optimal stationary policy when the state space is denumerably infinite. Throughout this chapter we have an MDC Δ with a denumerable state space and an approximating sequence (Δ_N). We will require that (1) the minimum average cost in Δ_N be constant, (2) the sequence of constant minimum average costs in (Δ_N) converge to the (constant) minimum average cost in Δ, and (3) any limit point of a certain sequence of optimal stationary policies for (Δ_N) be optimal for Δ.

It might seem natural to begin with one of the assumption sets from Chapter 7 for the existence of an optimal stationary policy and then add additional assumptions in order to carry out the computational program. However, this is not the approach we take. Recall that in Chapter 3 we introduced Assumption FH and in Chapter 4 we introduced Assumption DC, both related to properties of the approximating sequence. A similar approach is followed in this chapter. A set (AC) of assumptions is introduced specifically to guarantee that the computational program can be carried out. The results in this chapter are largely independent of the material in Chapter 7. Selected results from Chapter 7 will occasionally be called upon. If the reader has omitted Chapter 7, then these results may be scanned as they are needed.

In Section 8.1 the (AC) assumptions are introduced, and the major result of the chapter is proved. In Section 8.2 we discuss the verification of these assumptions.

In Section 8.3 we show how to verify the (AC) assumptions for several models, including the single-server queue with reject option, the single-server queue with controllable service rates, and the routing to parallel queues model. In Section *8.4 the routing model with a cost for changing the decision is treated. In Section 8.5 we give computational results for the single-server queue with controllable service rates, and in Section 8.6 computational results for the routing model are presented.

Section 8.7 presents a generalization of the (AC) assumptions. This material is useful in Chapter 9.

8.1 THE (AC) ASSUMPTIONS

Let Δ be an MDC with a denumerable state space S. The objective is to compute an average cost optimal stationary policy. This is done by computing optimal stationary policies in an approximating sequence and showing that any limit point of these policies is average cost optimal for Δ.

We now give a set (AC) of assumptions that allows this to be accomplished. Let us assume that we have an AS $(\Delta_N)_{N \geq N_0}$ for Δ. The (AC) assumptions are as follows:

(AC1). There exist a (finite) constant J^N and (finite) function r^N on S_N such that

$$J^N + r^N(i) = \min_a \left\{ C(i,a) + \sum_{j \in S_N} P_{ij}(a;N) r^N(j) \right\}, \quad i \in S_N, N \geq N_0. \tag{8.1}$$

(AC2). We have $\limsup_{N \to \infty} r^N(i) < \infty$ for $i \in S$.

(AC3). There exists a nonnegative (finite) constant Q such that $-Q \leq \liminf_{N \to \infty} r^N(i)$ for $i \in S$.

(AC4). We have $\limsup_{N \to \infty} J^N =: J^* < \infty$ and $J^* \leq J(i)$ for $i \in S$.

Here is the major result of the chapter. It utilizes Lemma 7.2.1.

Theorem 8.1.1. Assume that the (AC) assumptions hold. Then:

(i) The quantity $J^* = \lim_{N \to \infty} J^N$ is the minimum average cost in Δ.
(ii) Any limit point e^* of a sequence e^N of stationary policies realizing the minimum in (8.1) is average cost optimal for Δ.

Proof: It follows from (AC1) and Proposition 6.5.1(ii) that J^N is the constant minimum average cost in Δ_N and that any stationary policy realizing the minimum in (8.1) is average cost optimal for Δ_N.

Let e^N realize the minimum in (8.1). Fix a sequence N_x. By Proposition B.5 there exist a subsequence N_u of N_x and a stationary policy e^* such that $\lim_{u\to\infty} e^{N_u} = e^*$. This implies that $e^{N_u}(i) = e^*(i)$ for sufficiently large u (dependent on i).

By (AC4) there exist a subsequence N_v of N_u and a number J_0 such that $\lim_{v\to\infty} J^{N_v} = J_0 < \infty$. Let $w(i) = \liminf_{v\to\infty} r^{N_v}(i)$. It follows from (AC2–3) that w is a finite function bounded below by $-Q$.

For a fixed state $i \in S$ and sufficiently large v, (8.1) may be written

$$J^{N_v} + r^{N_v}(i) = C(i, e^*) + \sum_{j\in S_{N_v}} P_{ij}(e^*; N_v) r^{N_v}(j). \tag{8.2}$$

Take the limit infimum of both sides of (8.2), and employ Proposition A.2.5 to obtain

$$J_0 + w(i) \geq C(i, e^*) + \sum_j P_{ij}(e^*) w(j). \tag{8.3}$$

Since this argument may be carried out for each i, it is the case that (8.3) holds for $i \in S$.

It then follows from Lemma 7.2.1 that $J_{e*}(i) \leq J_0$ for all i. Using (AC4), we see that $J_{e*}(.) \leq J_0 \leq J^* \leq J(.) \leq J_{e*}(.)$, and hence these terms are all equal. This proves that e^* is average cost optimal with constant average cost $J_0 = J^*$.

Since the argument may be carried out for any initial sequence, it follows that the limit in (i) must hold. □

Suppose that an approximating sequence has been constructed for a model and that the (AC) assumptions have been verified for a particular sequence r^N. Let us assume that we can compute r^N, and hence J^N and the resulting average cost optimal policy realizing the minimum in (8.1). For N sufficiently large, it follows from Theorem 8.1.1 that $J^N \approx J$, where J is the minimum average cost in Δ, and that e^N is close to optimal for Δ.

In practice we will carry out this operation until J^N is varying by less than some tolerance and e^N is unchanging. Then we may be confident that an average cost optimal policy for Δ has been determined and that a very close approximation to the minimum average cost has been obtained. For some models a complete picture of the optimal policy may not be attainable, and we must be satisfied that it has been computed in the region of the state space S of most interest. This limitation is illustrated in Section 8.6.

8.2 VERIFICATION OF THE ASSUMPTIONS

We will employ the Value Iteration Algorithm 6.6.4 to calculate an average cost optimal stationary policy for Δ_N. Proposition 6.6.3 justifies the VIA and deals with a relative value function $r_n^N(i) = v_n^N(i) - v_n^N(x^N)$, where the *base point* x^N is an arbitrary element of S_N. We will verify the (AC) assumptions for a fixed base point x. After the discussion of the verification methods has been completed, it will be argued that the base point may be chosen arbitrarily (and may vary with N), and the computation will still yield the same average cost optimal stationary policy given in Theorem 8.1.1. (Section *8.4 treats an example for which the transformation in Proposition 6.6.6 is applied to the AS.)

Before beginning this development, the reader is advised to skim the material in Sections C.4 and C.5 of Appendix C. As you do this, feel free to omit the accompanying background results and focus solely on grasping the important notion of conformity. We now discuss how to relate the definitions in Section C.4 to Δ and (Δ_N).

Notice that any stationary policy d for Δ induces a stationary policy $d|N$ for Δ_N, where we have $P_{ij}(d|N) = P_{ij}(d(i); N)$, $i, j \in S_N$. A similar result holds for a randomized stationary policy. This means that any (randomized) stationary policy induces a Markov chain and accompanying approximating sequence for that MC as defined in Definition C.4.1. Now let d be a z standard policy for Δ as defined in Definition 7.5.1. Then (Δ_N) is *conforming at d* if the MC and AS corresponding to d satisfy Definition C.4.8. If d is a (randomized) stationary policy inducing an MC with a positive recurrent class R_d having finite average cost, then the AS is *conforming on R_d* if it satisfies Definition C.4.10. When first encountered these notions seem involved, but they are quite natural. Informally, conformity at d means that the AS is "well-behaved" with respect to the MC induced by d. Thus steady state probabilities and the average cost under $d|N$ converge to the steady state probabilities and average cost under d.

Here we give a template of four steps to validate the VIA and verify the (AC) assumptions.

Proposition 8.2.1. Let $(\Delta_N)_{N \geq N_0}$ be an AS for Δ, and let x be a distinguished state (we may assume that $x \in S_N$ for all N). Carrying out the following four step template justifies the use of the value iteration algorithm in Δ_N and verifies that the (AC) assumptions hold for the function $r^N(.) = \lim_{n \to \infty}(v_n^N(.) - v_n^N(x))$.

Step 1. Show that every stationary policy for Δ_N induces a unichain MC with aperiodic positive recurrent class containing x.

Step 2. Show that there exists an x standard policy d for Δ such that the AS is conforming at d.

Step 3. Do one of the following:

(i) Show that $v_n^N(i) \le v_n(i)$ for all n, N, and $i \in S_N$.

(ii) Show that $V_\alpha^N(i) \le V_\alpha(i)$ for all $\alpha \in (0,1)$, N, and $i \in S_N$.

(iii) Show that the minimum average cost in Δ is constant, and that there exists an average cost optimal stationary policy f inducing an MC with a positive recurrent class R_f such that the AS is conforming on R_f.

Step 4. Do one of the following:

(i) Show that $v_n^N(i) \ge v_n^N(x)$ for all n, N, and $i \in S_N$.

(ii) Show that $V_\alpha^N(i) \ge V_\alpha^N(x)$ for all $\alpha \in (0,1)$, N, and $i \in S_N$.

(iii) Show that there exists a nonempty finite set G such that v_n^N takes on a minimum in G for all n and N. Moreover there exists a stationary policy g inducing an MC with a positive recurrent class $R_g \supset G \cup \{x\}$ having finite average cost and such that the AS is conforming on R_g.

(iv) Show that there exists a nonempty finite set G such that V_α^N takes on a minimin in G, for all N and $\alpha \in (0,1)$. Moreover there exists a stationary policy g inducing an MC with a positive recurrent class $R_g \supset G \cup \{x\}$ having finite average cost and such that the AS is conforming on R_g.

Proof: Under Step 1 it follows from Proposition 6.4.1 that the minimum average cost in Δ_N is constant. Clearly Assumption OPA holds. Hence VIA 6.6.4 may be carried out in Δ_N. It follows from Proposition 6.6.3 that $r^N(.) = \lim_{n\to\infty}(v_n^N(.) - v_n^N(x))$ exists. This provides a solution to (8.1) and verifies that (AC1) holds. We show that (AC2–3) hold for the function r^N and that (AC4) is valid.

Step 2 enables us to verify (AC2). Note that $r^N(x) \equiv 0$, and hence we may assume that $i \neq x$. Note that $v_n^N \le v_m^N \le v_{\theta,m}^N$, where $m \ge n$ and θ is any m step policy for Δ_N. Define θ to follow $d|N$ until state x is reached, and then to follow the optimal finite horizon policy for n steps. Then $v_n^N(i) \le c_{ix}^N(d|N) + v_n^N(x)$, and hence $r^N(i) \le c_{ix}^N(d|N)$. From the conformity at d it follows that $\limsup_{N\to\infty} r^N(i) \le c_{ix}(d) < \infty$.

Step 3 enables us to verify (AC4). We first show that $J(.)$ is finite. Since d from Step 2 is x standard, it follows that $J(i) \le J_d < \infty$ for $i \in S$.

Now assume that Step 3(i) holds. Using Theorem 6.4.2(v), this implies that

$$\begin{aligned} J^N &= \lim_{n\to\infty} \frac{v_n^N(i)}{n} \\ &\le \limsup_{n\to\infty} \frac{v_n(i)}{n} \\ &\le \limsup_{n\to\infty} \frac{v_{\theta,n}(i)}{n} \\ &= J_\theta(i) \end{aligned} \tag{8.4}$$

for any policy θ for Δ. Then clearly $J^N \le J(i)$ for all $i \in S$, and hence $J^* \le J(.) < \infty$.

If Step 3(ii) holds, then the argument is analogous to that in (8.4) but uses Proposition 6.2.3 and then Proposition 6.1.1. We omit the proof.

Now assume that Step 3(iii) holds. In this case we have $J(i) = J_f(i) \equiv J$, for some constant $J < \infty$. This follows since f is optimal and the minimum average cost is constant. Fix $i \in R_f$. Then $J^N \le J^N_{f|N}(i)$. Taking the limit supremum of both sides and using the conformity on R_f yields $J^* \le J$. This completes the verification of (AC4).

Step 4 enables us to verify (AC3). If Step 4(i) holds, then $r^N(i) = \lim_{n\to\infty}(v^N_n(i) - v^N_n(x)) \ge 0$. This verifies (AC3) with $Q = 0$.

Assume that h^N_α from Theorem 6.4.2 is defined using distinguished state x. It then follows from Proposition 6.5.1(iii) and Step 1 that $r^N \equiv h^N$. So if Step 4(ii) holds, then we again have the validity of (AC3) with $Q = 0$.

Now assume that Step 4(iii) holds. From Step 1 it follows that $g|N$ induces a unichain MC with a positive recurrent class $W(N)$ containing x. We claim that for sufficiently large N, it is the case that $W(N) \supset G$. For $j \in G$ it is the case that x and j communicate in the MC induced by g. By the definition of an AS, it is the case that they communicate in the MC induced by $g|N$ for sufficiently large N, say $N \ge N_j$. Hence $j \in W(N)$ for $N \ge N_j$. Then for sufficiently large N, say $N \ge N^*$, we have $G \subset W(N)$. Note that the conformity was not necessary to obtain this result.

Let us assume that $N \ge N^*$ and observe that $c^N_{xj}(g|N) < \infty$ for $j \in G$. Moreover it follows from the conformity on R_g and Proposition C.4.6 that $c^N_{xj}(g|N) \to c_{xj}(g)$. Let $Q =: \max_{j\in G}\{c_{xj}(g)\}$.

For fixed $i \ne x$, n, and N, choose $j \in G$ such that $v^N_n(i) \ge v^N_n(j)$. Using reasoning similar to that in the verification of (AC2), we have

$$r^N_n(i) = (v^N_n(i) - v^N_n(j)) + r^N_n(j) \ge r^N_n(j) \ge -c^N_{xj}(g|N). \tag{8.5}$$

This implies that $\liminf_{N\to\infty} r^N(i) \ge -Q$, and hence (AC3) holds.

The first portion of the proof under Step 4(iv) is as under Step 4(iii). To finish the proof, observe that $r^N \equiv h^N$, and let Q be as in Step 4(iii). Assume that the initial state is x, and fix $j \in G$. We may follow $g|N$ until state j is reached and then follow an α discounted optimal policy in Δ_N. If T denotes the time to reach j, then this yields

$$\begin{aligned} V^N_\alpha(x) &\le c^N_{xj}(g|N) + E[\alpha^T]V^N_\alpha(j) \\ &\le c^N_{xj}(g|N) + V^N_\alpha(j). \end{aligned} \tag{8.6}$$

For fixed $i \ne x$, α, and N, choose $j \in G$ such that $V^N_\alpha(i) \ge V^N_\alpha(j)$. Then it follows from (8.6) that

$$h^N_\alpha(i) = (V^N_\alpha(i) - V^N_\alpha(j)) + h^N_\alpha(j) \ge h^N_\alpha(j) \ge -c^N_{xj}(g|N). \tag{8.7}$$

The proof is then completed as in Step 4(iii). □

The following special case of Proposition 8.2.1 arises frequently:

Corollary 8.2.2. Let Δ have state space $S = \{0, 1, 2, \ldots\}$, let (Δ_N) have state space $S_N = \{0, 1, \ldots, N\}$, and send the excess probability to N. Assume that the following hold:

(i) Every stationary policy for Δ_N induces a unichain MC with aperiodic positive recurrent class containing 0.
(ii) For $n, N \geq 1$ the value function $v_n^N(i)$ is increasing in $0 \leq i \leq N$.
(iii) For $n \geq 1$ the value function $v_n(i)$ is increasing in i.
(iv) There exists a 0 standard policy d for Δ such that $m_{i0}(d)$ and $c_{i0}(d)$ are increasing in $i \geq 1$.

Then the conclusions of Proposition 8.2.1 hold for the function $r^N(i) = \lim_{n \to \infty}(v_n^N(i) - v_n^N(0))$.

Proof: We show that the four-step procedure in Proposition 8.2.1 may be carried out for $x = 0$. Clearly Step 1 holds. To verify Step 2, note that (C.37–38) become the requirements that the mean first passage time and cost from $i \geq 1$ to 0 under d are increasing in i. Then Step 2 follows from Proposition C.5.3 and (iv). Note that (3.19), for $\alpha = 1$, becomes the requirement that $v_n(N) \leq v_N(r)$ for $r > N$. This is equivalent to (iii). Step 3(i) then follows from Proposition 3.3.4. Step 4(i) follows from (ii). □

Next we present a way of validating the four step procedure for an ATAS that sends excess probability to a finite set. It is based on the (BOR) assumptions from Section 7.5. If Chapter 7 has been omitted, then the proof of the next result should be skipped.

Proposition 8.2.3. Assume that the following hold:

(i) There exists a z standard policy d for Δ.
(ii) There exists $\epsilon > 0$ such that $D = \{i | C(i, a) \leq J_d + \epsilon \text{ for some } a\}$ is a finite set.
(iii) There exists a stationary policy g for Δ that induces an MC with a positive recurrent class $R_g \supset D \cup \{z\}$ with finite average cost.
(iv) (It will be shown that there then exists an average cost optimal stationary policy for Δ and that any optimal stationary policy induces an MC with at least one positive recurrent class.) If e is an optimal stationary policy for Δ, we assume that the MC induced by e has a single positive recurrent class, which contains z.

(v) The AS (Δ_N) is an ATAS that sends excess probability to $D \cup \{z\}$.

(vi) Every stationary policy for Δ_N induces a unichain MC with aperiodic positive recurrent class containing z.

Then the VIA and the (AC) assumptions hold for the function $r^N(i) = \lim_{n\to\infty}(v_n^N(i) - v_n^N(z))$.

******Proof:* We will show that the four-step template in Proposition 8.2.1 can be carried out with $x = z$. It follows from (vi) that Step 1 holds.

By Proposition C.5.2 and (v) it follows that the ATAS is conforming at d, and hence Step 2 holds.

Let us now show Step 3(iii). Note that (i–iii) imply that the (BOR) assumptions in Theorem 7.5.6 hold. The condition in (iii), which we denote (BOR3*), is slightly stronger than (BOR3). If (BOR3*) holds, then (BOR3) holds with $\theta_i \equiv g$.

It follows from Theorem 7.5.6 that (7.9) is an equality (the ACOE for Δ) and that any stationary policy e realizing (7.9) is average cost optimal for Δ. It follows from Theorem 7.5.6 and (iv) that e is z standard. It then follows from Proposition C.5.2 that the ATAS is conforming at e, and hence Step 3(iii) holds.

Finally we show that Step 4(iv) holds. We first show that the (BOR) assumptions, with (BOR3*), hold for Δ_N for sufficiently large N. It follows from (vi) that $d|N$ is a z standard policy for Δ_N. This verifies (BOR1).

Since the ATAS is conforming at d, it follows that $J^N_{d|N} \to J_d$. Hence, for N sufficiently large, we have $J^N_{d|N} \le J_d + \epsilon/2$. Then $D_N =: \{i \in S_N | C(i,a) \le J^N_{d|N} + \epsilon/2 \text{ for some } a\} \subset D$. The set D_N satisfies (BOR2).

To verify (BOR3*) for Δ_N, it is sufficient to show that $D \subset R^N_{g|N}$ for N sufficiently large. The proof is similar to that of Step 4(iii) in Proposition 8.2.1, and we omit it.

We have verified that (BOR) holds for Δ_N for sufficiently large N. It then follows from the proof of Theorem 7.5.6 (statement (*) applied to Δ_N) that V^N_α takes on a minimum in $D_N \subset D$. This together with (iii) and Proposition C.5.2 applied to g shows Step 4(iv). □

This completes our discussion of the verification of the (AC) assumptions. In certain examples it is not possible to employ either Proposition 8.2.1 or Proposition 8.2.3, and in these cases a modified approach must be used. Example 8.4.1 illustrates this situation.

Remark 8.2.4. Let us now discuss the base point for the computation, as we promised earlier. The results in Section 8.2 have verified the (AC) assumptions for a specific base point x. It has been shown that $r^N(i) = \lim_{n\to\infty}(v_n^N(i) - v_n^N(x))$ exists and that r^N satisfies (AC2–3). Now suppose that we wish to use another base point. More generally, suppose that we choose $x^N \in S_N$ so that the base point may be an arbitrary element of S_N and may vary with N. What happens then?

It follows from Proposition 6.6.3 that $w^N(i) = \lim_{n\to\infty}(v_n^N(i) - v_n^N(x^N))$ exists. Since $v_n^N(i) - v_n^N(x^N) = (v_n^N(i) - v_n^N(x)) + (v_n^N(x) - v_n^N(x^N))$, it follows that the limit of the last term must exist and equal a constant u^N. Hence we have $w^N(i) = r^N(i) + u^N$, and so the functions w^N and r^N differ by a constant (which may depend on N).

Hence the class of stationary policies realizing the minimum in (8.1) equals the class realizing the minimum in (8.1) with r^N replaced by w^N. It is proved in Theorem 8.1.1 that any limit point of such a sequence of stationary policies is average cost optimal for Δ. Hence this continues to be true even if the optimal policy e^N is computed using w^N. □

The important conclusion is that once the (AC) assumptions and the hypotheses of Proposition 6.6.3 have been verified, then the optimal stationary policies in the approximating sequence may be computed using any desired base point, and the base point may even vary with N.

8.3 EXAMPLES

In this section we show how to verify the (AC) assumptions in several models.

Example 8.3.1. This is Example 7.6.1. This is a single-buffer/single-server model with the option of rejecting arriving batches. Under action a the arriving batch is admitted, whereas under action r it is rejected. We operate under the basic assumptions of Example 7.6.1.

Let $S_N = \{0, 1, \ldots, N\}$. There is excess probability possible only under the admit action, and any such probability is mapped to N. This means that if the admission of a batch would cause a buffer overflow, then the probability of such an event is given to the full buffer state N.

We employ Corollary 8.2.2. Since p_0 and μ are positive, it is always possible for the system to transition downward in one slot. This means that any stationary policy for Δ_N is 0 standard. Since $P_{00}(r) = 1$ and $P_{00}(a) = p_0 > 0$, it is the case that every stationary policy has a single aperiodic positive recurrent class containing 0. This verifies (i).

Lemma 7.6.2 shows that (iii) holds. This is intuitively clear since, if the process begins in $i \geq 1$ and operates optimally for n steps, it cannot do better than if it begins in $i - 1$ and operates optimally for n steps. The same argument convinces us that this is also true for Δ_N, and hence that (ii) holds.

It remains to verify (iv). Let d be the policy that always rejects. It is shown in the proof of Proposition 7.6.3 that d is 0 standard and that $m_{i0}(d)$ and $c_{i0}(d)$ are increasing in $i \geq 1$.

Hence the conclusions of Corollary 8.2.2 are valid for this model. □

Example 8.3.2. This is Example 7.6.4. This is a single-buffer/single-server model with the actions being the allowable service rates. Arriving batches are always admitted. We operate under the basic assumptions from Example 7.6.4.

Let $S_N = \{0, 1, \dots, N\}$. If a batch arrives that would cause a buffer overflow, then the probability of that event is given to the full buffer state N. We again employ Corollary 8.2.2.

The actions are the (geometric) service rates, and under action a the packet at the head of the line is served at rate a. Since p_0 and a are positive, it is always possible for the system to transition downward in one slot. This means that any stationary policy for Δ_N is 0 standard. Since $P_{00} = p_0 > 0$, it is the case that every stationary policy has a single aperiodic positive recurrent class containing 0. This verifies (i).

Lemma 7.6.5 shows that (iii) holds. This is intuitively clear since, if the process begins in $i \geq 1$ and operates optimally for n steps, it cannot do better than if it begins in $i-1$ and operates optimally for n steps. The same argument convinces us that this is also true for Δ_N and hence that (ii) holds.

It remains to verify (iv). Let d be the policy that serves at maximum rate a_K. It is shown in Lemma 7.6.6 that d is standard with $R_d = [0, \infty)$. At most one packet may be served in a slot. This implies that $m_{i0}(d) = m_{ii-1}(d) + m_{i-1,0}(d)$, and hence $m_{i0}(d)$ is increasing in $i \geq 1$. A similar result is true for the first passage costs. Hence (iv) holds.

Hence the conclusions of Corollary 8.2.2 are valid for this model. □

Example 8.3.3. This is Example 7.6.8. This concerns the routing of batches of packets to two parallel queues. We assume that the basic assumptions in Example 7.6.8 hold with the exception that one (or both) holding costs may be bounded. In this example there is no cost for changing the routing decision. Example 8.4.1 treats this model when there is a cost for changing the routing decision.

The ATAS may be described as follows: Each buffer is limited to N customers. If a decision is made to route to buffer 1 (say) and the arrival of a batch of a certain size would cause a buffer overflow in buffer 1, then the probability of that event is assigned to the full buffer state at buffer 1.

We apply the four-step procedure in Proposition 8.2.1 with $\mathbf{x} = (0, 0)$. Since $p_0 > 0$ and the service rate at each buffer is positive, there is a positive probability of each buffer decreasing by 1 in a given slot (or remaining empty if currently empty). This means that any stationary policy for Δ_N is $\mathbf{x}$ standard. Since $P_{\mathbf{xx}}(1) = P_{\mathbf{xx}}(2) = p_0 > 0$, it is the case that any stationary policy has a single aperiodic positive recurrent class containing $\mathbf{x}$, and hence Step 1 holds.

The fixed splitting d from Lemma 7.6.9 is $\mathbf{x}$ standard (in fact induces an irreducible MC on S). Let us verify that (C.37–38) hold for the MC induced by d. It will then follow from Proposition C.5.3 that the ATAS is conforming at d, and this will complete Step 2.

Assume that d chooses 1 in state $\mathbf{i} = (i_1, i_2)$ (if it chooses 2 the argument is similar). Recall from Example 2.5.6 that we introduce a variable s, where $s = i_2$ if $i_2 = 0$ or there is no service completion at buffer 2, and $s = i_2 - 1$ if there is a service completion at buffer 2. Then (C.37) becomes $m_{(N,s)\mathbf{x}}(d) \leq m_{(r,s)\mathbf{x}}(d)$ for $r > N$. This holds if the expected first passage time is increasing in the first coordinate, with the second coordinate held fixed. This is intuitively clear. The same argument works for the expected first passage costs in (C.38).

We show that Step 3(i) holds by applying Proposition 3.3.4 with $\alpha = 1$. It is easily seen that (3.19) holds if v_n is increasing in one coordinate, with the other held fixed. This was proved in Problem 7.11. A similar proof shows that v_n^N is increasing in one coordinate, with the other held fixed. This implies that $v_n^N \geq v_n^N(\mathbf{x})$, and hence Step 4(i) holds.

Hence the conclusions of Proposition 8.2.1 are valid for this model. □

Example 8.3.4. In this model batches of packets arrive to a buffer, with $p_j = P$ (a batch of size j arrives in a slot) > 0 for $j \geq 0$.

The state of the system is the number $i \geq 0$ of packets in the buffer. There is a null action in state 0. When in state $i \geq 1$ the action set is $\{1, 2, \ldots, i\}$, where action k means that k packets are served perfectly in one slot. (Serving more than one packet at a time is known as *batch service*.)

There is an increasing holding cost $H(i)$ with $H(0) = 0$ and $\lim_{i \to \infty} H(i) = \infty$. There is a nonnegative service cost $B(k)$ that is increasing in k. We have $C(0) = 0$ and $C(i, k) = H(i) + B(k)$ for $1 \leq k \leq i$. The transition probabilities are given by $P_{0j} = p_j$ and $P_{ii-k+j}(k) = p_j$ for $j \geq 0$ and $1 \leq k \leq i$. Finally we assume that $\sum_{j \geq 1} p_j[H(j) + B(j)] < \infty$.

Consider an ATAS with $S_N = \{0, 1, \ldots, N\}$ that sends excess probability to 0. We employ Proposition 8.2.3 (with $z = 0$), and note that (v) holds.

Now consider an arbitrary stationary policy e for Δ. Since $p_0 > 0$ and at least one packet must be served in each slot, it follows that i leads to 0 under e. Since $p_i > 0$, it follows that 0 leads to i. Hence e induces an irreducible MC on S.

The same reasoning shows that any stationary policy for Δ_N induces an irreducible MC on S_N. Since $P_{00} = p_0 > 0$, it follows that the chain is aperiodic. This verifies (vi).

Let d be defined by $d(i) = i$ for $i \geq 1$ so that in each slot all the waiting packets are perfectly served. This policy induces an irreducible MC on S whose transition matrix has identical rows. By Remark C.2.7(ii) it will follow that d is 0 standard if the induced MC is positive recurrent with finite average cost. Because the rows are identical, we may view the expected time to return to 0 as the expectation of a geometric random variable with probability of success p_0. Hence $m_{00}(d) = 1/p_0$ and $\pi_0 = p_0$. Then it is easy to see from Proposition C.1.2(i) that $\pi_i(d) = p_i$. From Proposition C.2.1(i) we have $J_d = \sum_{j \geq 1} p_j[H(j) + B(j)] < \infty$. This verifies that (i) holds.

Since H is unbounded, it is clear that (ii) holds, and in fact $D = [0, i^*]$ for some i^*. Since $R_d = S$, it follows that we may take $g = d$ in (iii).

It remains to verify (iv). We have shown above that any stationary policy for Δ induces an irreducible MC on S. Since an optimal stationary policy induces a MC with at least one positive recurrent class, it follows that (iv) holds.

Hence the conclusions of Proposition 8.2.3 are valid for this model. □

Notice that this proof makes crucial use of two facts: (1) There is a positive probability of a batch of any size arriving in any slot, and (2) at least one packet must be served in any slot. Problem 8.2 explores this example when (1) is weakened.

*8.4 ANOTHER EXAMPLE

The example in this section is the routing problem of Example 8.3.3 except that a cost is allowed for changing the routing decision. Proposition 8.2.1 is not directly applicable, but its basic approach remains valid in modified form. Another complicating factor is that the VIA may not hold for the approximating sequence but will hold for the transformed version of the AS. The reasoning is somewhat more involved and may be omitted on first reading.

Example 8.4.1. This is the routing problem of Example 8.3.3 except that a cost is allowed for changing the routing decision. The states are $[(i_1, i_2), k^*]$, where $\mathbf{i} = (i_1, i_2)$ is the current buffer level and k^* is the previous routing decision.

The ATAS is defined in Example 2.5.6. Recall that the content of each buffer is limited to N customers and that the informational tag k^* is carried along.

Let $x = [(0, 0), 1]$ and $y = [(0, 0), 2]$, and consider Δ_N. Since p_0 and the service rates are positive, it is clear that for any initial state and under any stationary policy e, one of x or y may be reached. Hence e induces a MC with at most two positive recurrent classes. There are four possibilities for decisions made in $\{x, y\}$. If $e(x) = e(y) = 1$, then y leads to x and $P_{xx}(e) > 0$. Hence e induces a MC with a single aperiodic positive recurrent class. Similar reasoning holds if $e(x) = e(y) = 2$. If $e(x) = 1$ and $e(y) = 2$, then $P_{xx}(e) > 0$ and $P_{yy}(e) > 0$, and hence e induces a MC with one or two aperiodic positive recurrent classes. However, if $e(x) = 2$ and $e(y) = 1$, then there is a single positive recurrent class containing $\{x, y\}$, and there is the possibility that this class is periodic.

For this reason we must effect the aperiodicity transformation on Δ_N discussed in Section 6.6. Let us assume that this transformation has been carried out yielding Δ_{N*}.

Recall Problem 7.11(iv). Its solution should show that the fixed splitting d from Lemma 7.6.9 induces an irreducible positive recurrent MC with finite average cost on the state space S. Note that $d|N$ may be used to verify Proposition 6.4.1(v), and hence the minimum average cost in Δ_N is constant. (This was proved for a stationary policy but is also valid for a randomized stationary policy.)

Then it follows from Proposition 6.6.6 that the minimum average cost in Δ_{N*} is given by τJ^N and that the VIA may be carried out in Δ_{N*}, yielding a solution to the ACOE for Δ_N. It is the case that $r^{N*}(.) = \lim_{n \to \infty} (v_n^{N*}(.) - v_n^{N*}(x))$ is a solution to (6.37). This verifies (AC1).

To verify that (AC4) holds, we first show that (3.19) holds. Assume that the process is in state $[\mathbf{i}, 1]$ and that action 1 is chosen. Then (3.19) becomes the requirement that $v_n([(N,s),1]) \le v_n([(r,s),1])$, where $r > N$ and s is the auxillary variable introduced in Example 8.3.3. This is clearly valid. Now assume that the process is in state $[\mathbf{i},1]$ and that action 2 is chosen. Then (3.19) becomes the requirement that $v_n([(s,N),2]) \le v_n([(s,r),2])$, where $r > N$ and s is the auxillary variable. This is clearly valid. Similar reasoning holds if the process is in state $[\mathbf{i},2]$.

Hence it follows from Proposition 3.3.4 that $v_n^N \le v_n$. The verification of (AC4) then follows exactly as in (8.4).

It remains to verify that (AC2–3) hold for r^{N*}. We first show that (Δ_N) is conforming at d by verifying that (C.37–38) hold and then applying Proposition C.5.3. If d chooses 1, then (C.37) becomes $m_{[(N,s),1]x}(d) \le m_{[(r,s),1]x}(d)$ for $r > N$. This holds if the expected first passage times to x are increasing in the first buffer content, with the second buffer content held fixed. This is intuitively clear. If d chooses 2, then (C.37) becomes $m_{[(s,N),2]x}(d) \le m_{[(s,r),2]x}(d)$ for $r > N$. This is true by the same reasoning. Notice that to effect the first passage, we make decisions randomly according to the fixed splitting d until we reach an empty system at the same time that the previous decision was 1. Similar results hold for the expected first passage costs. This verifies (C.37–38), and hence (Δ_N) is conforming at d.

We cannot claim that $(\Delta_N{}^*)$ is "conforming at d" because the restriction of d to $\Delta_N{}^*$ does not induce an AS for the Markov chain induced by d, as defined in Definition C.4.1. Nevertheless, let us see what can be deduced from the fact that (Δ_N) is conforming at d.

Recall that $d|N$ induces a positive recurrent MC on Δ_N (and on $\Delta_N{}^*$). Hence we need not worry about multiple classes or transient states. Since the steady state probabilities associated with $d|N$ are identical for Δ_N and $\Delta_N{}^*$, it is clear that the convergence of the steady-state probabilities behaves properly. Moreover $J_{d|N}^{N\,*} = \tau J_{d|N}^{N} \to \tau J_d$.

We need to examine the convergence of the mean first passage times and costs. Problem *8.3 asks you to prove that

$$m_{ij}^{N*}(d|N) = \begin{cases} \dfrac{1}{\tau}\, m_{ij}^{N}(d|N), & j \ne i, \\ m_{ij}^{N}(d|N), & j = i, \end{cases}$$

$$c_{ij}^{N*}(d|N) = \begin{cases} c_{ij}^{N}(d|N), & j \ne i, \\ \tau c_{ij}^{N}(d|N), & j = i. \end{cases} \tag{8.8}$$

Let us see how (8.8) may be used to complete the proof. The verification of (AC2) follows as in the proof of Proposition 8.2.1. To verify (AC3), note that it is intuitively clear that

$$\begin{aligned} v_n^{N*}([\mathbf{i},1]) &\ge v_n^{N*}(x) \\ v_n^{N*}([\mathbf{i},2]) &\ge v_n^{N*}(y), \qquad \text{all } \mathbf{i}, n \ge 1. \end{aligned} \tag{8.9}$$

This implies that $r_n^{N*}([\mathbf{i}, 1]) \geq 0$. Moreover we have

$$\begin{aligned} r_n^{N*}([\mathbf{i}, 2]) &= (v_n^{N*}([\mathbf{i}, 2]) - v_n^{N*}(y)) + (v_n^{N*}(y) - v_n^{N*}(x)) \\ &\geq v_n^{N*}(y) - v_n^{N*}(x) \\ &\geq -c_{xy}^{N*}(d|N) \\ &= -c_{xy}^{N}(d|N). \end{aligned} \tag{8.10}$$

Since the last term converges to $-c_{xy}(d)$, it follows that (AC3) holds with $Q = c_{xy}(d)$. □

8.5 SERVICE RATE CONTROL QUEUE

In this section we give computational results for a special case of Example 8.3.2. Recall that this model is a single-server queue with service rate control. We compute an average cost optimal stationary policy under the assumption that the packet arrival process is Bernoulli. That is, there is a probability p of a single packet arriving in any slot and a probability $1 - p$ of no arrival, where $0 < p < 1$. The holding cost is given by $H(i) = Hi$, where H is a positive constant. This is ProgramThree.

Under the assumption that $p < a_K$, the basic assumptions are valid, and it follows from Proposition 7.6.7 that any optimal stationary policy e is standard with $R_e = [0, \infty)$. Moreover, if e realizes the ACOE (7.48) and breaks ties by choosing to serve at the lowest optimal rate, then $e(i)$ is increasing in i and eventually chooses a_K.

So it is likely (unless there are ties) that the optimal policy computed using (AC) will be increasing in i and eventually choose a_K. Our computational results bear this out. The optimal policy may be given as a sequence of $K - 1$ intervals, with the first interval corresponding to service at rate a_1, the second to service at rate a_2, and so on. The interval at which it is optimal to serve at maximum rate is then obvious and may be omitted.

The expressions for the VIA 6.6.4 are given by

$$\begin{aligned} w_n(0) &= (1-p)u_n(0) + pu_n(1) \\ w_n(i) &= Hi + \min_a \{C(a) + a(1-p)u_n(i-1) \\ &\quad + [(1-a)(1-p) + ap]u_n(i) + (1-a)pu_n(i+1)\}, \qquad 1 \leq i \leq N-1, \\ w_n(N) &= HN + \min_a \{C(a) + a(1-p)u_n(N-1) \\ &\quad + [(1-a)(1-p) + ap + (1-a)p]u_n(N)\} \\ u_{n+1}(i) &= w_n(i) - w_n(0), \qquad 0 \leq i \leq N. \end{aligned} \tag{8.11}$$

The second and third equations in (8.11) may be evaluated in the same loop by introducing an auxillary variable that equals $i + 1$ for $1 \leq i \leq N - 1$ and equals N for $i = N$.

We would like to have a *benchmark policy* to compare with the optimal policy. Assume that rate a satisfies $p < a$. Then the policy $d(a)$ that always serves at rate a has finite average cost and can be implemented with no buffer observation. This is called *open loop control*. Our benchmark policy d serves at the rate a that minimizes $J_{d(a)}$. That is, under d we serve at the constant rate that yields $J_d = \min_{a>p}\{J_{d(a)}\}$.

In the case of non-Bernoulli arrivals, the stability condition is $\lambda < a$, and we can calculate J_d by employing the same program that calculates the optimal policy but reducing the actions to the single one a. Or a separate program to do this efficiently can be given. However, in the case of Bernoulli arrivals, it is possible to give a closed form expression for $J_{d(a)}$.

Proposition 8.5.1. Assume that a satisfies $p < a$, and let $d(a)$ be the policy that always serves at rate a. Then

$$J_{d(a)} = \frac{Hp(1-p)}{a-p} + \frac{pC(a)}{a}. \tag{8.12}$$

Proof: Let $r = (1-a)p/[a(1-p)]$. The steady state probabilities of the MC induced by $d(a)$ are given by $\pi_0 = 1 - (p/a)$ and $\pi_i = \pi_0 r^i/(1-a)$ for $i \geq 1$. This can be shown by verifying that Proposition C.1.2(i) holds for these values. Then the expression in (8.12) follows after some algebra. Problem 8.4 asks you to fill in the details. □

Remark 8.5.2. It is intuitively clear (and may be proved by induction on (8.11)) that if H and $C(a)$ are multiplied by a positive constant, then the optimal average cost is multiplied by that constant, and the optimal policy is unchanged. For this reason we assume that $H = 1$ in all our scenarios. We may then examine the effect of a cost of service small relative to 1, as well as the effect of a large cost of service. Whether a service rate option is less than p or greater than p will be seen to be a crucial factor. In all scenarios we used the weaker convergence criterion (Version 1) of the VIA. □

Remark 8.5.3. Consider the situation in which there are just two service rates. In this case the service rate cost is linear in the rates, and we have $C(a) = Ca + C^*$, where it is easily seen that

$$C = \frac{C(a_2) - C(a_1)}{a_2 - a_1},$$

$$C^* = \frac{C(a_1)a_2 - C(a_2)a_1}{a_2 - a_1}. \tag{8.13}$$

We know from Proposition 7.6.7(ii) that if $C^* \geq -1$, then the benchmark policy $d(a_2)$ is optimal. This can be intuitively explained by observing that in this case the cost of higher service is not a great deal more than for slower service, and so it always pays to serve at the higher rate. However, if $C^* < -1$, then higher-rate service costs a great deal more than lower-rate service, and it may be optimal to serve for a while at the lower rate. We will use this result to check the program. □

Scenarios 8.5.4. Here $K = 2$. The results are summarized in Table 8.1. A dash indicates that the entries in that box are identical with the corresponding box in the previous column. The row labeled J_d gives the value of a yielding the benchmark policy and the average cost under that policy. The row labeled J contains the approximation generated by the program. Proposition 7.6.7(iii) implies that the optimal policy eventually serves at maximum rate, and hence the policy may be indicated by a single interval that gives the buffer content for which it is optimal to serve at the slower rate. Note that $\varnothing$ means that it is optimal to serve at maximum rate. Because the program printout is given in four columns, we choose N divisible by 4. For most of the scenarios we selected $N = 96$ and $\epsilon = 0.00005$, and then $N = 120$ and $\epsilon = 0.000005$. It was always the case that the optimal policy was immediately indicated and unchanging for large N. In fact, in these examples, the optimal policy is typically determined for much smaller values of N. However, determining an approximation to J accurate to three decimal places necessitates the larger N and smaller ϵ.

In Scenario 1 it is the case that $C^* \geq -1$, and hence we know that it is optimal to serve at maximum rate. This is confirmed by the program. In the rest of the scenarios we have $C^* < -1$, and it turns out to be optimal to initially serve at the slower rate.

In Scenario 2 the system is stable under both rates. It is optimal to serve at the slower rate when the buffer content is no more than 4. In Scenario 3 the system is unstable under the slower (free) rate, and it is only optimal to serve at this rate when there is a single packet in the buffer. Scenario 4 examines this system when the faster rate costs twice as much as under Scenario 3. The content under which it is optimal to serve at the slower rate only increases from 1 to 2. Under Scenario 5 the system is stable under both rates and the higher rate costs 20 times the lower rate. In this case it is optimal to serve at the slower rate for a buffer content of no more than 5. Scenario 6 examines this system when the costs of both rates from Scenario 5 are multiplied by a factor of 5. In this case it is optimal to serve at the slower rate for buffer content of no more than 18. Scenarios 7 and 8 have large packet arrival rates.

In conclusion we see that if the queue is unstable under a given rate, then this rate will be used sparingly, even if it is free. □

Scenarios 8.5.5. Here $K = 3$. The results are summarized in Table 8.2. The optimal policy may be given as two intervals, with the first indicating the buffer content level at which the controller should serve at slowest rate, and the

Table 8.1 Results for Scenarios 8.5.4

Scenario	1	2	3	4	5	6	7	8
p	0.4	0.4	0.7	0.7	0.7	0.7	0.9	0.9
Service rates	0.5 0.6	—	0.6 0.9	—	0.75 0.9	—	0.85 0.92	0.9 0.95
Costs	2.0 2.5	1.0 5.0	0.0 5.0	0.0 10.0	0.5 10.0	2.5 50.0	1.0 5.0	0.0 10.0
C^*	−0.5	−19.0	−10.0	−20.0	−47.0	−235.0	−47.57	−180.0
J_d	$a = 0.6$ 2.8667	$a = 0.5$ 3.2000	$a = 0.9$ 4.9389	$a = 0.9$ 8.8278	$a = 0.75$ 4.6667	$a = 0.75$ 6.5333	$a = 0.92$ 9.3913	$a = 0.95$ 11.2737
J	2.867	3.028	4.249	6.650	3.866	6.505	9.374	6.546
Savings	0.0	0.172	0.690	2.178	0.801	0.028	0.017	4.728
Optimal policy	Ø	[1, 4]	[1]	[1, 2]	[1, 5]	[1, 18]	[1]	[1, 4]

Table 8.2 Results for Scenarios 8.5.5

Scenario	1	2	3	4	5	6	7	8
p	0.3	0.2	0.2	0.5	0.5	0.7	0.8	0.9
Service rates	0.2 0.4 0.8	0.1 0.4 0.7	—	0.3 0.5 0.9	—	0.8 0.9 0.99	0.7 0.85 0.95	0.92 0.95 0.99
Costs	0.9 1.3 2.1	0.01 0.5 5.0	0.0 0.01 10.0	0.0 0.5 25.0	0.0 0.1 50.0	0.1 5.0 50.0	0.0 10.0 25.0	1.0 5.0 10.0
N	48	48 64	—	—	—	100 500 1000	80 100	64 80
J_d	$a = 0.8$ 1.2075	$a = 0.4$ 1.0500	$a = 0.4$ 0.8050	$a = 0.9$ 14.5139	$a = 0.9$ 28.4028	$a = 0.8$ 2.1875	$a = 0.85$ 12.6118	$a = 0.92$ 5.4783
J	1.2075	1.028	0.805	6.015	8.003	2.143	11.577	4.457
Savings	0.0	0.022	0.0	8.499	20.399	0.045	1.035	1.021
Optimal policy	ØØ	Ø[1, 3]	Ø[1, 7]	Ø[1, 4]	Ø[1, 7]	[1, 5] [6, >1000]	[1] [2, 8]	[1, 5] Ø

second the level at which the controller should serve at the middle rate. Thus ∅∅ means that it is optimal to serve at maximum rate.

In Scenario 1 we have $C(a) = 2a + 0.5$, and hence according to the theory it is optimal to serve at maximum rate. This is confirmed by the program. In Scenario 2 the costs are moderate, and the queue is unstable under the slowest service rate. The optimal policy never uses the slowest rate and switches from the middle to the fastest rate for buffer content of 4 or more. Scenario 3 examines the effect of making the slowest rate free, of drastically reducing the cost of the middle rate, and doubling the cost of the fastest rate. The optimal policy still does not employ the slowest rate; the content at which it is optimal to switch to the fastest rate moves up modestly from 4 to 8. In this case the minimum average cost and the average cost under the benchmark policy are identical to three decimal places.

Scenario 4 has two inexpensive rates yielding unstable queues and a highly costly fastest rate. Scenario 5 examines the effect of reducing the cost of the middle rate and increasing the cost of the fastest rate. The optimal policy is modestly changed.

The queue under Scenario 6 is stable under all the rates. The fastest rate is very costly compared to the others. We know that at some point it is optimal to switch to this rate, but this point was not located for $N = 1000$. In the normal range of operation it is optimal to serve at the second fastest rate for content of 6 or more.

The queue under Scenario 7 is unstable under the free slowest rate, and it is optimal to serve at this rate when the buffer content is 1 and to switch from the middle to the fastest rate when it reaches 9. See Fig. 8.1.

The queue under Scenario 8 is stable under all three rates. In this interesting example it is never optimal to use the middle rate. □

8.6 ROUTING TO PARALLEL QUEUES

In this section we give computational results for a special case of Example 8.3.3. Recall that this concerns the routing of batches of packets to one of two parallel servers. We compute an optimal policy under the assumption that the packet arrival process is Bernoulli (p), as in Section 8.5. The holding cost for $\mathbf{i} = (i_1, i_2)$ is $H_1 i_1 + H_2 i_2$, where H_1 and H_2 are positive constants. This is ProgramFour.

Under the assumption that $p < \mu_1 + \mu_2$ the basic assumptions are valid, and it follows from Proposition 7.6.10 that the (CAV*) assumptions hold and any optimal stationary policy is positive recurrent at $\mathbf{x} = (0, 0)$. Then from Theorem 7.5.6 we see that the optimal stationary policy realizing the ACOE is $\mathbf{x}$ standard.

Under the stability condition we have $\rho =: p/(\mu_1 + \mu_2) < 1$. A system with small ρ is called *lightly loaded*, one with moderate ρ is called *moderately loaded*, and one with ρ close to 1 is called *heavily loaded.*

Let us develop the equation for the VIA 6.6.4. A couple of notational devices

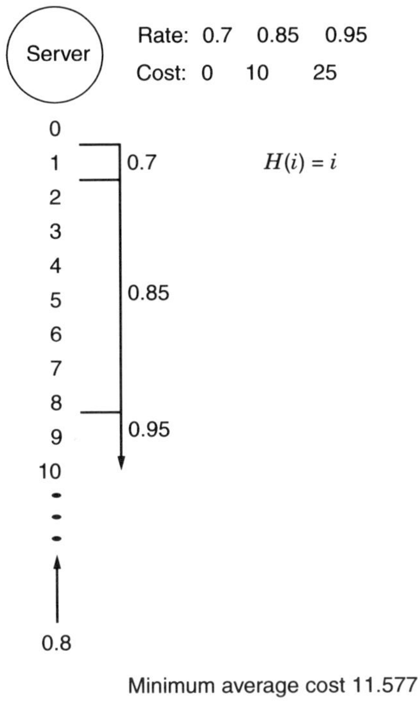

Figure 8.1 Scenario 7 from Table 8.2.

will facilitate this. Let q_1 equal $i_1 + 1$ if $0 \le i_1 \le N - 1$ and equal N if $i_1 = N$. The variable q_2 is defined similarly. Let s_1 equal $i_1 - 1$ if $0 < i_1 \le N$ and equal 0 if $i_1 = 0$. The variable s_2 is defined similarly.

We now develop some pieces that will be combined to form the expression in Step 2 of the VIA. These pieces are constructed to hold for all states **i**.

Let

$$y_n(\mathbf{i}) = (1 - \mu_1)(1 - \mu_2)u_n(i_1, i_2) + \mu_1(1 - \mu_2)u_n(s_1, i_2) + (1 - \mu_1)\mu_2 u_n(i_1, s_2) + \mu_1\mu_2 u_n(s_1, s_2). \tag{8.14}$$

This is what is expected to happen if there is no arrival. It is independent of the routing decision.

Let

$$z_n^1(\mathbf{i}) = (1 - \mu_1)(1 - \mu_2)u_n(q_1, i_2) + \mu_1(1 - \mu_2)u_n(s_1 + 1, i_2) + (1 - \mu_1)\mu_2 u_n(q_1, s_2) + \mu_1\mu_2 u_n(s_1 + 1, s_2). \tag{8.15}$$

This is what is expected to happen if routing decision 1 is chosen and there is an arrival. The expression z_n^2 is defined analogously and represents what is expected to happen if routing decision 2 is chosen and there is an arrival. The reader should check that (8.14–15) indeed hold for every state.

The VIA equations become

$$\begin{aligned} w_n(\mathbf{i}) &= H_1 i_1 + H_2 i_2 + (1-p)y_n(\mathbf{i}) + p\min\{z_n^1(\mathbf{i}), z_n^2(\mathbf{i})\}, \\ u_{n+1}(\mathbf{i}) &= w_n(\mathbf{i}) - w_n(\mathbf{x}). \end{aligned} \tag{8.16}$$

We now construct a benchmark open-loop policy to compare with the optimal policy. It is a naive benchmark in that we can actually do better with open-loop control (see the Bibliographic Notes). We employ it because it has the virtue of being easily understood and its average cost is readily computed.

Recall that in Lemma 7.6.9 we showed that there exists a fixed splitting inducing a standard MC on S. The *optimal fixed splitting* is the fixed splitting with minimum average cost and this is our benchmark.

Proposition 8.6.1. Let $\epsilon =: \mu_1 + \mu_2 - p > 0$. The average cost under the optimal fixed splitting d^* is specified by the following cases:

Case 1: If $H_1 = H_2$ $(=H)$ and $\mu_1 = \mu_2$, then $J_d* = Hp(2-p)/\epsilon$.

Now let $\beta_1 = \sqrt{\mu_1(1-\mu_1)}$ and $\beta_2 = \sqrt{\mu_2(1-\mu_2)}$. Define

$$F(x) = \frac{H_1(\mu_1 - x)[1-(\mu_1 - x)]}{x} + \frac{H_2(\mu_2 + x - \epsilon)[1-(\mu_2 + x - \epsilon)]}{\epsilon - x}, \qquad 0 < x < \epsilon. \tag{8.17}$$

Case 2: If $H_1 = H_2$ and $\mu_1 \neq \mu_2$, then $J_{d*} = F[\epsilon\beta_1/(\beta_1 + \beta_2)]$.

Case 3: If $H_1 \neq H_2$, then find x^*, with $0 < x^* < \epsilon$, satisfying

$$H_1\left[1 + \left(\frac{\beta_1}{x*}\right)^2\right] = H_2\left[1 + \left(\frac{\beta_2}{\epsilon - x^*}\right)^2\right]. \tag{8.18}$$

then $J_{d*} = F(x^*)$.

Proof: Let $d(q)$ be the fixed splitting that sends a packet to buffer 1 with probability q and to buffer 2 with probability $1 - q$. This splitting has finite average cost if $pq < \mu_1$ and $p(1-q) < \mu_2$. Observe that each buffer acts as an independent single-server queue with fixed service rate. The average cost under d^* is the sum of the average cost for each buffer. From Proposition 8.5.1 it follows that

$$J_{d(q)} = \frac{H_1 pq(1-pq)}{\mu_1 - pq} + \frac{H_2 p(1-q)[1-p(1-q)]}{\mu_2 - p(1-q)}. \tag{8.19}$$

We sketch the rest of the proof, with the details left as Problem 8.7. If we express (8.19) in terms of μ_1, μ_2, ϵ, and the unknown $x = \mu_1 - pq$, then we obtain (8.17). The left side is called $F(x)$. The stability requirements become $0 < x < \epsilon$.

To minimize $F(x)$, we solve $F'(x) = 0$. After some tedious algebra this reduces to (8.18). Thus the value x^* yielding the minimum is the solution of (8.18) and $J_{d*} = F(x^*)$.

It is easy to see that in Cases 1 and 2 we obtain the stated results. To implement the solution under Case 3, we can solve (8.18) using bisection or another method for finding roots. □

It is easy to see that if both H_1 and H_2 are multiplied by a positive constant U, then J is multiplied by U and the optimal policy is unchanged. For this reason we may assume that $H_1 = 1$ in all our runs. We employ Version 1 of the VIA.

Checking Scenarios 8.6.2. In the first scenario we set $H_1 = 1$, $H_2 = 0$, $p = 0.6$, $\mu_1 = 0.8$, and $\mu_2 = 0.7$. Since it costs nothing to be in the second buffer, the optimal policy should always choose 2, and the minimum average cost should be 0. This is confirmed by the program. In the second scenario H_1, H_2, and p are as above, and $\mu_1 = 0.5$, and $\mu_2 = 0.4$. Since it costs nothing to be in the second buffer, the optimal policy should always choose 2, and the minimum average cost should be 0. This is confirmed by the program. Notice that in this case the second queue is unstable.

In the third scenario we let $H_1 = 2$, $H_2 = 1$, $p = 0.7$, $\mu_1 = 0.01$, and $\mu_2 = 0.9$. Since the holding cost in the second buffer is smaller than that in the first buffer and since the service rate in the first buffer is very small, we would expect the optimal policy to almost always choose 2 and $J \approx 1.05$ from (8.12). This is confirmed by the program. □

Scenarios 8.6.3. In these scenarios we set $H_1 = H_2 = 1$ and $\mu_1 = \mu_2$ (= μ). Here J represents the minimum average number of packets in the system. It can also be taken as a measure of the minimum average total system delay. Recall that an arriving packet is not "counted" in the system until the slot following its arrival (in this model the arrival slot is devoted to routing, and hence the packet is not available for service). It is intuitively clear that the optimal policy routes an arriving packet to the shortest queue, and this is confirmed by the program. Hence our interest does not lie in computing the optimal policy but rather in comparing J with J_{d*}. The comparison shows the reduction in the average number in the system gained by observing the system and implementing the optimal policy compared with exercising open loop control using the best fixed splitting.

Table 8.3 Results for Scenarios 8.6.3

Scenario	1	2	3	4	5
p	0.3	0.6	0.6	0.8	0.8
μ	0.7	—	0.4	0.9	0.5
ρ	0.21	0.43	0.75	0.44	0.8
J_{d*}	0.4636	1.05	4.2	0.96	4.8
J	0.4336	0.8969	2.4517	0.8951	2.5486
Savings	0.03	0.1531	1.7483	0.0649	2.2514

The results are in Table 8.3. Under Scenario 1 we have a lightly loaded system in which the optimal policy effects a reduction in the average number in the system of $0.03/0.4636 = 6\%$. Under Scenario 2 we have a moderately loaded system, and the reduction in the average number in the system is 15%. Under Scenario 3 we have a fairly heavily loaded system in which the average number of packets is reduced by 42%. Scenario 4 represents a moderately loaded system, and the average number of packets is reduced by 7%. Scenario 5 is a fairly heavily loaded system with a reduction of 47%.

For these examples we chose $N = 39$ with a tolerance of 5×10^{-9} and confirmed with $N = 47$ and a tolerance of 5×10^{-10}. The value of J is actually determined quite accurately for much smaller values of N.

Consider a fixed state $\mathbf{i}$. If one of the coordinates of $\mathbf{i}$ is close to the boundary N, then the calculated optimal decision may be incorrect. The reason for this is the weak convergence criterion in Version 1 of VIA 6.6.4. There are two ways to mitigate this. The first way is to use Version 2, which will increase the run time. The second way is to increase N. As N is increased, a given state will receed from the boundary region, and the calculated optimal policy in that state will pull in. The second way requires more memory and also a modest increase in the run time. An additional factor in this particular case is that some of the values in the minimization in (8.16) may be equal, causing an ambiguity. □

Scenarios 8.6.4. Table 8.4 presents results for the more general case in which the holding costs and/or the service rates are unequal. The first three scenarios have holding costs equal to 1 (and hence we are finding the minimum average number in the system) but unequal service rates. Scenario 1 is a lightly loaded system. The reduction in the average number in the system is 7%.

Let us explicate the optimal policy given for Scenario 1. See Fig. 8.2. Clearly the controller will favor buffer 1; however, when this buffer reaches a certain level, then the controller will switch to buffer 2. The entries indicate the switching points for fixed levels of buffer 1. For example, if $i_1 = 6$ or 7, then the controller will route an arriving packet to buffer 2 if $0 \le i_2 \le 4$. The other entries are intepreted similarly. This defines a *switching curve*, and it could be given graphically. For the discrete situation in which the optimal policy will be

Table 8.4 Results for Scenarios 8.6.4

Scenario	1	2	3	4
p	0.3	0.5	0.8	0.9
μ_1	0.7	0.6	0.7	0.8
μ_2	0.5	0.4	0.4	0.8
ρ	0.25	0.5	0.73	0.63
H_1	1.0	1.0	1.0	1.0
H_2	1.0	1.0	1.0	2.0
J_{d*}	0.520	1.420	2.897	2.001
J	0.482	1.090	1.792	1.508
Savings	0.038	0.330	1.105	0.493
Optimal policy	(1, 0) (2, 1) (3–4, 2) (5, 3) (6–7, 4) (8, 5) (9–10, 6) (11, 7) (12, 8) (13–14, 9) (15, 10) (16–17, 11) (18, 12) (19, 13) (20, 14)	(1, 0) (2, 1) (3–4, 2) (5, 3), (6–7, 4) (8, 5) (9–10, 6) (11, 7) (12–13, 8) (14–15, 9) (16, 10) (17–18, 11) (19, 12) (20, 13)	(2, 1) (3–4, 2) (5, 3) (6–7, 4) (8–9, 5) (10, 6) (11–12, 7) (13–14, 8) (15, 9) (16–17, 10) (18–19, 11) (20, 12)	($2 \leq i_1 \leq 13$, 1) ($14 \leq i_1 \leq 30$, 2)

implemented through table lookup, it is more efficient to give the policy as we have done in Table 8.4.

We might conjecture that the optimal policy operates by routing an arriving packet to the buffer that minimizes its expected system time. This is the *individually optimal* policy, since it is what the packet would choose to do if it had the freedom to route itself. However, the optimal policy does not operate quite this way. Assume that a packet arrives to find the system in state (1, 0). If it is routed to buffer 1, then its expected system time (less the arrival slot) is $(0.7)(1/0.7) + (0.3)(2/0.7) = 1.86$. This is found by conditioning on what happens to the packet in buffer 1 during the arrival slot. If it is routed to buffer 2, then its expected system time is $1/0.5 = 2$. Hence the packet would send itself to 1, whereas the optimal policy sends it to 2. This shows that the optimal policy (which may be regarded as *socially optimal*) is not the same as the individually optimal policy. However, for many of the entires on the switching curve, the socially optimal and individually optimal decisions do coincide. (For the case of equal service rates, the socially optimal and individually optimal policies coincide.)

For these scenarios we chose $N = 47$ except for the last one, for which $N = 59$. Because of the weak convergence criterion we can only be confident of the optimal policy away from the boundaries and, as a rough rule of thumb, for states with coordinates not more than $N/2$. This can be mitigated as pre-

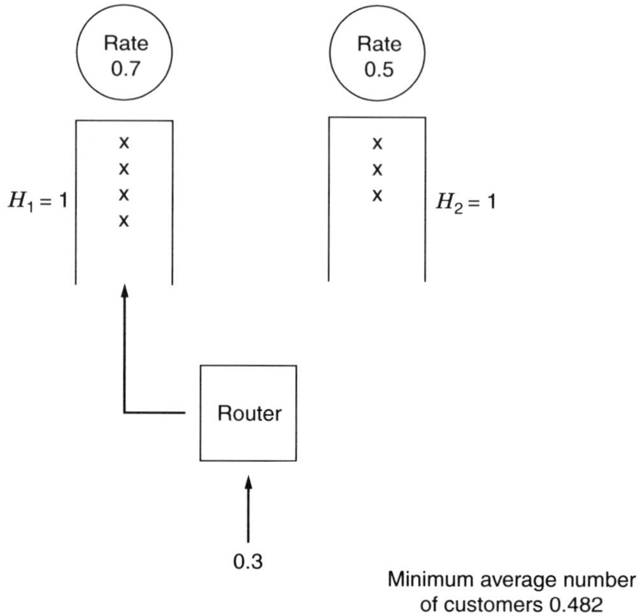

Figure 8.2 Scenario 1 from Table 8.4.

viously discussed, either through changing the convergence criterion or by increasing N.

However, notice that there is another way to look at this. For Scenario 1 the minimum average number in the system is less than 0.5, so it will be extremely rare for either buffer content to be above 20. If the optimal policy is given as in Table 8.4, then on those rare occasions in which the content of buffer 1 exceeds 20 we could simply implement the best fixed splitting. The resulting policy should be quite close to optimal.

Scenario 2 is a moderately loaded system. The reduction in the average number in the system is 23%. Scenario 3 is a somewhat heavily loaded system, and the reduction in the average number in the system is 38%. In Scenario 4 the service rates are equal but it costs twice as much to hold packets in the second buffer. See Fig. 8.3. The optimal policy strongly favors the first buffer. For example, if $i_1 = 20$, then an arriving packet will be routed to the second buffer if and only if its content is less than or equal to 2. The reduction in the average holding cost is 25%. □

Remark 8.6.5. When the average number in the system is being minimized, these scenarios allow us to give some very rough guidelines on the percentage reduction that can be expected from employing the optimal policy. When the system is lightly loaded, reductions around 5–10% may be effected. When the system is moderately loaded, reductions around 15–25% may be effected, and

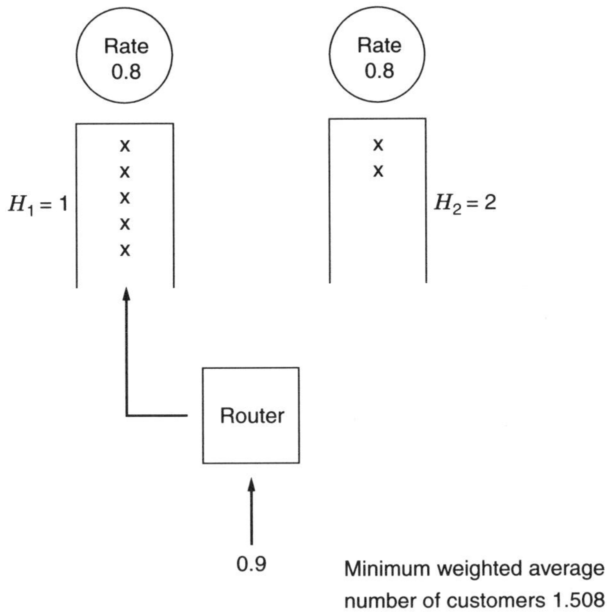

Figure 8.3 Scenario 4 from Table 8.4.

when the system is somewhat heavily loaded, reductions around 30–50% may be effected. □

8.7 WEAKENING THE (AC) ASSUMPTIONS

For some models it is not possible to verify the (AC) assumptions. In these situations it is useful to have a weaker set of assumptions under which the conclusions of Theorem 8.1.1 remain valid. We will not attempt to find the absolutely weakest conditions under which the conclusions of Theorem 8.1.1 hold. Rather, we give a useful set (WAC) of assumptions under which they hold. It will be clear from the proof of Proposition 8.7.1 how to further weaken (WAC) if necessary.

The idea is to weaken (AC3) by allowing Q to be a function. This necessitates some additional assumptions. We have (WAC1), (WAC2), and (WAC4) identical to their (AC) counterparts. The new assumption (WAC3) will be given in two parts.

(WAC3$_1$). There exists a nonnegative (finite) function Q on S such that $-Q(i) \le \liminf_{N \to \infty} r^N(i) =: u(i)$ for $i \in S$.

(WAC3$_2$). Let e be a stationary policy for Δ and $X_0 = i$ an initial state.

Then:

(i) $\lim_{N \to \infty} \sum_{j \in S_n} P_{ij}(e; N)Q(j) = \sum_j P_{ij}(e)Q(j) < \infty$,
(ii) $-\infty < E_e[u(X_n)]$ for $n \geq 1$,
(iii) $\liminf_{n \to \infty} E_e[u(X_n)]/n \geq 0$.

Suppose that (AC) holds. If we let $Q(.) \equiv Q$ from (AC3), then it is easy to see that (WAC3) holds. In this case, both summations in (WAC3_2)(i) equal Q. Hence (AC) $\Rightarrow$ (WAC), and the latter is a weaker set of assumptions. Here is the existence result under (WAC).

Proposition 8.7.1. Assume that the (WAC) assumptions hold. Then the conclusions of Theorem 8.1.1 are valid.

Proof: We proceed as in the proof of Theorem 8.1.1 noting that $w \geq u$, up to (8.2), which may be written

$$J^{N_v} + r^{N_v}(i) + \sum_{j \in S_{N_v}} P_{ij}(e^*; N_v)Q(j)$$
$$= C(i, e^*) + \sum_{j \in S_{N_v}} P_{ij}(e^*; N_v)\{r^{N_v}(j) + Q(j)\}. \tag{8.20}$$

Note that the term added to both sides is finite, since it is a summation over a finite set. We then take the limit infimum of both sides and employ Proposition A.1.8 and (WAC3_2)(i) to obtain (8.3).

We now apply the proof technique from Lemma 7.2.1. This requires that $-\infty < E_{e*}[w(X_n)] < \infty$ for all n. The right inequality follows as in the proof of Lemma 7.2.1, while the left inequality follows from (WAC3_2)(ii) and the fact that $w \geq u$.

We may then proceed as in the proof of Lemma 7.2.1 where (7.7) becomes

$$\frac{v_{e*,n}(i)}{n} \leq J_0 + \frac{w(i) - E_{e*}[w(X_n)]}{n}. \tag{8.21}$$

Taking the limit supremum of both sides and using (WAC3_2)(iii) yields $J_{e*}(i) \leq J_0$. The proof is then completed as before. □

BIBLIOGRAPHIC NOTES

The approximating sequence method was originally developed for computing average cost optimal policies, and the (AC) assumptions were introduced in Sennott (1997a) with further results given in Sennott (1997b). The original version of (AC) was based on the relative value function h_α^N from Theorem 6.4.2. We would like to thank Dr. Eitan Altman for the suggestion that we give (AC) in terms of a general solution to the ACOE in Δ_N, which is the version given here.

Thomas and Stengos (1985) give a method for the computation of optimal average cost policies in denumerable state space MDCs. Their method assumes that the costs are bounded and in addition assumes a condition guaranteeing the existence of a bounded solution to the ACOE. It is clear from the results in Chapter 7 that this is a very limiting assumption that fails to hold in many models of interest. While this assumption renders a direct comparison of the methods quite difficult, it appears that one of the value iteration schemes is the same as our ATAS that sends the excess probability to a distinguished state. The paper also gives some approximate policy iteration algorithms.

Van Dijk (1991) discusses an MDP 1 and a related MDP 2. Under certain assumptions it is possible to give a bound on the difference between the minimum average costs in 1 and 2. An example of Van Dijk is related to an ATAS that maps excess probability in a state i to a state that is a function of i. The method is applied to a particular network model, and some computations are presented. The quantity that is computed is the bound, and the minimum average cost and optimal policy are not indicated. The ideas in this paper may give an avenue to develop bounds on the convergence of the minimum average cost in Δ_N to the minimum average cost in Δ. We have not treated this topic, and it remains a fruitful direction for further exploration.

The material in Sections 8.2–4 appears in a somewhat different form in Sennott (1997a, b). The ideas behind the notion of conformity are detailed in the Bibliographic Notes to Appendix C. The example in Section 8.5 appears in Sennott (1997a), and the routing example from Section 8.6 appears in Sennott (1997b). The assumptions given in Section 8.7 further generalize the assumptions in Sennott (1997a). The ideas parallel the development in Section 7.7.

There has been a large amount of work on the model of routing customers to parallel queues. Most of this work attempts to characterize optimal policies, develop bounds, or find good suboptimal policies. To the best of our knowledge there has been no attempt to calculate optimal policies or values. We will highlight a few results.

Much of this work considers the model in continuous time and assumes that the servers are exponential and the customer arrival process is Poisson. If there are finitely many servers serving at the same rate, then Winston (1977) proves that the policy of sending an arriving customer to the shortest queue maximizes roughly the discounted number of service completions in any finite interval $[0, T]$. Weber (1978) extends this result to a general arrival process and servers with nondecreasing hazard rates.

Hajek (1984) is a seminal paper treating a related model with two stations and proving that the optimal policy is described by a switching function.

For the case of Poisson arrivals and unequal rate exponential servers, Krishnan (1987) introduces a heuristic based on the optimal fixed splitting. This method performs a calculation based on the current state and the optimal splitting and chooses an action based on that calculation that will give better performance than the optimal fixed splitting.

Stidham and Weber (1993) is a survey stating results on the routing problem and related models with many additional references.

There is a line of development that considers open-loop routing that performs better than the optimal fixed splitting. Assume that there are K parallel buffers. The idea is to construct a deterministic sequence of integers k, with $1 \le k \le K$, such that if a customer arrives and the value of the sequence is k, then that customer is routed to server k. The sequence can be constructed so that the average number in the system is less than under a standard random implementation of the optimal fixed splitting. Hajek (1985) initiated this line of research and showed how to construct the sequence for $K = 2$. Rosberg (1985) develops a certain sequence for $K \ge 2$. See also Arian and Yevy (1992). Milito and Fernandez-Gaucherand (1995) examine the problem for a fixed number of arrivals. Shanthikumar and Xu (1997) examine a related problem.

PROBLEMS

8.1. Consider the model in Problem 7.14, but with no cost for changing the decision. Let the ATAS be defined as in Example 8.3.3. Verify that the hypotheses of Proposition 8.2.1 are satisfied.

8.2. Consider Example 8.3.4 modified so that some p_j may be 0. In particular, assume that $0 < p_0$ and $\sup\{j | p_j > 0\} = \infty$. Verify that the hypotheses of Proposition 8.2.3 still hold.

***8.3.** Verify (8.8). *Hint:* For $i \ne j$, what is ${}_j u_{ik}^N(d|N)^*$?

8.4. Fill in the details in the proof of Proposition 8.5.1.

8.5. Run ProgramThree for the following scenarios, and discuss the results. Be sure to set N and NUMACT (=K) appropriately for each run.

(a) $p = 0.6$, $a_1 = 0.65$, $a_2 = 0.9$, $C(a_1) = 1.95$, and $C(a_2) = 2.7$.

(b) $p = 0.8$, $a_1 = 0.8$, $a_2 = 0.85$, $C(a_1) = 0$, and $C(a_2) = 20$.

(c) $p = 0.5$, $a_1 = 0.48$, $a_2 = 0.52$, $a_3 = 0.8$, $C(a_1) = 0$, $C(a_2) = 0.5$, and $C(a_3) = 10$.

(d) $p = 0.7$, $a_1 = 0.7$, $a_2 = 0.8$, $a_3 = 0.9$, $a_4 = 0.99$, $C(a_1) = 0$, $C(a_2) = 1$, $C(a_3) = 5$, and $C(a_4) = 15$.

8.6. Make up some scenarios of your own for ProgramThree, and discuss the results.

8.7. Fill in the details in the proof of Proposition 8.6.1.

8.8. Consider the routing model under the conditions in Scenarios 8.6.3. Recall that in this case J is the minimum average number of packets in the system, a quantity that is related to the total average system delay. Let D be the delay suffered by a randomly arriving packet (*including* the arrival slot) under the optimal policy e (which routes an arriving packet to the shorter buffer and, if the buffers are equal, then routes it to buffer 1, say). We develop an expression for $E[D]$. Let J_0 be the average number of packets in the shorter buffer, and let U be the probability that at least one buffer is empty (both under the policy e in steady state). Prove that $E[D] = U + (1 + 2J_0)/\mu \le 1 + (1 + J)/\mu$. *Hint:* Show that the probability that an arriving packet finds the system in state $\mathbf{i}$ in steady state is equal to $\pi_{\mathbf{i}}(e)$.

8.9. Run ProgramFour for the following scenarios, and discuss the results:
(a) $p = 0.4,\ \mu_1 = \mu_2 = 0.4,\ H_1 = H_2 = 1.$
(b) $p = 0.7,\ \mu_1 = 0.7,\ \mu_2 = 0.3,\ H_1 = H_2 = 1.$
(c) $p = 0.9,\ \mu_1 = 0.6,\ \mu_2 = 0.5,\ H_1 = H_2 = 1.$
(d) $p = 0.6,\ \mu_1 = 0.7,\ \mu_2 = 0.7,\ H_1 = 1,\ H_2 = 1.5.$

8.10. Consider a system (see Fig. 1.5) with two independent geometric servers, with server 1 serving at rate μ_1 and server 2 at rate μ_2, where $0 < \mu_1 < \mu_2 < 1$. Batches of packets arrive to an infinite capacity buffer with $p_j = P$ (a batch of size j arrives in any slot), where $0 < p_0$ and $\lambda > 0$ is the mean batch size. The arrival of batches is independent of the service times. An arriving batch is "counted" in the buffer at the beginning of the slot following its arrival.

At that time, if the buffer is nonempty and server 2 is free, then the packet at the head of the line is instantaneously routed to server 2 and begins service. If server 2 is busy but server 1 is free, then the controller has two choices: a = route the packet to server 1, and b = allow server 1 to remain idle. Last consider the situation in which there are at least two packets in the system and both servers are free. Then the packet behind the one routed to server 2 may be instantaneously routed to server 1 (action a) or held in the buffer (action b). This takes place simultaneously with the routing of the head packet. It is helpful to make a sketch showing the possibilities.

It is desired to compute a policy that minimizes the total average number of packets in the system. This problem will guide you through setting up an MDC to model this system and verifying that the (AC) assumptions hold for a suitable approximating sequence.

(a) Let $\mathbf{s}$ be an ordered pair of 0's and 1's indicating whether or not a server is busy, with 1 indicating "busy." An appropriate state space S consists of all pairs $(i, \mathbf{s})$. What does i represent?

(b) Which of the states have a single action (and what is it), and which have action set $\{a, b\}$?

(c) What is the cost in state $(i, \mathbf{s})$?

(d) Develop the transition probabilities.

This completes the modeling of this system as an MDC Δ. Define an ATAS (Δ_N) by letting $S_N = \{(i, \mathbf{s}) | i \leq N\}$ and sending the excess probability to N. Let us verify that the hypotheses of Proposition 8.2.1 hold.

(e) Show that any stationary policy e for Δ_N is $(0, \mathbf{0})$ standard and has an aperiodic positive recurrent class.

(f) Argue informally that Steps 3(i) and 4(i) hold.

(g) Assume that $\lambda < \mu_1 + \mu_2$, and let d be the policy that always chooses a. Prove that d induces an irreducible positive recurrent MC on S. *Hint:* Use Corollary C.1.6 with test function y(i, **s**) = i + # in service.

***(h)** Assume the results in (g) and in addition that the second moment $\lambda^{(2)}$ of the arrival process is finite. Prove that d is standard. *Hint:* Use Corollary C.2.4 with test function $r(i, \mathbf{s}) = K(i + \#\text{ in service})^2$ for some positive constant K.

(i) Argue that (C37–38) hold for $(0, \mathbf{0})$ and the MC induced by d.

This problem was treated, in the continuous time framework, by Lin and Kumar (1984) who proved that the optimal policy is of *threshold* type; that is, server 1 will idle until the buffer content reaches a certain level. A procedure for calculating the threshold is also given. Also see Shenker and Weinrib (1989). An advantage of our approach is that it generalizes to more than two servers. The optimal policy can then be found computationally. However, the dimension of the state space and the number of possible transitions both increase rapidly with the number of servers. Various observations on the obvious behavior of an optimal policy can be made to somewhat reduce the number of actions one needs to consider. Notice that in the problem formulation above we require a packet to be sent to server 2 if it is free. This must be true of the optimal policy, since we wish to minimize the total average number in the system and there is no penalty for employing the faster server. Building this in as a requirement (rather than proving it) simplifies the analysis.

CHAPTER 9

Optimization Under Actions at Selected Epochs

In this chapter we consider systems in which actions are available only at selected slots, known as *epochs*, rather than at every slot. This appears to be a dramatically new situation that the previous theory would be inadequate to handle. However, it will be seen shortly that these systems may be modeled as MDCs and hence that the previously developed theory applies. For this reason little new theory is needed, and this chapter concentrates on showing how the method works in examples.

Section 9.1 discusses a modeling dichotomy, namely whether a random quantity such as a service time is adequately determined by a single sample or must be repeatedly sampled as the service evolves. Section 9.2 deals with the theory behind repeated samples.

Section 9.3 presents models involving service control of a single-server queue, and Section 9.4 presents models involving arrival control of a single-server queue. Sections 9.5 and 9.6 verify that the computation of an average cost optimal policy may be carried out for two of the service control examples. Sections 9.7 and 9.8 present computational results for two of the models.

9.1 SINGLE- AND MULTIPLE-SAMPLE MODELS

Let us assume that we wish to repair a machine, and let Y denote the repair time. Here Y is a discrete random variable on the positive integers $\{1, 2, \dots\}$. It may have a bounded or an unbounded distribution.

As an example, consider the situation in which there are three possible failure modes. These modes are repaired by replacing certain boards, and as soon as the appropriate mode is determined, then it is a matter of replacing the boards for that mode, which takes a fixed amount of time. Assume that mode 1 repair takes 5 units of time, mode 2 repair takes 6 units, and mode 3 takes 9 units. Assume further that the probability of a mode 1 failure is 0.25, the probability of a mode 2 failure is

0.35, and the probability of a mode 3 failure is 0.40. Then the repair time Y has distribution $P(Y = 5) = 0.25$, $P(Y = 6) = 0.35$, and $P(Y = 9) = 0.40$.

In this case it is reasonable to assume that a single sample from the distribution of Y is sufficient to remove all ambiguity. If a sample from the distribution yields $Y = 5$, then we can reasonably assume that it will take exactly 5 units of time to repair the machine. This is known as a *single-sample* (SS) model. The idea behind an SS model is that one observation of the underlying distribution suffices to remove all ambiguity and from then on the situation behaves deterministically.

Now consider another machine. Assume that it has several failure modes. The situation is further complicated by the fact that as repairs are being made, it may be discovered that further work is necessary. An initial determination of a failure mode may not be sufficient to predict how long it will take to actually repair the machine. Let us assume that careful records kept over a period of time indicate that the total repair time Y is well-modeled by a truncated Poisson distribution with mean $\lambda = 3$ hours. This means that

$$\begin{aligned} P(Y = y) &= \frac{[e^{-3}/(1 - e^{-3})]3^y}{y!} \\ &= \frac{0.0524(3)^y}{y!}, \qquad y \geq 1. \end{aligned} \tag{9.1}$$

The factor in the denominator comes from the fact that the truncated Poisson distribution is not allowed to assume the value 0.

The service takes at least one slot (one hour equals one slot). Assume that we have been repairing the machine for one hour. There are two possibilities: Either the repair is finished, or it is not finished. The probability that it is finished is $P(Y = 1) = 0.1572$. The probability that it is unfinished is $P(Y > 1) = 0.8428$. We may view this as our first sample. Namely we sampled to see whether or not we were finished in one hour.

Now assume that we are not finished. Let Y_1 be the remaining (*residual*) repair time, and observe that it is at least one hour. Because uncertainty remains concerning how long the residual repair will take, it is reasonable to let Y_1 be governed by the conditional distribution

$$\begin{aligned} P(Y_1 = y) &= P(Y = y + 1 | Y > 1) \\ &= \frac{P(Y = y + 1)}{P(Y > 1)} \\ &= \frac{(0.0524/0.8428)3^{y+1}}{(y + 1)!} \\ &= \frac{0.0622(3)^{y+1}}{(y + 1)!}, \qquad y \geq 1. \end{aligned} \tag{9.2}$$

The residual repair time is a random variable Y_1 governed by the distribution in (9.2). Remember that this represents the remaining repair time.

We now repeat this procedure. Continue to repair the machine for another hour. At the end of this time, either the repair is finished (having taken a total of two hours) or it is not finished. The probability that it is finished is $P(Y_1 = 1) = 0.2799$. (Observe that this is not the unconditioned probability $P(Y = 2) = 0.2358$.) The probability that it is not finished is $P(Y_1 > 1) = 0.7201$.

Assuming that we are still not finished, let us repeat the argument. Let Y_2 be the residual repair time, and observe that it is at least one hour. We let Y_2 be governed by the conditional distribution

$$\begin{aligned} P(Y_2 = y) &= P(Y = y + 2 | Y > 2) \\ &= \frac{P(Y = y + 2)}{P(Y > 2)} \\ &= \frac{(0.0524/0.607)3^{y+2}}{(y+2)!} \\ &= \frac{0.0863(3)^{y+2}}{(y+2)!}, \qquad y \geq 1. \end{aligned} \tag{9.3}$$

This process can be continued indefinitely. A model of this type is known as a *multiple-sample* (MS) model.

It is important to realize that the SS and MS model types are independent of whether the underlying distribution is bounded or unbounded. We now illustrate this claim.

To illustrate an MS model with a bounded distribution, assume that Y is uniformly distributed over $\{1, 2, 3, 4\}$. Then under the MS model it is easy to see that Y_1 is uniformly distributed on $\{1, 2, 3\}$, that Y_2 is uniformly distributed on $\{1, 2\}$, and finally that $P(Y_3 = 1) = 1$ because, if the service has not terminated after three slots, then it must terminate in the fourth slot.

To illustrate an SS model with an unbounded distribution, consider the truncated Poisson distribution in (9.1). We may take a single sample from this distribution. It is likely to be around the mean, and in fact $P(1 \leq Y \leq 5) = 0.9118$. Let us say that the sample yields $Y = 4$. Then it may be appropriate in certain circumstances to assume that the repair time is deterministically four hours.

If you are given the task of modeling a system, should you use an SS model or an MS model? This depends entirely on the physical situation and which one would be perceived as more appropriate under the particular circumstances. It is the case that an MS model will be somewhat more complicated mathematically. However, this minor complication may be deemed worth the extra effort to achieve good results from the model. It is also entirely possible that a system may be most appropriately modeled with some distributions being SS and others being MS.

9.2 PROPERTIES OF AN MS DISTRIBUTION

Assume that the distribution of a random quantity is treated using the MS model. In this case a certain restriction on the distribution will prove useful if it is desired to verify the (AC) assumptions. This and related matters are discussed in this section.

Let Y be a random variable representing the quantity being modeled, and let its distribution be given by $u_y = P(Y = y)$ for $y = 1, 2, 3, \ldots$. Then $F(y) = P(Y \leq y)$ is its cumulative distribution, and the complement of the cumulative is $F^*(y) = P(Y > y) = 1 - F(y)$. Note that these quantities are defined for $y = 0$ with $F(0) = 0$ and $F^*(0) = 1$.

It is helpful to have a specific situation in mind to motivate the material. We will assume that Y represents the length of service of a customer. However, keep in mind that these concepts are general and apply to other situations. In the service case $F^*(y)$ is the probability that the service lasts more than y slots, i.e., the probability that it lasts at least $y + 1$ slots.

Here is an expression for the moments of Y involving F^*.

Proposition 9.2.1. We have $E[Y] = \sum_{y=0}^{\infty} F^*(y)$ and

$$E[Y^k] = 1 + \sum_{z=0}^{k-1} \binom{k}{z} \left[\sum_{y=1}^{\infty} y^z F^*(y) \right], \qquad k \geq 2. \tag{9.4}$$

Proof: Note that

$$\begin{aligned} E[Y] &= \sum_{y=1}^{\infty} y u_y \\ &= \sum_{y=1}^{\infty} y(F^*(y-1) - F^*(y)) \\ &= \sum_{y=0}^{\infty} F^*(y), \end{aligned} \tag{9.5}$$

where we recall that $F^*(0) = 1$. Writing out a few terms of the summation in the second line gives the third line.

To prove (9.4), observe that

$$\begin{aligned}
E[Y^k] &= \sum_{y=1}^{\infty} y^k u_y \\
&= \sum_{y=1}^{\infty} y^k (F^*(y-1) - F^*(y)) \\
&= F^*(0) + \sum_{y=1}^{\infty} [(y+1)^k - y^k] F^*(y) \\
&= 1 + \sum_{y=1}^{\infty} \left(\sum_{z=0}^{k-1} \binom{k}{z} y^z \right) F^*(y). \qquad (9.6)
\end{aligned}$$

Then (9.4) follows by interchanging the order of summation (valid since all terms are nonnegative). □

Remark 9.2.2. It is clear from (9.4) that $E[Y^k] < \infty$ if and only if $\sum y^{k-1} F^*(y) < \infty$. □

Now let us assume that the service has lasted for s slots and that *it is not completed.* Then Y_s is a random variable denoting the residual (remaining) service time. Generalizing the argument given in Section 9.1, we see that its distribution is given by

$$\begin{aligned}
P(Y_s = y) &= P(Y = s + y | Y > s) \\
&= \frac{P(Y = s + y)}{P(Y > s)} \\
&= \frac{u_{s+y}}{F^*(s)}, \qquad y \geq 1. \qquad (9.7)
\end{aligned}$$

Notice that the distribution given in (9.7) for $s = 0$ coincides with the distribution of Y. Hence we set $Y_0 = Y$, and assume that (9.7) applies for $s \geq 0$.

From (9.7) it follows that the complement of the cumulative distribution for Y_s is given by

$$F_s^*(y) = P(Y_s > y)$$
$$= \frac{\sum_{w=y+1}^{\infty} u_{s+w}}{F^*(s)}$$
$$= \frac{F^*(s+y)}{F^*(s)}. \tag{9.8}$$

Using (9.8) and Proposition 9.2.1 (applied to the distribution for Y_s), it is immediate that

$$E[Y_s] = \frac{\sum_{y=s}^{\infty} F^*(y)}{F^*(s)},$$

$$E[Y_s^k] = 1 + \frac{1}{F^*(s)} \sum_{z=0}^{k-1} \binom{k}{z} \left[\sum_{y=1}^{\infty} y^z F^*(s+y), \right], \qquad k \geq 2. \tag{9.9}$$

Notice that for $s = 0$ the expressions in (9.9) reduce to the results in Proposition 9.2.1.

The original development of these concepts had to do with reliability theory, and in this case Y represents the lifetime of a component. For this reason $E[Y_s]$ is known as the *mean residual lifetime*.

Proposition 9.2.3. Fix a positive integer k. Then $E[Y^k] < \infty$ implies that $E[Y_s^k] < \infty$ for all $s \geq 0$.

Proof: This is Problem 9.1. □

Here is an important property that may be possessed by the mean residual lifetimes.

Definition 9.2.4. Assume that there exists a (finite) constant U such that $E[Y_s] \leq U$ for $s \geq 0$. Then the distribution of Y has *bounded mean residual lifetimes* (BMRL). We denote this by BMRL-U. □

In a service time distribution that is not BMRL, if the service does not terminate as time goes on, the expected additional service required grows without bound. It is clear that any service with this property is undesirable, and hence the BMRL assumption is fairly natural. Nevertheless, it does entail the following strong consequence.

Proposition 9.2.5. If the distribution of Y is BMRL, then Y has finite moments of all orders.

Proof: Let U be the bound in Definition 9.2.4. Taking the reciprocal of the first equation in (9.9) and manipulating yields

$$1 - \frac{F^*(s)}{\sum_{y=s}^{\infty} F^*(y)} \leq 1 - \frac{1}{U}. \tag{9.10}$$

Letting $w(s) = \sum_{y=s}^{\infty} F^*(y)$, it is easily seen that this becomes

$$\frac{w(s+1)}{w(s)} \leq 1 - \frac{1}{U}. \tag{9.11}$$

Observe that $w(0) = E[Y]$. Applying (9.11) inductively, we see that $w(1) \leq E[Y](1 - 1/U)$, $w(2) \leq w(1)(1 - 1/U) \leq E[Y](1 - 1/U)^2$, and in general $w(s) \leq E[Y](1 - 1/U)^s$. Since $w(s) \geq F^*(s)$, this yields

$$F^*(y) \leq E[Y]\left(1 - \frac{1}{U}\right)^y, \qquad y \geq 0. \tag{9.12}$$

Then for $k \geq 2$ we have

$$\sum_{y=1}^{\infty} y^{k-1} F^*(y) \leq E[Y] \sum_{y=1}^{\infty} y^{k-1}\left(1 - \frac{1}{U}\right)^y < \infty, \tag{9.13}$$

and hence it follows from Remark 9.2.2 that all moments of Y are finite. □

Problem 9.3 asks you to show that any finite (bounded) distribution is BMRL. The next result shows that common infinite distributions are BMRL.

Proposition 9.2.6. The following distributions are BMRL:

(i) Geometric
(ii) Negative binomial
(iii) Truncated Poisson

Proof: Assume that Y has a geo(μ) distribution, where $0 < \mu < 1$. Then Y represents the number of repeated independent Bernoulli trials until the first success is achieved, where P(success) $= \mu$. Then $F^*(y) = P$(failure in first y trials) $= (1-\mu)^y$, and it is easily seen from (9.9) that $E[Y_s] = E[Y] = 1/\mu$. In this case the mean residual lifetimes are constant, which is what we would expect from the memoryless property of the geometric.

Assume that Y has a neg bin(μ, r) distribution, where $0 < \mu < 1$ and $r \geq 2$.

Then Y represents the number of repeated independent Bernoulli trials until exactly r successes are achieved. If $Y > s$, then s trials have been observed without achieving r successes. It is clear that the expected additional time to achieve r successes is bounded by the unconditional mean $E[Y] = r/\mu$.

Assume that Y has a trun Pois(λ) distribution, where $\lambda > 0$. Then

$$P(Y = y) = \left(\frac{e^{-\lambda}}{1 - e^{-\lambda}} \right) \frac{\lambda^y}{y!}, \qquad y \geq 1. \tag{9.14}$$

Using (9.7) and the definition of $E[Y_s]$ gives

$$\begin{aligned} E[Y_s] &= \frac{\sum_{y=1}^{\infty} y\lambda^{s+y}/(s+y)!}{\sum_{y=1}^{\infty} \lambda^{s+y}/(s+y)!} \\ &= \frac{1 + \dfrac{2\lambda}{(s+2)} + \dfrac{3\lambda^2}{(s+2)(s+3)} + \dfrac{4\lambda^3}{(s+2)(s+3)(s+4)} + \dots}{1 + \dfrac{\lambda}{(s+2)} + \dfrac{\lambda^2}{(s+2)(s+3)} + \dfrac{\lambda^3}{(s+2)(s+3)(s+4)} + \dots} \\ &\leq 1 + \lambda + \frac{\lambda^2}{2!} + \frac{\lambda^3}{3!} + \dots \\ &= e^{\lambda}. \end{aligned} \tag{9.15}$$

The second line follows by factoring out and canceling the common term $\lambda^{s+1}/(s+1)!$. Focus on the second line. Its denominator is bounded below by 1; hence its reciprocal is bounded above by 1. Its numerator is bounded by the sequence in the third line, which is the power series for e^{λ}. □

Remark 9.2.7. Another proof of Proposition 9.2.6 is usually given. This is based on a stronger concept than BMRL, namely that of increasing failure rates (IFR). It is the case that IFR $\Rightarrow$ BMRL and distributions with IFR possess many nice properties. For more on this concept, see the references in the Bibliographic Notes. We have not chosen to develop this topic further here. □

The distributions in Proposition 9.2.6 are the commonest infinite distributions on $\{1, 2, \dots\}$. However, there is a wealth of others whose properties for modeling have been little explored. Johnson and Kotz (1969) and Johnson et al. (1997) contain a great deal of material on discrete distributions.

9.3 SERVICE CONTROL OF THE SINGLE-SERVER QUEUE

In this section we show various ways to model the service time control of a discrete time single-server queue. Some additional models are given in the problems.

Example 9.3.1. Batch Arrivals with SS Service Control. In each slot there is a probability p_j of a batch of j customers arriving for $j \geq 0$, and the arrivals are independent slot to slot. An arrival is counted in the buffer and available for service at the beginning of the slot following its arrival.

There is a finite set A of actions with $a \in A$ corresponding to a particular service time distribution. If Y_a denotes the service time under action a, then we let $u_y(a) = P(Y_a = y)$, $y \geq 1$. Also F_a denotes the cumulative distribution function of the service time, and F_a^* the complement of the cumulative. It is assumed that a new service distribution may be chosen only at the beginning of a service so that the whole service must be completed under the chosen distribution.

The epochs of decision are the slots in which a new service is to begin. Under the SS model, when action a is chosen, then the service time at that particular decision epoch is determined by a single sample taken from the distribution F_a. Given the sample value, the service proceeds "lock step" until it is completed.

Let us see how to model this system as an MDC. The state space S consists of the following states: State 0 means that the buffer is empty; there are no decisions in this state. State $i \geq 1$ means that there are i in the buffer, and a new service is to be initiated; the action set is A. The *service-in-progress* states are necessary to represent progress during a service of length at least 2. In this case (i, s), for $s \geq 1$, means that there are currently i in the buffer and that a service is ongoing and has s slots remaining until completion. Note that in the service-in-progress states there are no actions (null action). (In state $(i, 1)$ we know that the service will be completed in the current slot, and hence we might be tempted to want to choose the next service distribution at this time. However, this is not allowed, since it is desirable to base the decision in part on how many new customers enter during the final slot of service.)

We assume that there is a cost $C(a)$ for choosing action a (incurred at the beginning of the service) and a cost rate $d(s)$ for having an ongoing service with s slots remaining until completion. In addition a holding cost $H(i)$ is incurred when there are i customers in the buffer (including the one currently in service), where $H(0) = 0$. For example, if the system is in state (8, 2), then we know that there are currently 8 customers in the buffer and that 2 slots remain in a service (this does not tell us how many were in the buffer when the service began). Since there is no action in state (i, s), we can commit a slight abuse of notation and denote the cost as $C(i, s)$. The cost structure is given by $C(0) = 0$ and

$$\begin{aligned} C(i,a) &= H(i) + C(a), \\ C(i,s) &= H(i) + d(s), \qquad i \geq 1. \end{aligned} \tag{9.16}$$

Note that under a, a service of length 1 incurs a total cost of $C(a)$, whereas a service of length $y \geq 2$ incurs a total cost of $C(a) + \sum_{s=1}^{y-1} d(s)$.

The transition probabilities are given by

$$\begin{aligned} &P_{0j} = p_j, \\ &\begin{cases} P_{i,i-1+j}(a) = p_j u_1(a), \\ P_{i(i+j,y-1)}(a) = p_j u_y(a), \qquad y \geq 2, \end{cases} \\ &\begin{cases} P_{(i,s)(i+j,s-1)} = p_j, \qquad s \geq 2, \\ P_{(i,1)i-1+j} = p_j. \end{cases} \end{aligned} \tag{9.17}$$

This completes the specification of this model as an MDC. We may then optimize it. Either the infinite horizon discounted or the average cost criterion is perhaps more suitable than the finite horizon criterion. □

Example 9.3.2. Batch Arrivals with MS Service Control. The arrival process and service time distributions are as in Example 9.3.1. However, when action a is chosen, then service is begun under F_a as explained in Section 9.1. As service continues, its additional time is determined from the residual lifetime distributions. So at the beginning of the service, it is not known how long the service will take.

To model this as an MDC, let S be as in the previous example except that the service-in-progress states are (i,a,s) for $a \in A$ and $s \geq 1$. The state (i,a,s) means that there are currently i in the buffer, that a service is ongoing under F_a, and that the service has lasted for s slots and is *not completed.* If the distribution governed by a is finite with maximum value B_a, then we have $s \leq B_a - 1$.

Let us assume that the cost $C(a)$ is incurred when action a is chosen and that a cost rate of $d(a)$ is incurred for each unit of time (after the first slot) that service is ongoing. A holding cost $H(i)$ is incurred when there are i customers in the buffer (including the one currently in service), where $H(0) = 0$. The cost structure is given by $C(0) = 0$ and

$$\begin{aligned} C(i,a) &= H(i) + C(a), \\ C(i,a,s) &= H(i) + d(a), \qquad i \geq 1. \end{aligned} \tag{9.18}$$

The transition probabilities are given by

$$P_{0j} = p_j,$$

$$\begin{cases} P_{i,i-1+j}(a) = p_j u_1(a), \\ P_{i(i+j,a,1)}(a) = p_j F_a^*(1), \end{cases}$$

$$\begin{cases} P_{(i,a,s)(i+j,a,s+1)} = p_j \left(\dfrac{F_a^*(s+1)}{F_a^*(s)} \right), \\ P_{(i,a,s)i-1+j} = p_j \left(\dfrac{u_{s+1}(a)}{F_a^*(s)} \right). \end{cases} \tag{9.19}$$

The last line is the probability that the service is completed in the next slot, given that it has been ongoing for s slots and has not been completed, and this follows from (9.7). The second to the last line is the probability that the service will not be completed in $s + 1$ slots, given that it has not been completed in s slots, and this follows from (9.8).

We have modeled this as an MDC and may now proceed to optimize it. □

Example 9.3.3. Batch Arrivals with MS Service Control and Intervention. In this variant of the last model we specify a positive integer U as a cutoff value. If the process reaches state (i, a, U), then the customer currently in service is ejected from the system, and a penalty cost is incurred. This mechanism imposes a firm limit on the amount of service provided to any customer. Problem 9.5 asks you to model this as an MDC. □

Example 9.3.4. Priority Batch Arrivals with Uncontrolled MS Service. There are priority and nonpriority classes of customers. There is a probability p_j (respectively, q_j) of a batch of j priority (respectively, nonpriority) customers arriving in any slot. The arrival processes are independent slot to slot and class to class.

A choice of service time distribution might be included in the decision options, but to keep the model simple, let us assume that the service time of any customer is governed by a single distribution with cumulative distribution F and complement F^*. When a service is completed and the system is nonempty, then the server has a decision to make. The actions are a = serve a priority customer, b = serve a nonpriority customer, and c = idle. The server is not allowed to idle when both classes of customers are present.

The state space S consists of the following states: State (0, 0) means that the buffer is empty; there are no decisions in this state. State (i, x), with at least one coordinate positive, means that there are i priority and x nonpriority customers present and a new service may begin. If i, $x \geq 1$, then the action set is $\{a, b\}$, since a new service must be initiated. If $i = 0$, then the action set is $\{b, c\}$ (see Fig. 9.1), while if $x = 0$, then the action set is $\{a, c\}$. State (i, x, a, s) is a service-in-progress state such that a priority customer is being served and the service

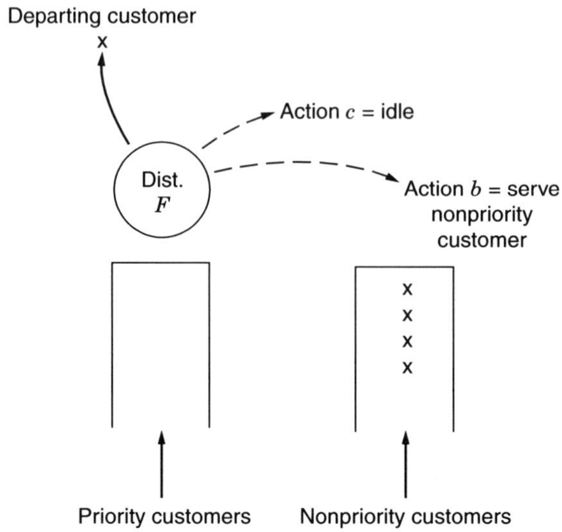

Figure 9.1 Example 9.3.4 with just emptied priority buffer.

has been ongoing for s slots and is not completed. State (i, x, b, s) has a similar interpretation with a nonpriority customer being served. There are no actions in these states.

Assume that there is no cost for service. There are nonnegative holding costs $H(i)$ (respectively, $W(x)$) for holding i priority (respectively, x nonpriority) customers in the system. We assume that $H(0) = W(0) = 0$ and that $H(i) > W(i)$ for $i \geq 1$.

A few of the transition probabilities are

$$P_{(0,x)(i,x+y)}(c) = p_i q_y, \qquad x \geq 1,$$

$$P_{(i,x)(i+j,x+y,a,1)}(a) = p_j q_y F^*(1), \qquad i \geq 1,$$

$$P_{(i,x,b,s)(i+j,x-1+y)} = p_j q_y \left(\frac{u_{s+1}}{F^*(s)} \right), \qquad x \geq 1. \tag{9.20}$$

It is clear how to obtain the remaining probabilities. This completes the specification of the model as an MDC. □

Example 9.3.5. Markov Modulated Batch Arrivals with MS Service Control. This is a generalization of Example 9.3.2 with a more complicated arrival process.

Consider an irreducible Markov chain on the finite state space $\{1, 2, \ldots, K\}$ with transition probabilities Q_{kk^*}. When the MC is in state k, then the batch arrival process is governed by the distribution $(p_j(k))_{j\geq 0}$. A state of the MC is known as a *phase*. As the chain moves from phase to phase the batch arrival process changes.

The service time model is as in Example 9.3.2. Notice that in this model the service times are being controlled and the arrival process is uncontrolled. In modeling this system, let us consider various options. Under option 1 the phase is known at the beginning of each slot, and Problem 9.6 asks you to model the system under this assumption.

Option 2 assumes that the system has been operating for a long time (and hence that the MC has reached steady state) and that no information on the phases is utilized. In this case it is reasonable to assume that P(system is in phase k) = π_k, where π_k is the steady state probability associated with phase k. Convince yourself that under option 2 the MDC model is exactly as in Example 9.3.2 with P(a batch of size j arrives) = $\sum_k p_j(k)\pi_k$.

Option 3 assumes that the phase is unknown and attempts to make a statistical inference concerning the current phase, given the historical data on customer arrivals. This approach will not be treated here. □

9.4 ARRIVAL CONTROL OF THE SINGLE-SERVER QUEUE

In this section we discuss several models dealing with the control of arrivals (often known as *flow control*) to a single-server queue. To avoid obscuring the ideas, we will assume that the service time is uncontrolled. Obviously one could control both simultaneously but at the expense of increased model complexity.

In the models in Section 9.3, it was assumed that the customers were essentially identical and that the service time characteristics were attached to the server. Controlling the service involved adjusting the characteristics of the server rather than any inherent properties of the arriving customers.

Now we switch to the point of view that the customer brings work into the system and that the controller has the option of adjusting the work load in various ways. Notice that one perspective or the other may apply in a given application.

Example 9.4.1. SS Arrival Distribution Control with Fixed Service Rate. In this example we think of the customers as packets and assume that the service rate is fixed at one packet per slot. An action a from the finite set A corresponds to a particular phase a of arrivals and the length of the phase is governed by a distribution with cumulative distribution function F_a, complement F_a^*, and probability function $u(a)$. As long as this phase is operative, then P(a batch of size j arrives in a slot) = $p_j(a)$. Under the SS variant of this model, a single sample is taken from the phase distribution, and this determines how long the phase lasts. Notice that the control is exercised on the *arrival statistics* rather than on individual arrivals. A new action may be selected when a phase terminates.

To model this as an MDC, the state space S consists of the following states. State $i \geq 0$ means that there are currently i packets in the buffer and a phase has just ended; the action set is A. The phase-in-progress state (i, a, s) means that there are $i \geq 0$ in the buffer and that phase a (of length at least 2) has $s \geq 1$ slots remaining until completion.

There is a nonnegative holding cost $H(i)$ on the buffer content, where $H(0) = 0$. In addition we assume that a nonnegative reward $R(a)$ is earned when phase a is chosen, and a nonnegative reward rate of $r(a, s)$ is earned when there are s slots remaining of phase a. We assume that there exists a (finite) constant B that is an upper bound for the rewards and reward rates. This assumption enables the rewards to be incorporated into the cost structure as negative costs. To accomplish this, the net cost is incremented by B to make it nonnegative. If the system is optimized under the average cost criterion, then $J - B$ will be the minimum average cost and hence $B - J$ the maximum average reward.

Under these assumptions the costs are given by

$$
\begin{aligned}
C(i, a) &= H(i) - R(a) + B, \\
C(i, a, s) &= H(i) - r(a, s) + B. \qquad (9.21)
\end{aligned}
$$

Compare this cost structure with that in Example 9.3.1. In the latter the server is choosing a service distribution and hence must pay for it with presumably "on average faster" service costing more. We assumed that the cost rate depended only on the remaining service time and not on the particular service action choice. This makes sense because once the service length is determined, then the service proceeds lock step until it is completed. In Example 9.4.1 we are thinking of competing customer classes, and there may be a reward associated with allowing a certain class of packets into the system. It makes sense to allow the reward rate to depend on both the class identity and the remaining length of the phase, since the arrival statistics from that class are operative in *each* slot of the phase.

To develop the transition probabilities, it is helpful to introduce $i^* =: (i-1)^+$. This quantity equals $i - 1$ for $i \geq 1$ and 0 for $i = 0$. Recall that the service rate is 1 packet per slot. Then the transition probabilities are given by

$$
\begin{cases}
P_{i,i^*+j}(a) = p_j(a)u_1(a), \\
P_{i(i^*+j,a,y-1)}(a) = p_j(a)u_y(a), \qquad y \geq 2,
\end{cases}
$$

$$
\begin{cases}
P_{(i,a,s)(i^*+j,a,s-1)} = p_j(a), \qquad s \geq 2, \\
P_{(i,a,1)i^*+j} = p_j(a).
\end{cases} \qquad (9.22)
$$

□

Example 9.4.2. MS Arrival Distribution Control with Fixed Service Rate. This is as in Example 9.4.1 except that the length of each phase is determined under the MS model. Problem 9.8 asks you to model this as an MDC. □

Example 9.4.3. Semi-Markov Modulated Batch Arrivals with Fixed Service and Reject Option. Assume first that we have an irreducible Markov chain as in Example 9.3.5. This chain will be the mechanism underlying the phase process, which works as follows: If the MC is in state k, we assume that a sample is taken from the transition probability distribution Q to determine the next state, say it is k^*. Then associated with the pair (k, k^*) is a cumulative distribution function $F(k, k^*)$ and probability function $u(k, k^*)$ that determines how long the current phase will last. Let us employ the SS model for the phase length. As long as the distribution $F(k, k^*)$ is operative, then P(a batch of size j arrives in a slot) $= p_j(k, k^*)$. This is known as a *semi-Markov modulated batch arrival process*. The above description embodies full generality. As a special case it might happen that the phase length distribution depends only on the state k rather than on the pair (k, k^*).

Here is one possible method of control. Assume that when a new phase is to begin, the controller may choose action a = accept all incoming batches under that phase or b = reject all incoming batches under that phase. This decision is made with knowledge of the current MC state k but before the next state or the sampled length of the resulting phase are revealed.

There is a nonnegative holding cost H on packets in the buffer. In addition there is a fixed penalty cost G for choosing b and a cost rate $g(s)$ incurred when there are s slots remaining in a rejection phase. Let us assume a fixed service rate of one packet per slot.

To model this as an MDC, we observe that the state space S consists of the following states: State (i, k) means there are i packets in the buffer, a new phase is to begin, and that phase is determined by state k in the MC. The action set is $\{a, b\}$. State (i, k, k^*, I, s) means that there are i packets in the buffer and that $s \geq 1$ slots remain of an ongoing phase associated with (k, k^*). Here I is an indicator variable with $I = 0$ meaning that a was chosen at the beginning of the phase and $I = 1$ meaning that b was chosen. Notice that if $I = 1$, then we know that no new packets will enter the system during the whole phase. If the phase identity depends only on k, then the next MC state k^* may be omitted from the state description.

The costs are given by

$$\begin{aligned} C(i, k, a) &= H(i), \\ C(i, k, b) &= H(i) + G, \\ C(i, k, k^*, I, s) &= H(i) + Ig(s). \end{aligned} \tag{9.23}$$

Recall that the service rate is one packet per slot, and let i^* be an auxillary variable as in Example 9.4.1. The transition probabilities are given by

$$\begin{cases} P_{(i,k)(i^*+j,k^*)}(a) = Q_{kk^*}p_j(k, k^*)u_1(k, k^*), \\ P_{(i,k)(i^*+j,k,k^*,0,y-1)}(a) = Q_{kk^*}p_j(k, k^*)u_y(k, k^*), \qquad y \geq 2, \end{cases}$$

$$\begin{cases} P_{(i,k)(i^*,k^*)}(b) = Q_{kk^*}u_1(k, k^*) \\ P_{(i,k)(i^*,k,k^*,1,y-1)}(b) = Q_{kk^*}u_y(k, k^*), \qquad y \geq 2, \end{cases}$$

$$
\begin{cases} P_{(i,k,k^*,0,s)(i^*+j,k,k^*,0,s-1)} = p_j(k,k^*), & s \geq 2, \\ P_{(i,k,k^*,0,1)(i^*+j,k^*)} = p_j(k,k^*), \end{cases}
$$

$$
\begin{cases} P_{(i,k,k^*,1,s)(i^*,k,k^*,1,s-1)} = 1, & s \geq 2, \\ P_{(i,k,k^*,1,1)(i^*,k^*)} = 1. \end{cases} \tag{9.24}
$$

This completes the specification of this model as an MDC. □

9.5 AVERAGE COST OPTIMIZATION OF EXAMPLE 9.3.1

Once a system has been modeled as an MDC, then it may be optimized. We focus on optimization under the average cost criterion. To employ the approximating sequence method requires verification of the (AC) assumptions (or the weaker (WAC) assumptions discussed in Section 8.7). In this section we show how to verify the (WAC) assumptions for Example 9.3.1.

Let λ be the mean and $\lambda^{(2)}$ the second moment of the batch size. Let τ_a be the mean service time under action a and $\tau_a^{(2)}$ the second moment of the service time. We operate under the following basic assumptions:

(BA1). We have $\lambda^{(2)} < \infty$.

(BA2). For some a^* it is the case that $\lambda\tau_{a^*} < 1$ and $\tau_{a^*}^{(2)} < \infty$.

(BA3). We have $\tau_a < \infty$ for all a.

(BA4). There exist a and $y \geq 2$ such that $u_y(a) > 0$.

(BA5). The holding cost is given by $H(i) = Hi$ for some positive constant H. The cost rate $d(s) \equiv 0$.

Note that (BA5) is assumed for convenience. The result could be proved under more general conditions, but at the expense of increased complexity. Since $\tau_{a^*} \geq 1$, it follows from (BA2) that $\lambda < 1$, and hence $p_0 > 0$. There is no loss of generality in assuming that $p_0 < 1$, since otherwise no customers will ever arrive. The condition in (BA4) rules out another triviality. If it fails to hold, then the service time under every action is exactly one slot, and this situation is of no interest.

Remark 9.5.1. In this section and the next we will be calculating a number of expected times and costs. Suppose that i is a state in an MDC, and we need to calculate the expected first passage time (or cost) from i to a distinguished state. What is usually important is whether this quantity is a constant, a linear function of i, a quadratic function of i, and so on, rather than the specific parameters involved. In what follows, all the "U" functions are assumed to be finite. We say a function is U_0 if it is a constant. It is $U_1(i)$ (respectively, $U_2(i)$) if it is a linear (respectively, quadratic) function of i.

This notation gives us a lot of flexibility. A function of i and s is $U_1(i, s)$ if it is a linear function of i and s. Such a function can have i, s, and constant terms. It is $U_2(i, s)$ if it is a quadratic function of i and s. Such a function can have i^2, s^2, and is terms as well as linear terms. The function is $U_1(i)U_1(s)$ if it is a product of a linear function of i and a linear function of s. Other combinations can be created in an obvious way. □

Lemma 9.5.2. Assume that the (BA) assumptions hold, and let $X(s)$ be the number of customers arriving in s slots. Then

$$E[X(s)] = \lambda s = U_1(s),$$
$$E[(X(s))^2] = \lambda^{(2)} s + \lambda^2 s(s-1) = U_2(s). \tag{9.25}$$

Proof: Let X_k be the number of customers arriving in slot k. Then $X(s) = \sum_{k=1}^{s} X_k$. Using the linearity of the expectation and the fact that $E[X_k] = \lambda$ yields the first line of (9.25).

We have $\text{var}[X_k] = E[(X_k)^2] - (E[X_k])^2 = \lambda^{(2)} - \lambda^2$. Moreover the variance of $X(s)$ is linear because the summands are independent. Using these facts and some algebra yields the second line of (9.25). □

We now show that there exists a standard policy.

Lemma 9.5.3. Assume that the (BA) assumptions hold, and let d be the stationary policy that always chooses a^*. Then d is 0 standard.

Proof: Since $p_0 > 0$, there is a path from any nonzero state to 0 in the MC induced by d (indeed, in the MC induced by any stationary policy). Hence this MC has a single communicating class R containing 0.

We consider three cases concerning the service time distribution under a^*: In Case 1 either this distribution is unbounded, or it is bounded and its maximum value B is the largest possible service time under any action. In Case 2 the maximum value of this distribution is B, and a service under at least one other action may be longer than B. Moreover either ($B \geq 2$) or ($B = 1$ and $p_0 + p_1 < 1$). In Case 3 we have $B = 1$ and $p_0 + p_1 = 1$.

Under Case 1 the policy d induces an irreducible Markov chain on S (why?), and hence $R = S$.

In Case 2 a queue can always build up, and hence the states $D = \{0, 1, 2, \ldots\} \subset R$. Note that states (i, s) for $1 \leq s \leq B-1$ are also in R, whereas states (i, s) for $s \geq B$ are transient.

We first argue informally that under either Case 1 or Case 2, the class R is positive recurrent with finite average cost. Assume that a service has just commenced under a^*. The expected number of customers arriving during that service is $\sum u_y(a^*)(\lambda y) = \lambda \tau_{a^*} < 1$. Since one customer is served and the expected number of arrivals is less than 1, on average when the service is finished the buffer will contain fewer customers than when the service began. Eventually, in finite expected time, the MC will reach 0.

Let m be the expected time to go from state 1 to state 0. Starting in state i, the chain must transition to $i-1$, then $i-2$, and so on, to reach 0. Each of these first passages is a statistical replica of the one from 1 to 0. Hence $m_{i0}(d) = im$. Then $m_{00}(d) = 1 + \sum p_j(jm) = 1 + m\lambda < \infty$. This shows that R is positive recurrent.

We now upper bound the expected cost of a service initiated in i. Let $P_j(y)$ be the probability that exactly j customers arrive in y slots. Then

$$
\begin{aligned}
E_d[\text{cost of service in } i] &\le C(a^*) + H \sum_y u_y(a^*) \sum_j P_j(y)(i+j)y \\
&= C(a^*) + H \sum_y y u_y(a^*)(i + \lambda y) \\
&= C(a^*) + H[i\tau_{a^*} + \lambda \tau_{a^*}^{(2)}] \\
&= U_1(i).
\end{aligned}
\tag{9.26}
$$

The first line of (9.26) follows by assuming that the customers arriving during the service are charged a holding cost throughout the length of the service. Line two follows from the first line of (9.25). Note that the terms in the third line are finite by (BA2). It is then possible to argue (we omit the details) that $c_{i0}(d) = U_2(i)$. Thus $c_{00}(d) = \sum p_j U_2(j) < \infty$ by (BA1).

This completes the informal proof that R is positive recurrent with finite average cost. Problem 9.9 outlines a rigorous proof of the positive recurrence.

This shows that d is 0 standard in Case 1. To complete the proof under Case 2, we need to deal with the transient service-in-progress states. Let us develop expressions for the expected time and cost to reach D rather than the larger set R. If the process starts in transient state (i, s), then in s steps it will reach D. Hence $m_{(i,s)R}(d) = U_1(s)$. Moreover, using reasoning similar to that in (9.26), we have

$$
\begin{aligned}
c_{(i,s)R}(d) &\le H \sum_j P_j(s)(i+j)s \\
&= H[is + \lambda s^2] \\
&= U_1(i)U_1(s) + U_2(s).
\end{aligned}
\tag{9.27}
$$

This completes the proof in Case 2.

It remains to consider Case 3. In this case a queue cannot build up, and it is easy to see that $R = \{0, 1\}$. All other states of S are transient. We refer to Case 3 as the *Special Case*.

Since R is finite, it is a positive recurrent class with finite average cost. It may easily be seen that $m_{i0}(d) = i/p_0 = U_1(i)$ and $c_{i0}(d) = [C(a^*)i + Hi(i+1)/2]/p_0 = U_2(i)$.

Using reasoning similar to that above, we have

$$\begin{aligned} m_{(i,s)0}(d) &\leq s + \sum_j P_j(s) m_{i+j,0}(d) \\ &= s + \sum_j P_j(s) U_1(i+j) \\ &= U_1(i,s). \end{aligned} \tag{9.28}$$

Moreover

$$\begin{aligned} c_{(i,s)0}(d) &\leq \sum_j P_j(s)[H(i+j)s + c_{i+j,0}(d)] \\ &= H(is + \lambda s^2) + \sum_j P_j(s) U_2(i+j) \\ &= U_2(i,s). \end{aligned} \tag{9.29}$$

Note that (9.29) utilizes the second line of (9.25). Thus d is 0 standard in the Special Case. The proof is completed. □

We now consider the verification of the assumptions for average cost computation. If all the service time distributions are bounded, then it is possible to verify the (AC) assumptions given in Section 8.1. However, if at least one distribution is unbounded, then we must employ the (WAC) assumptions from Section 8.7. (These only differ from (AC) in that (WAC3) is weaker than (AC3).) The (WAC) approach is more general and works for bounded or unbounded distributions. For this reason we show how to verify the (WAC) assumptions.

First choose and fix a sequence $M(N)$ of positive integers such that $M(N) \to \infty$ as $N \to \infty$. Then let $S_N = \{i | 0 \leq i \leq N\} \cup \{(i,s) | 1 \leq i \leq N, 1 \leq s \leq M(N) - 1\}$. This means that the buffer is not allowed to contain more than N customers and no service can last more than $M(N)$ slots. If a batch arrives that would cause a buffer overflow, then the probability of that event is given to the corresponding full buffer state. So, if the system is in state $(i,s) \in S_N$, then the probability of a batch of more than $N - i$ customers is given to state $(N, s-1)$. The probability of a sampled service time greater than $M(N)$ is given to a service time of $M(N)$. So, for example, if the system is in state i, $1 \le i < N$, then the probability that the service time is greater than $M(N)$ and that a single customer arrives is given to state $(i+1, M(N)-1)$. Other possibilities are handled in the obvious way. This defines (Δ_N).

The next proof utilizes Section 7.7 and may be omitted if desired.

Proposition 9.5.4. Assume that the (BA) assumptions hold for Example 9.3.1, and let the AS be as above. Then the VIA is valid for (Δ_N) and the (WAC) assumptions hold for the function $r^N(.) = \lim_{n\to\infty} (v_n^N(.) - v_n^N(0))$.

**Proof:* We follow the four-step template in Proposition 8.2.1 (with $x = 0$) with the exception of Step 4 which verifies (AC3). Instead, we will directly verify (WAC3) from Section 8.7.

Since $p_0 > 0$, it is easy to see that under any stationary policy there is a path from any nonzero state in Δ_N to state 0. Moreover we have $P_{00} = p_0 > 0$. Hence the induced MC is unichain with aperiodic positive recurrent class containing 0, and Step 1 holds.

Let d be the 0 standard policy in Lemma 9.5.3. If we can show that (C.37–38) hold, then Step 2 will follow from Proposition C.5.3. Consideration of a few cases will make this clear. If the process is in state i, where $1 \le i \le N$, then it may transition to state $(r, y-1)$, where $r > N$ and $y > M(N)$. The probability associated with this is given to $(N, M(N)-1)$. It is clearly the case that $m_{(N,M(N)-1)0}(d) \le m_{(r,y-1)0}(d)$, since during the shorter service time fewer customers are expected to enter the system. Similar arguments may be given for other cases and for the first passage costs. Hence (C.37–38) hold.

We verify that Step 3(iii) holds. Let us first show that the (H) assumptions from Section 7.7 hold for Δ (with distinguished state 0). This will give us the existence of an average cost optimal stationary policy. It follows from Lemma 9.5.1 and Proposition 7.5.3 that (H1–2) hold. It remains to verify (H3–5).

Let us first explain where we are going and then show how to get there. It follows from (H1) that $(1-\alpha)V_\alpha(0)$ is bounded in α. Let Z be an upper bound for this quantity.

We will first show that

$$L(i) = c_{01}(d) \quad \text{and} \quad L(i,s) = c_{01}(d) + sZ, \qquad i \ge 1, \tag{9.30}$$

works in (H3). For the function in (9.30) it is easy to see that (H4) holds. Check it out! Property (BA3) is needed.

Now define

$$W =: \max_{1 \le s} \left\{ c_{01}(d) + sZ - \frac{H\lambda s(s-1)}{2} \right\}. \tag{9.31}$$

Note that $c_{01}(d) + Z \le W < \infty$, where the finiteness follows since the subtracted quantity is a quadratic in s. Let h be any limit function (Definition 7.2.2(i).) We will show that

$$h \ge -W. \tag{9.32}$$

Since h is bounded below by a constant, the validity of (H5) will follow.

So to complete the verification of (H), we show that (9.30) and (9.32) hold. It is helpful to define the function z_α, for $\alpha \in (0, 1)$, by $z_\alpha(1) \equiv 0$ and

$$z_\alpha(s) = (s-1)\alpha^{s-1} + (s-2)\alpha^{s-2} + \ldots + (1)\alpha, \qquad s \ge 2, \tag{9.33}$$

and note that

$$\lim_{\alpha \to 1^-} z_\alpha(s) = \frac{s(s-1)}{2}. \tag{9.34}$$

To proceed, observe that

$$\begin{aligned} V_\alpha(i) &\ge V_\alpha(1), \\ V_\alpha(i, s) &\ge V_\alpha(1, s), \qquad i \ge 1. \end{aligned} \tag{9.35}$$

If the process is in state $i \ge 1$, then the situation is probabilistically identical to the situation in state 1 except that the holding cost is greater. Hence the first line of (9.35) is clear. The reasoning for the second line is similar.

Using (9.35), we see that it is only necessary to verify (9.30) and (9.32) for states with a buffer content of 1. Using reasoning similar to that in (8.6) but applied to Δ yields $h_\alpha(1) \ge -c_{01}(d)$, and this verifies the first part of (9.30). It also shows that $h(1) \ge -c_{01}(d) \ge -W$ and hence verifies (9.32) for decision epochs.

We now obtain an expression for $h_\alpha(1, s)$. Let $P_j(s)$ be the probability that exactly j customers arrive in the remaining s slots of the service. Iterating the discount optimality equation (4.9), it may be seen that

$$
\begin{aligned}
V_\alpha(1,s) &= E[\text{discounted holding cost for } s \text{ slots}] + \alpha^s \sum_j P_j(s)V_\alpha(j) \\
&\geq H\lambda[(\alpha + \alpha^2 + \cdots + \alpha^{s-1}) + (\alpha^2 + \ldots + \alpha^{s-1}) + \ldots + \alpha^{s-1}] \\
&\quad + \alpha^s \sum_j P_j(s)V_\alpha(j) \\
&\geq H\lambda z_\alpha(s) + \alpha^s[(p_0)^s V_\alpha(0) + (1 - (p_0)^s)V_\alpha(1)], \qquad s \geq 1. \qquad (9.36)
\end{aligned}
$$

To obtain the second line, the holding cost due to the customer in service has been discarded; the first term on the right is the expected holding cost of the customers arriving during the service. The third line follows from (9.33) and (9.35). It is useful for the reader to check the validity of (9.36) for $s = 1, 2$.

Subtracting $V_\alpha(0)$ from both sides of (9.36) and performing some algebraic manipulation yields

$$
\begin{aligned}
h_\alpha(1,s) &\geq H\lambda z_\alpha(s) + \alpha^s(1 - (p_0)^s)h_\alpha(1) - \left(\frac{1-\alpha^s}{1-\alpha}\right)[(1-\alpha)V_\alpha(0)] \\
&\geq H\lambda z_\alpha(s) - c_{01}(d) - sZ. \qquad (9.37)
\end{aligned}
$$

Since $z_\alpha(s) \geq 0$, it follows that $h_\alpha(1,s) \geq -c_{01}(d) - sZ$, and this verifies the second statement of (9.30).

It follows from (9.37) and (9.34) that

$$
h(1,s) \geq \frac{H\lambda s(s-1)}{2} - c_{01}(d) - sZ. \qquad (9.38)
$$

It then follows from (9.31) that (9.32) holds. This completes the verification of the (H) assumptions.

From Proposition 7.7.2 there exists an average cost optimal stationary policy f for Δ and the minimum average cost is a finite constant. In the MC induced by f there is a single communicating class R_f that contains 0. Since $p_0 < 1$ and every service lasts at least one slot, we must have $1 \in R_f$. (These statements are true for any stationary policy.) Let us consider various cases concerning the service time distribution under $f(1)$.

Under Case 1, this distribution is unbounded, and then $R_f = S$. Since f is average cost optimal and the holding cost is unbounded as the number of customers increases, it is intuitively clear that the chain is positive recurrent. Hence f is 0 standard.

Under Case 2, the service time has maximum value B, where either $(B \geq 2)$ or $(B = 1$ and $p_0 + p_1 < 1)$. In Case 2 a queue may build up, and we have

$i \in R_f$ for $i \geq 0$. The argument that R_f is positive recurrent is as in Case 1. There may be transient service-in-progress states if there are service time distributions yielding longer services than under f. Using reasoning similar to that in Lemma 9.5.3, it is easy to see that the expected time and cost to reach a decision epoch from one of the transient states are finite. Thus f is 0 standard.

Under Case 3, the Special Case, the service time is one slot and $p_0 + p_1 = 1$. Then $R_f = \{0, 1\}$. We do not need to argue that f is 0 standard in this case.

Using the fact that f is 0 standard under Cases 1 and 2, it may be shown just as was argued above for d, that (C.37–38) hold for the MC induced by f. The validity of this argument does not require the policy to always choose the same service time distribution. Instead, it relies crucially on two facts: first that exactly one customer is serviced at a time, and second that in a given decision state i, the action $f(i)$ is constant (which is true for any stationary policy). It then follows from Proposition C.5.3 that (Δ_N) is conforming at f.

Now assume we are in the Special Case. Then it may be seen directly that $\pi_0 = p_0$, $\pi_1 = p_1$, and $J = p_1(C(f(1)) + H)$. These same results hold for Δ_N for $N \geq 1$, and hence (Δ_N) is conforming on R_f. Hence in all three cases we have conformity and Step 3(iii) holds.

It remains to verify (WAC3) from Section 8.7. We have already shown that r^N satisfies (8.1). It follows from Proposition 6.5.1(iii) that

$$r^N(.) = \lim_{\alpha \to 1^-} h_\alpha^N(.), \qquad h_\alpha^N(.) = V_\alpha^N(.) - V_\alpha^N(0). \tag{9.39}$$

Using (9.39) and following the method in the verification of (H3-5) above verifies (WAC3$_1$) with the function Q being a constant. The rest of (WAC3) follows immediately. The details are ommitted. □

9.6 AVERAGE COST OPTIMIZATION OF EXAMPLE 9.3.2

In this section we show how to verify the (AC) assumptions for Example 9.3.2. Recall that Y_a denotes the service time under action a. We let $Y_{a,s}$ be the residual service time under a, given that the service has been ongoing for s slots and is not completed. Recall that τ_a is the mean service time and $\tau_a^{(2)}$ the second moment of the service time distribution under a. Let $\tau_{a,s} = E[Y_{a,s}]$ be the mean residual service time and $\tau_{a,s}^{(2)}$ the second moment of the residual service time distribution.

We operate under the following basic assumptions:

(BA1). We have $\lambda^{(2)} < \infty$.

(BA2). For some a^* it is the case that $\lambda\tau_{a^*} < 1$.

(BA3). There exists a (finite) constant U such that $\tau_{a,s} \leq U$ for all a and s.

(BA4). There exist a and $y \geq 2$ such that $u_y(a) > 0$.

(BA5). The holding cost is given by $H(i) = Hi$ for some positive constant H. The cost rate $d(a) \equiv 0$.

We may argue as in Section 9.5 that $0 < p_0 < 1$. Note that (BA3) says that the service time distributions are all BMRL. The purpose of (BA4) is to rule out the trivial case in which every service takes exactly one slot. The conditions in (BA5) are assumed for convenience. The result can be proved under more general conditions.

We first show that there exists a standard policy.

Lemma 9.6.1. Assume that the (BA) assumptions hold, and let d be the stationary policy that always chooses a^*. Then d is 0 standard.

Proof: Note that for $a \neq a^*$, states of the form (i, a, s) are transient under the MC induced by d. To see if there are additional transient states, we need to consider two cases. In Case 1, one of three situations holds: the distribution under a^* is unbounded, or it is bounded with maximum value $B \geq 2$, or $B = 1$ and $p_0 + p_1 < 1$. In any of these situations a queue can build up and the remaining states of S form a single communicating class R.

In Case 1 an informal argument that R is positive recurrent with finite average cost may be given in a similar manner to the proof of Lemma 9.5.3. We omit the reasoning. A formal proof of the positive recurrence is outlined in Problem 9.10.

To complete the proof in Case 1, let $D = \{0, 1, 2, \ldots\}$, and assume that the process is in transient state (i, a, s). Note that reaching D and reaching R are equivalent. We now show that the expected time and cost to reach D are finite. Note that for *fixed s*, we may consider $\tau_{a,s}$ and $\tau_{a,s}^{(2)}$ to be finite constants. The mean residual service times are finite by (BA3), and the second moments are finite by Propositions 9.2.3 and 9.2.5.

Remember the "U" notation from Remark 9.5.1. Do not confuse it with the bound U in (BA3). Observe that $m_{(i,a,s)D}(d) = \tau_{a,s} < \infty$ by (BA3). Conditioning on the length of the residual service time yields

$$\begin{aligned} c_{(i,a,s)D}(d) &\leq H \sum_k P(Y_{a,s} = k) \sum_j P_j(k)(i + j)k \\ &= H[i\tau_{a,s} + \lambda\tau_{a,s}^{(2)}] \\ &= U_1(i). \end{aligned} \tag{9.40}$$

The first line follows by assuming that the holding cost is incurred on all the customer arrivals for the whole length of the residual service time. The second line follows from the first line of (9.25) and the finiteness is argued above. This proves that d is 0 standard in Case 1.

In Case 2, the Special Case, we have $B = 1$ and $p_0 + p_1 = 1$. In this case a queue cannot built up. Then $R = \{0, 1\}$, and the remaining states are transient. It is clear that $m_{i0}(d)$ and $c_{i0}(d)$ are as given in the proof of Lemma 9.5.3. By using reasoning similar to that above and in Lemma 9.5.3, we can show that $m_{(i,a,s)0}(d) = U_1(i)$ and $c_{(i,a,s)0}(d) = U_2(i)$. This completes the proof in the Special Case. □

To define (Δ_N) for this example, choose and fix a sequence $M(N)$ of positive integers such that $M(N) \to \infty$ as $N \to \infty$. Let $S_N = \{i | 0 \le i \le N\} \cup \{(i, a, s) | 1 \le i \le N, \text{ all } a, 1 \le s \le M(N) - 1\}$. This means that the buffer is not allowed to contain more than N customers. If the system is in state (i, a, s), where $1 \le s \le M(N) - 2$, then just as in Δ, a sample is taken from the appropriate residual service time distribution to see whether the service finishes in the next slot or not. If the system is in state $(i, a, M(N) - 1)$, then it is declared finished in the next slot.

So, if the system is in state (i, a, s), $i \le N$ and $s \le M(N) - 2$, then the probability of a batch of more than $N - i$ customers, and a continuing service is given to state $(N, a, s, +1)$. If the system is in state $(i, a, M(N) - 1)$, $i \le N$, then the probability of a single customer arriving is given to state i (recall that the service is declared finished in the next slot). Other possibilities are handled in the obvious way.

Here is the main result.

Proposition 9.6.2. Assume that the (BA) assumptions hold for Example 9.3.2, and let the AS be as above. Then the VIA is valid for (Δ_N) and the (AC) assumptions hold for the function $r^N(.) = \lim_{n \to \infty} (v_n^N(.) - v_n^N(0))$.

Proof: We follow the four-step template in Proposition 8.2.1 with $x = 0$, except that (AC3) will be verified directly rather than through Step 4.

Since $p_0 > 0$, it is easy to see that under any stationary policy there is a path from any nonzero state in Δ_N to state 0. Moreover we have $P_{00} = p_0 > 0$. Hence the induced MC is unichain with aperiodic positive recurrent class containing 0. This completes Step 1.

Let us give an informal argument that (C.37-38) hold for the MC induced by d. It will then follow from Proposition C.5.3 that the AS is conforming at d, and hence Step 2 holds. Assume that the process is in state i with $1 \le i \le N$. It may transition in the next slot to $(r, a^*, 1)$ where $r > N$. The probability of this event is given to the state $(N, a^*, 1)$. It is clear that $m_{(N,a^*,1)0}(d) \le m_{(r,a^*,1)0}(d)$. This argument relies on the fact that to reach 0 every customer must be served, one at a time. If the process is in state (i, a, s) with $1 \le i \le N$ and $1 \le s \le M(N) - 2$, then it may transition in the next slot to $r > N$. The probability of this event is given to the state N, and it is clear that $m_{N0}(d) \le m_{r0}(d)$. If the process is in state $(i, a, M(N) - 1)$ with $1 \le i \le N$, then it may transition in the next slot to $(i + j, a, M(N))$ with

$i - 1 + j > N$. The probability of this event is given to the state N, and it is clear that $m_{N0}(d) \le m_{(i+j,a,M(N))0}(d)$. The argument for the other cases and for the first passage costs is similar and hence (C.37–C38) hold.

Let us now verify that Step 3(iii) holds. We first verify that the (SEN) assumptions from Section 7.2 hold for Δ (with distiguished state 0). This will give us the existence of an average cost optimal stationary policy with constant minimum average cost. It follows from Lemma 9.6.1 and Proposition 7.5.3 that (SEN1–2) hold. It remains to verify (SEN3).

Observe that

$$\begin{aligned} V_\alpha(i) &\ge V_\alpha(1), \\ V_\alpha(i,a,s) &\ge V_\alpha(1,a,s), \qquad i \ge 1. \end{aligned} \tag{9.41}$$

If the process is in state $i \ge 1$, then the situation is probabilistically identical to the situation in state 1 except that the holding cost is greater. Hence the first line of (9.41) is clear. The reasoning for the second line is similar.

Using (9.41), we see that it is only necessary to verify (SEN3) for states with a buffer content of 1. Using reasoning similar to that in (8.6) but applied to Δ yields $h_\alpha(1) \ge -c_{01}(d)$.

Now assume that the process starts in $(1, a, s)$, and let $P_j(k)$ be the probability that exactly j customers arrive during k slots of a service. Then

$$\begin{aligned} V_\alpha(1,a,s) &\ge \sum_{k=1}^{\infty} \alpha^k P(Y_{a,s} = k) \sum_j P_j(k) V_\alpha(j) \\ &\ge \sum_{k=1}^{\infty} \alpha^k P(Y_{a,s} = k)[(p_0)^k V_\alpha(0) + (1 - (p_0)^k) V_\alpha(1)]. \end{aligned} \tag{9.42}$$

Subtracting $V_\alpha(0)$ from both sides yields

$$\begin{aligned} h_\alpha(1,a,s) &\ge h_\alpha(1) \sum_{k=1}^{\infty} \alpha^k P(Y_{a,s} = k)(1 - (p_0)^k) \\ &\qquad - [(1-\alpha)V_\alpha(0)] \sum_{k=1}^{\infty} \left(\frac{1-\alpha^k}{1-\alpha} \right) P(Y_{a,s} = k) \\ &\ge -c_{01}(d) - Z \sum_{k=1}^{\infty} k P(Y_{a,s} = k) \\ &\ge -c_{01}(d) - ZU, \end{aligned} \tag{9.43}$$

where Z is an upper bound for $(1-\alpha)V_\alpha(0)$, and U is from (BA3). Thus (SEN3) holds.

It follows from Theorem 7.2.3 that there exists an average cost optimal stationary policy f for Δ and the minimum average cost is a finite constant J. There is a path from any nonzero state to state 0 under the MC induced by f, which implies that there is a single communicating class R_f containing 0. Since $p_0 < 1$ and every service lasts at least one slot, we must have $1 \in R_f$. Let us consider two cases concerning the service time distribution under $f(1)$.

Under Case 1, one of three situations holds: this distribution is unbounded; or it is bounded with maximum value $B \geq 2$; or $B = 1$ and $p_0 + p_1 < 1$. In any of these situations a queue can build up, and R_f contains all the decision epochs. Note that states of the form (i, a, s) are transient if the action a did not arise under f. The reasoning in the proof of Lemma 9.6.1 shows that the expected time and cost of a first passage from a state (i, a, s) to a decision epoch are finite. Since f is average cost optimal and since the holding cost is unbounded, it is intuitively clear that R_f must be positive recurrent. Hence f is 0 standard.

It may be argued that (C.37–38) hold for the MC induced by f. The validity of this argument does not require the policy to always choose the same service time distribution. Instead, it relies crucially on two facts: first that exactly one customer is serviced at a time, and second that in a given decision state i, the action $f(i)$ is constant (which is true for any stationary policy). It then follows from Proposition C.5.3 that (Δ_N) is conforming at f.

Under Case 2, we have $B = 1$ and $p_0 + p_1 = 1$. In this case a queue cannot build up, and $R = \{0, 1\}$. Then it may be seen directly that $\pi_0 = p_0$, $\pi_1 = p_1$, and $J = p_1(C(f(1)) + H)$. These same results hold for (Δ_N) for $N \geq 1$, and hence (Δ_N) is conforming on R_f. Hence in both cases we have conformity, and Step 3(iii) holds.

The verification of Steps 1, 2, and 3 shows that (AC1), (AC2), and (AC4) hold. We now give a direct proof that (AC3) holds. We have already shown that r^N satisfies (8.1). It follows from Proposition 6.5.1(iii) that (9.39) is valid for this example. The counterpart of (9.41) holds in (Δ_N), and hence it is sufficient to verify (AC3) for a buffer content of 1.

Using reasoning similar to that in (8.6) yields, for sufficiently large N, that $h_\alpha^N(1) \geq -c_{01}^N(d|N)$. Then using (9.39) and the conformity of the AS at d, we obtain $\liminf_{N\to\infty} r^N(1) \geq -c_{01}(d)$. This verifies (AC3) for decision epochs.

Now consider service-in-progress states. We may mimic the argument in (9.42) in (Δ_N) to obtain, for $1 \leq s \leq M(N) - 1$,

$$V_\alpha^N(1, a, s) \geq \sum_{k=1}^{M(N)-s-1} \alpha^k P(Y_{a,s} = k) \left\{ \sum_{j=0}^{N-1} P_j(k) V_\alpha^N(j) + \sum_{j=N}^{\infty} P_j(k) V_\alpha^N(N) \right\}$$

$$+ \alpha^{M(N)-s} \sum_{k=M(N)-s}^{\infty} P(Y_{a,s} = k) \left\{ \sum_{j=0}^{N-1} P_j(M(N) - s) V_\alpha^N(j) \right.$$

$$+ \sum_{j=N}^{\infty} P_j(M(N) - s)V_\alpha^N(N)\Bigg\}$$

$$\geq \sum_{k=1}^{M(N)-s-1} \alpha^k P(Y_{a,s} = k)[(p_0)^k V_\alpha^N(0) + (1 - (p_0)^k)V_\alpha^N(1)]$$

$$+ \alpha^{M(N)-s} \sum_{k=M(N)-s}^{\infty} P(Y_{a,s} = k)[(p_0)^{M(N)-s}V_\alpha^N(0)$$

$$+ (1 - (p_0)^{M(N)-s})V_\alpha^N(1)]. \tag{9.44}$$

We now subtract $V_\alpha^N(0)$ from both sides. Using reasoning similar to that in (9.43) yields

$$h_\alpha^N(1, a, s) \geq -c_{01}^N(d|N)$$

$$- [(1 - \alpha)V_\alpha^N(0)]\left\{\sum_{k=1}^{M(N)-s-1} kP(Y_{a,s} = k)\right.$$

$$\left. + (M(N) - s) \sum_{k=M(N)-s}^{\infty} P(Y_{a,s} = k)\right\}$$

$$\geq -c_{01}^N(d|N) - [(1 - \alpha)V_\alpha^N(0)]\tau_{a,s}$$

$$\geq -c_{01}^N(d|N) - [(1 - \alpha)V_\alpha^N(0)]U. \tag{9.45}$$

This yields $r^N(1, a, s) = \lim_{\alpha \to 1^-} h_\alpha^N(1, a, s) \geq -c_{01}^N(d|N) - J^N U$. Recall that we have verified (AC4), and hence $\limsup_{N \to \infty} J^N \leq J$. Then taking the limit infimum of both sides yields $\liminf_{N \to \infty} r^N(1, a, s) \geq -c_{01}(d) - JU$. This verifies (AC3) with $Q = c_{01}(d) + JU$. □

9.7 COMPUTATION UNDER DETERMINISTIC SERVICE TIMES

Let us consider a single-server queue in which the actions are choices of deterministic service times. Assume that action k corresponds to a service time of exactly k units. If $k = 1$, then the SS and MS models of this service time coincide. If $k \geq 2$, then the SS model yields a single sample of k units. An examination of (9.17) and (9.19) shows that the SS and MS models with deterministic service times are essentially equivalent, except that the SS model looks at remaining service time while the MS model looks at elapsed service time.

Let us adopt the point of view of Example 9.3.2 and look at elapsed service time. We compute an optimal policy under the assumption that the customer arrival process is Bernoulli (p), with $0 < p < 1$, and actions $k = 1, 2, 3$. The holding cost is given by $H(i) = Hi$, where H is a positive constant. This is ProgramFive.

The basic assumptions of Section 9.6 are valid and hence the conclusions of Proposition 9.6.2 hold. If it is the case that $C(1) > C(2) > C(3)$, then we would expect that an optimal policy f would satisfy $f(i)$ is decreasing in i, and this is born out by the program, with one exception. Hence we may give the optimal policy as two intervals, where the first interval indicates buffer content for which it is optimal to serve in three slots, and the second interval buffer content for which it is optimal to serve in two slots. In the remaining states it is optimal to serve in one slot. For example [1, 5] $\varnothing$ means that it is optimal to serve in three slots when the buffer content is no greater than 5 and optimal to serve in one slot for content greater than 5.

Given the AS as defined in Section 9.6 (note that there is no need to truncate the service time), let us develop the expressions for the VIA 6.6.4. Let i^* equal $i + 1$ if $1 \le i < N$ and equal N if $i = N$. Then

$$\begin{aligned}
w_n(0) &= (1-p)u_n(0) + pu_n(1),\\
w_n(i) &= Hi + \min\{C(1) + (1-p)u_n(i-1) + pu_n(i),\\
&\qquad C(2) + (1-p)u_n(i,2,1) + pu_n(i^*,2,1),\\
&\qquad C(3) + (1-p)u_n(i,3,1) + pu_n(i^*,3,1)\}\\
w_n(i,2,1) = w_n(i,3,2) &= Hi + (1-p)u_n(i-1) + pu_n(i)\\
w_n(i,3,1) &= Hi + (1-p)u_n(i,3,2) + pu_n(i^*,3,2), \qquad 1 \le i \le N,\\
u_{n+1}(.) &= w_n(.) - w_n(0).
\end{aligned} \tag{9.46}$$

Remark 9.7.1. It is intuitively clear, and may be proved by induction on (9.46), that if H and $C(.)$ are multiplied by a positive constant, then the optimal average cost is multiplied by that constant, and the optimal policy is unchanged. For this reason we assume that $H = 1$ in all our scenarios. In all scenarios we used the weaker convergence criterion (Version 1) of the VIA. □

Whether the queue is stable or unstable under a given action turns out to be crucial, as we would expect from the examples in Chapter 8. The policy that always chooses action k induces a stable MC if $pk < 1$. We have (a) stable under $\{1, 2, 3\}$ if $p < \frac{1}{3}$, (b) stable only under $\{1, 2\}$ if $\frac{1}{3} \le p < \frac{1}{2}$, and (c) stable only under $\{1\}$ if $p \ge \frac{1}{2}$.

The policy d that always serves in one slot is our benchmark. Then $R_d = \{0, 1\}$, and we may easily see that $\pi_0 = 1 - p$, $\pi_1 = p$, and $J_d = p(C(1) + H)$. Under condition (b) it is the case that the queue is also stable under the policy that always serves in two slots, and this policy might yield a lower average cost than the benchmark. A similar comment holds under (a). In these cases we could give a separate program to calculate the average cost under these policies for a potentially better benchmark. We have not chosen to do this here.

As a check on the program, we let $p = 0.3$, $H = 1$, and $C(.) \equiv 2$. It is clear that d should be optimal with $J = 0.3\ (2 + 1) = 0.9$, and this is born out by the program.

Remark 9.7.2. Let us make the intuitively plausible assumption that an optimal policy f eventually chooses $k = 1$. In this case R_f is finite, and a natural limit is imposed on the buffer content. For example, assume that $f = [1, 2]\ [3, 7]$. If there are 8 or more in the queue, then customers are served in one slot, and since no more than one customer can arrive in a given slot, the queue cannot build up. The states $0 \le i \le 8$, together with the appropriate service in progress states, form R_f. □

Scenarios 9.7.3. Some scenarios are given in Table 9.1. For each of them, except for Scenarios 7 and 8, we let $\epsilon = 0.0000005$ and $N = 80$ and confirmed with $N = 100$.

Scenarios 1 and 2 fall under the stability case (a). Note that in Scenario 2 we have $C(2) = 4\ C(3)$, and it is never optimal to serve in 3 slots. Scenarios 3 and 4 fall under the stability case (b). The optimal policy uses $k = 2$ quite selectively and $k = 3$ only when there is one customer in the queue.

The remaining scenarios fall under the stability case (c) so that the queue is unstable under $k = 2, 3$. In Scenario 5 the controller switches to $k = 1$ for buffer content of 3 or more, even though $C(1) = 10\ C(2)$. A similar comment holds for Scenario 8. In this case the program output is unclear for $N = 100$. Increasing to $N = 200$ yields the optimal policy unambiguously.

The program is not well-behaved in Scenario 7, in the sense that $f(i)$ is not decreasing in i, as we conjectured. Note that $C(1) > 6C(2)$ and that p is fairly

Table 9.1 Results for Scenarios 9.7.3

Scenario	1	2	3	4	5	6	7	8
p	0.1	0.2	0.4	0.4	0.6	0.8	0.8	0.9
Costs	15.0	10.0	15.0	20.0	50.0	25.0	40.0	30.0
	5.0	1.0	5.0	10.0	5.0	15.0	6.0	10.0
	0.5	0.25	0.5	0.1	0.1	0.5	0.2	0.1
J_d	1.6	2.2	6.4	8.4	30.6	20.8	32.8	27.9
J	0.393	0.667	3.483	4.576	15.528	19.755	25.667	25.213
Savings	1.207	1.533	2.917	3.824	15.072	1.045	7.133	2.687
Optimal policy	[1, 3] [4, 7]	∅ [1, 5]	[1] [2, 3]	[1] [2]	∅ [1, 2]	[1] ∅	∅ [1, 38]?	∅ [1]

large. On the basis of $N = 340$ we have conjectured the policy given in Table 9.1. Several comments are in order. If there is a nonunique optimal policy, then this might give rise to ambiguous output under any convergence condition. If there is a unique optimal policy, then the ambiguity might be resolved in one of several ways. First, a substantial increase in N might resolve it. Second, a smaller N might be sufficient under Version 2 of the VIA. Third, if neither of these remedies works, then we might conjecture that a policy e choosing $k = 2$ in states $[1, L]$ is optimal or close to optimal. We could then develop a program to calculate J_e for various values of L to get close to $J \approx 25.667$. This mystery is left for the interested reader to resolve! □

9.8 COMPUTATION UNDER GEOMETRIC SERVICE TIMES

Let us consider Example 9.3.2 with geometric service times. This is a single-server queue with Bernoulli (p) customer arrivals and with the actions being geometric rates $\{a_1, a_2, \ldots, a_K\}$, where $0 < a_1 < a_2 < \ldots < a_K < 1$. When a service rate choice is made for a customer about to enter service, then a single cost is incurred, and the chosen rate must be used until the service of that customer is completed. This may be contrasted to the example in Section 8.5 in which a new rate may be chosen (and cost incurred) in each slot of an ongoing service. In the present case we envisage a situation in which the *quality of service* (i.e., the service rate) given to a particular customer is to remain constant throughout the service of that customer. This is ProgramSix.

Let the AS be defined as in Section 9.6, where no service can last more than L slots. Let $i^* = i + 1$ if $1 \le i < N$ and equal N if $i = N$. Then the expressions for the VIA 6.6.4 are

$$
\begin{aligned}
w_n(0) &= (1-p)u_n(0) + pu_n(1),\\
w_n(i) &= Hi + \min_a\{C(a) + (1-p)au_n(i-1) + pau_n(i)\\
&\quad + (1-p)(1-a)u_n(i,a,1) + p(1-a)u_n(i^*,a,1)\},\\
w_n(i,a,s) &= Hi + (1-p)au_n(i-1) + pau_n(i)\\
&\quad + (1-p)(1-a)u_n(i,a,s+1)\\
&\quad + p(1-a)u_n(i^*,a,s+1), \qquad 1 \le s < L,\\
w_n(i,a,L) &= Hi + (1-p)u_n(i-1) + pu_n(i), \qquad 1 \le i \le N,\\
u_{n+1}(.) &= w_n(.) - w_n(0).
\end{aligned}
\tag{9.47}
$$

If H and $C(a)$ are multiplied by a positive constant, then the optimal average cost is multiplied by that constant and the optimal policy is unchanged. For this reason we assume that $H = 1$ in all our scenarios.

Let $d(a)$ be the policy that always serves at rate a for $a > p$. The benchmark policy d serves at the constant rate that realizes $J_d = \min_{a>p}\{J_{d(a)}\}$. Note that the benchmark is defined as in Section 8.5. The expression in (8.12) is not valid because the cost for service is charged only once, at the beginning of the service, rather than during each slot of the service. The next result shows how to modify (8.12) in this situation.

Proposition 9.8.1. Assume that a satisfies $p < a$. Let $d(a)$ be the policy that always serves at rate a, where the cost of service $C(a)$ is charged once at the beginning of the service. Then

$$J_{d(a)} = \frac{Hp(1-p)}{a-p} + pC(a). \tag{9.48}$$

Proof: Note that this MC and the one in Proposition 8.5.1 operate exactly the same and hence the steady state probabilities are identical. The first term of (8.12) is the expected holding cost and this remains the same. The second term is the expected service cost, which equals P(service is taking place) $C(a)$ = $(1-\pi_0)\, C(a) = (p/a)C(a)$. In the present case, this is modified to E[service cost] = P(service is taking place) $(1/E$[length of a service]) $C(a) = (p/a)aC(a)$ = $pC(a)$, and hence (9.48) holds. □

Scenarios 9.8.2. Here $K = 3$. The results are summarized in Table 9.2 and may be interpreted in a manner similar to Table 8.2. For these scenarios we chose $\epsilon = 0.00000005$, $N = 68$, and $L = 8$, which was the maximum allowable under the stack size restriction. The values of J and the optimal policies are, at a minimum, quite close to optimal. Optimality should be confirmed with larger values of N and L. Scenario 1 is a checking scenario and the results were as expected.

Table 9.2 Results for Scenarios 9.8.2

Scenario	1	2	3	4	5	6	7
p	0.6	0.3	0.5	0.6	0.7	0.8	0.9
Service	0.6	0.4	0.3	0.55	0.75	0.7	0.92
rates	0.7	0.6	0.5	0.8	0.8	0.85	0.95
	0.8	0.8	0.9	0.9	0.85	0.95	0.99
Costs	2.0	1.0	0.0	0.25	0.1	0.0	1.0
	2.0	5.0	0.1	10.0	2.0	10.0	5.0
	2.0	25.0	50.0	15.0	10.0	25.0	10.0
J_d	$a = 0.8$	$a = 0.6$	$a = 0.9$	$a = 0.8$	$a = 0.8$	$a = 0.85$	$a = 0.92$
	2.4	2.2	25.625	7.2	3.5	11.2	5.4
J	2.400	1.83	7.51	5.54	3.16	10.48	4.38
Savings	0.0	0.37	18.12	1.66	0.34	0.72	1.02
Optimal	∅∅	[1]	∅ [1, 6]	[1, 2]	[1, 2]	[1]	[1, 4]
policy		[2, 23]		[3, 7]	[3, 13]	[2, 35]	[5]

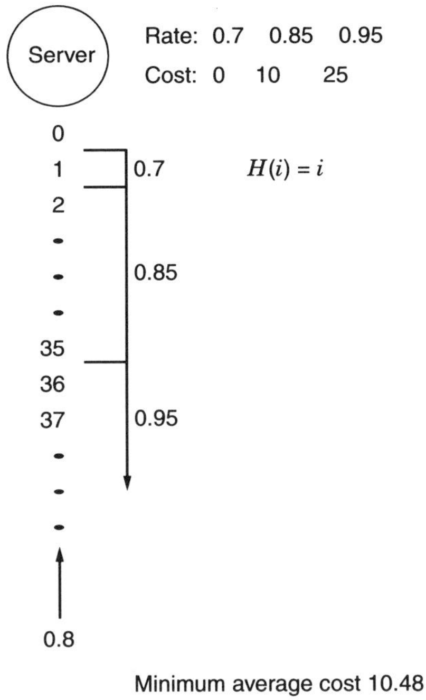

Figure 9.2 Scenario 6 from Table 9.2.

It is instructive to compare the model in this section with that in Section 8.5, when the respective parameters are equal (arrival probability, service rates, and service rate costs). Let us call this one Model New (the service rate can only be changed at the beginning of a service, and the service charge is incurred once during the service) and the other one Model Old (the service rate can be chosen during each slot, and the service charge is incurred during each slot of service).

Let us compare Scenario 3 in Table 9.2 and Scenario 5 in Table 8.2. We see that the optimal policy in Model New is slightly more conservative than that in Model Old. The former policy switches to the maximum service rate at the buffer content of 7, whereas the latter switches at a buffer content of 8. This behavior occurs because the queue can only be stabilized under the maximum rate and because the controller in Model New is "locked in" and cannot adjust the service rate until a service is finished. Hence it will tend to act more conservatively. Note that the minimum average cost of 7.51 is somewhat less than that of 8.003. This is because the cost of service is incurred only once rather than throughout the service. However, the difference is perhaps less than we would expect. This reflects the fact that most services finish in one or two slots.

For another comparison, consider Scenario 6 in Table 9.2 (see Fig. 9.2) and Scenario 7 in Table 8.2. In contrast to the previous case, the optimal policy in Model New is vastly different from that in Model Old. The reader might consider why this makes sense before reading further. Here is an explanation. Note that the queue is stable under both higher rates but the highest rate costs 2.5 times the middle rate. In Model New there is a trade-off between keeping the queue stabilized and not serving so fast that a service finishes quickly so that another begins and a new charge is incurred. In other words, the controller in Model New has a much greater incentive to choose the middle rate than the controller in Model Old. A service under rate 0.85 has a 15% chance of not finishing in one slot. During the second and subsequent slots of this service, no service charge in incurred. A service under rate 0.95 has only a 5% chance of not finishing in one slot, and hence it is less advantageous, at least for smaller buffer contents. Eventually the holding cost consideration prevails and forces a switch to the highest rate.

Note that the optimal policies for Scenario 7 in Table 9.2 and Scenario 8 in Table 8.2 differ only for a buffer content of 5. The slight difference between the optimal policies is more difficult to explain but is due to the same factors. □

BIBLIOGRAPHIC NOTES

Some of the material in this chapter is based on an unpublished paper *Service Control of Discrete-Time Single-Server Queues*, and the author would like to express her gratitude to the anonymous referees of this paper. Their helpful suggestions led to substantial improvements in portions of this chapter.

The author would also like to thank Dr. Ken Berk, a fellow of the American Statistical Society, for pointing out the difference between an SS model and an MS model and for emphasizing that care must be exercised to choose the formulation most appropriate in a given situation.

Some of the material in Section 9.2 is found in Barlow and Proschan (1965), Wolff (1989), and Ross (1996).

Although the vast majority of the literature on queueing systems deals with queues in continuous time, research on (uncontrolled) discrete time queueing systems has been steadily increasing in recent years. We mention only two references. Bruneel and Kim (1993) contains a comprehensive treatment of the GI/G/1 queue. This is as in Example 9.3.1 with a single-service distribution. Bruneel and Wuyts (1994) contains an analysis of the discrete time multiserver queueing system with constant service times.

Bournas, Beutler, and Teneketzis (1992) treat a discrete time flow control model. In this model there are several transmitters (queues with infinite buffers) competing for a single channel (server). The service is organized in phases of fixed length T. At the beginning of a phase the actions consist of the various allocations of slots within the phase for use by the various queues to service packets residing in their buffers. The objective is to show that there exists a stationary policy (allocation) minimizing the expected average packet waiting time. A more difficult problem is to incorporate a choice of phase length T into the decision process, together with a phase setup cost.

PROBLEMS

9.1. Prove Proposition 9.2.3.

9.2. Give a distribution with a finite mean that is not BMRL. *Hint:* Look at Proposition 9.2.5.

9.3. Verify that every distribution on $\{1, 2, \ldots, K\}$, where K is a finite positive integer, is BMRL.

9.4. What happens to (9.16) if the customer currently being served in an ongoing service does not incur a holding cost?

9.5. Model Example 9.3.3 as an MDC.

9.6. Model Example 9.3.5 as an MDC under the assumption that the phase of the chain is known at the beginning of each slot.

9.7. In this model assume that the batch arrival process is as in Example 9.3.1 and that service occurs in an MS fashion under a fixed distribution as in Example 9.3.4. If the buffer is nonempty in the slot following a service (or other activity) completion, then the server may choose action a = serve the next customer, or b = leave the queue to perform other tasks. Perhaps unfortunately, choosing b is referred to as *taking a vacation* and this type of model is a *vacation model*. However, taking a vacation does not connote idleness, since the server is free to perform other tasks elsewhere! Let us assume that the server must take a vacation when the buffer is empty.

The length of a vacation is determined in an MS fashion by a distribution G. Assume that there is a cost $H(i)$ for holding i customers in the buffer and a fixed reward of R at the beginning of each vacation. Develop this model as an MDC.

9.8. Model Example 9.4.2 as an MDC.

9.9. Give a rigorous proof of the positive recurrence of the class R in Lemma 9.5.3. This may be done by employing Corollary C.1.6 with test function $y(i) = Yi$, and $y(i, s) = s + Y(\lambda s + i - 1)$. Show that for an appropriate choice of the positive constant Y, we can make the drift in (C.10) identically equal to -1 for nonzero states. It is possible to prove that R has finite average cost by employing Corollary C.2.4 with an appropriate quadratic test function, but the argument is much more complicated.

9.10. Give a rigorous proof of the positive recurrence of the class R in Lemma 9.6.1. This may be done by employing Corollary C.1.6 with test function $y(i) = Yi$, and $y(i, a^*, s) = \tau_{a^*,s} + Y(\lambda\tau_{a^*,s} + i - 1)$. Show that for an appropriate choice of the positive constant Y, we can make the drift in (C.10) identically equal to -1 for nonzero states. It is possible to prove that R has finite average cost by employing Corollary C.2.4 with an appropriate quadratic test function, but the argument is much more complicated.

9.11. Run ProgramFive for the following scenarios. *Note:* The three parameters labeled "C" are the values of $C(1)$, $C(2)$, and $C(3)$, respectively.

(a) $H = 0.4$, $p = 0.25$, $C = 3, 1, 0.5$.

For the remainder of the scenarios, $H = 1$.

(b) $p = 0.25$, $C = 7.5, 2.5, 1.25$.
(c) $p = 0.1$, $C = 6, 5, 0.5$.
(d) $p = 0.2$, $C = 25, 10, 1$.
(e) $p = 0.3$, $C = 20, 8, 0.1$
(f) $p = 0.4$, $C = 15, 5, 0.1$.
(g) $p = 0.8$, $C = 20, 3, 0.1$.
(h) $p = 0.9$, $C = 40, 15, 0.1$.
(i) $p = 0.9$, $C = 30, 20, 0.1$.

For each scenario determine J and an optimal policy, and discuss your conclusions.

9.12. Run ProgramSix for the following scenarios. *Note:* The three parameters labeled "a" are the three service rates, and those labeled "C" are their respective costs.

(a) $H = 0.5$, $p = 0.4$, $a = 0.3, 0.6, 0.8$, $C = 0.5, 2, 6$.

For the remainder of the scenarios, $H = 1$.

(b) $p = 0.4$, $a = 0.3, 0.6, 0.8$, $C = 1, 4, 12$.
(c) $p = 0.2$, $a = 0.15, 0.3, 0.7$, $C = 0.1, 5, 15$.
(d) $p = 0.5$, $a = 0.6, 0.7, 0.8$, $C = 1, 20, 40$.
(e) $p = 0.6$, $a = 0.6, 0.8, 0.9$, $C = 0.1, 5, 15$.
(f) $p = 0.65$, $a = 0.6, 0.8, 0.9$, $C = 0.1, 5, 15$.

(g) $p = 0.8$, $a = 0.82, 0.9, 0.95$, $C = 0.01, 5, 10$.

(h) $p = 0.9$, $a = 0.92, 0.95, 0.98$, $C = 0.1, 1, 4$.

For each scenario determine J, J_d, and an optimal policy, and discuss your conclusions. Run some of the scenarios using ProgramFour and compare the optimal policies.

9.13. Consider Example 9.4.1. The basic assumptions (with obvious notation) are as follows:

(BA1). There exists a (finite) constant M such that $F_a(M) = 1$ for all a.

(BA2). For all a, either $u_1(a) > 0$ or there exist relatively prime integers y, z such that $u_y(a) > 0$ and $u_z(a) > 0$.

(BA3). The holding cost is Hi, where H is a positive constant. Moreover we have $r(a,s) \equiv r(a)$ for all a.

(BA4). There exists a^* such that $p_0(a^*) = 1$. Moreover $R(a^*) = r(a^*) = 0$.

(BA5). For all a, $p_0(a) > 0$ and $\lambda_a^{(2)} < \infty$.

(BA6). There exists $a\hat{}$ such that $p_0(a\hat{}) + p_1(a\hat{}) < 1$.

Note that (BA1) says that no phase can last more than M slots. Under (BA2) phase lengths are "aperiodic." Under (BA4) there exists a phase with no packet arrivals; during this phase no rewards are earned. Under (BA6) we avoid the trivial situation in which no more than a single packet can arrive in any slot under any phase. In this situation a queue cannot build up.

(a) Let d be the stationary policy that always chooses a^*. Show that d is 0 standard. *Hint:* Note that R_d is finite.

Define the AS so that the buffer cannot contain more than N packets. If a batch arrives that would cause a buffer overflow, then the probability of that event is given to the corresponding full buffer state. For example, if the system is in state (i, a, s) for $1 \le i \le N$, then the probability of $j > N-i+1$ packets arriving is given to state $(N, a, s-1)$.

Show that the (VIA) is valid for (Δ_N) and the (AC) assumptions hold for the function $r^N(.) = \lim_{n\to\infty} (v_n^N(.) - v_n^N(0))$. Follow the general procedure in Proposition 8.2.1.

(b) Note that (BA2) and (BA5) are needed to verify Step 1. If (BA2) is eliminated, then the aperiodicity transformation may be effected.

It should be possible to verify the constancy of the minimum average cost in (Δ_N) under weaker conditions than (BA5), but we do not explore this here.

(c) Argue that Step 2 holds for the policy d. *Hint:* It is only possible to exit S_N from a transient state (i, a, s), $a \neq a^*$.

(d) Argue informally that Step 3(ii) holds.

(e) Argue informally that $V_\alpha^N(i) \geq V_\alpha^N(0)$ and $V_\alpha^N(i, a, s) \geq V_\alpha^N(0, a, s)$. Use this to show that (AC3) holds.

CHAPTER 10

Average Cost Optimization of Continuous Time Processes

In this final chapter we show how to compute average cost optimal policies in a certain class of processes operating in continuous time.

In Section 10.1 we review the exponential distribution and the Poisson process. All the necessary background is contained in this section. The continuous time processes we deal with have countable state spaces, as before. If the process is in a given state and a certain action is taken, then the time until the next transition is exponentially distributed with a parameter dependent on the state and action. The theory may be extended to allow more general transition times, but for brevity and simplicity we restrict ourselves to the exponential case.

Section 10.2 formalizes the definition of a continuous time Markov decision chain (CTMDC). As an example, this section develops a CTMDC modeling the service rate control of an M/M/1 queue. This is the most famous queueing system. It consists of a single server, serving at exponential rate, with arrivals occurring according to a Poisson process. Here it is allowed to control the service rate. A new rate may be chosen when a service is to begin or when a new customer enters the system.

Section 10.3 discusses average cost optimization of the CTMDC. Under an assumption requiring the mean transition times to be bounded above and away from zero (which holds in practical models), it is possible to replace the CTMDC by a (discrete time) auxillary MDC. We may then bring the previously developed approximating sequence method into play to compute an average cost optimal stationary policy for the MDC. Under a reasonable assumption this policy is also optimal for the CTMDC. Hence, modulo the verification of the assumption, we have rigorously computed an average cost optimal stationary policy for the original continuous time process.

Section 10.4 gives computational results for the service rate control of an M/M/1 queue. In Section 10.5 we consider a system with arrivals according to a Poisson process and a pool of K identical exponential servers. The actions consist in choosing how many of these servers to turn on. This system is called an M/M/K queue with dynamic service pool. Computational results for this model are given.

In Section 10.6 we consider a polling model, as in Fig. 1.7. Customers arrive to each station according to a Poisson process. Both the service time of a customer and the walking times are exponentially distributed. If the server is currently at a station, then a decision to initiate a walk may be made under any of these conditions: The server has just arrived at the station, a service has just been completed, or a new customer has arrived somewhere in the system. The computation of an average cost optimal policy is illustrated.

10.1 EXPONENTIAL DISTRIBUTIONS AND THE POISSON PROCESS

In this section we discuss the structure of service times and customer arrivals in the continuous time models that are to be optimized. Let X be a random quantity. For specificity we will think of X as a service time, but other interpretations are also important. We say that X has an exponential (μ) distribution if its cumulative distribution function is

$$F_X(t) = P(X \leq t) = 1 - e^{-\mu t}, \qquad t \geq 0. \tag{10.1}$$

Here μ is a positive parameter known as the *rate of service*. The complement of the cumulative is $P(X > t) = e^{-\mu t}$, $t \geq 0$. The density is $f_X(t) = \mu e^{-\mu t}$, $t \geq 0$.

It may be shown that $E[X] = 1/\mu$. For example, if $\mu = 2$ customers per minute, then each service lasts on average 0.5 minute, and the server is serving at the rate of 2 customers per minute, on average.

Definition 10.1.1. Let $r(\delta)$ be a function of the positive number δ. Then r is $o(\delta)$ (read "little oh of delta") if $\lim_{\delta \to 0+} r(\delta)/\delta = 0$. This means that r is small relative to δ when δ is small. For example, $r(\delta) = \delta^2$ is $o(\delta)$ as is $r(\delta) = \delta^3$. However, $r(\delta) = \delta$ is not $o(\delta)$ nor is $r(\delta) = \delta^{1/2}$. □

Here are some important properties of the exponential distribution.

Proposition 10.1.2. Let X have an exp (μ) distribution. Then

$$P(X > x + y | X > y) = P(X > x), \qquad x, y > 0, \tag{10.2}$$

$$P(X \leq \delta) = \mu\delta + o(\delta). \tag{10.3}$$

As an aside, note that (10.2) is the famous *memoryless* property of the exponential distribution. It says that if a service was not completed in y time units, then the probability that it will be uncompleted after an *additional* x units is the same as the unconditional probability that the service lasts for at least x units.

In other words, a customer undergoing service according to an exponential distribution receives no credit for the amount of service rendered, if the service is still uncompleted. This is undoubtedly a limiting assumption. However, it holds (or approximately holds) for some important situations, such as the length of a telephone conversation. The assumption of exponential service times simplifies the mathematics considerably, since the model does not have to take into account the amount of service rendered.

Equation (10.3) says that for a small interval of time, the probabiliy that the service is completed in that amount of time is approximately proportional to the length of the interval, with proportionality constant μ.

Proof: To prove (10.2) note that

$$\begin{aligned} P(X > x + y | X > y) &= \frac{P(X > x + y, X > y)}{P(X > y)} \\ &= \frac{P(X > x + y)}{P(X > y)} \\ &= \frac{e^{-\mu(x+y)}}{e^{-\mu y}} \\ &= e^{-\mu x} \\ &= P(X > x). \end{aligned} \tag{10.4}$$

The first line follows from the definition of conditional probability. The second line is clear. The other lines follow from (10.1).

It follows from (10.1) that $P(X \leq \delta) = \mu\delta + [1 - \mu\delta - e^{-\mu\delta}]$. Hence to prove (10.3), it is sufficient to show that the expression in brackets is $o(\delta)$. Employing L'Hopital's rule yields

$$\lim_{\delta \to 0} \frac{1 - \mu\delta - e^{-\mu\delta}}{\delta} = \lim_{\delta \to 0} \frac{-\mu + \mu e^{-\mu\delta}}{1} = 0. \tag{10.5}$$

This proves (10.3). □

Consider a situation with server 1 and server 2, serving independently. The service time of a customer serviced by i follows an $\exp(\mu_i)$ distribution for $i = 1, 2$. We wish to obtain the probability that server 1 finishes first, as well as the probability of both services finishing in a given interval, and the distribution of the time until the first service completion. (Recall that there is a zero probability of two independent exponential distributions taking on the same value.)

Proposition 10.1.3. Let X_1 and X_2 be independent exponentially distributed random variables, with parameters μ_1 and μ_2, respectively. Then

$$P(X_1 \le \delta, X_2 \le \delta) = o(\delta), \tag{10.6}$$

$$P(X_1 < X_2) = \frac{\mu_1}{\mu_1 + \mu_2}. \tag{10.7}$$

Let $Y = \min(X_1, X_2)$. Then Y has an $\exp(\mu_1 + \mu_2)$ distribution.

Equation (10.6) says that if two independent exponential services are underway, then the probability that both will finish within a small interval of time is negligible. The last statement says that the time until the first service completion is also exponentially distributed, with a rate equal to the sum of the rates.

Proof: Note that

$$\begin{aligned} P(X_1 \le \delta, X_2 \le \delta) &= P(X_1 \le \delta)P(X_2 \le \delta) \\ &= (1 - e^{-\mu_1 \delta})(1 - e^{-\mu_2 \delta}). \end{aligned} \tag{10.8}$$

The first line follows from the independence of X_1 and X_2 and the second line follows from (10.1). It may be shown that this expression is $o(\delta)$ by using L'Hopital's rule as in (10.5). This proves (10.6).

To prove (10.7), we condition on the value of X_1 and use the law of total probability for continuous random variables to obtain

$$\begin{aligned} P(X_1 < X_2) &= \int_0^\infty P(X_1 < X_2 | X_1 = x) f_{X_1}(x)\, dx \\ &= \int_0^\infty P(X_2 > x) f_{X_1}(x)\, dx \\ &= \int_0^\infty e^{-\mu_2 x}(\mu_1 e^{-\mu_1 x}\, dx) \\ &= \left. \frac{-\mu_1 e^{-(\mu_1 + \mu_2)x}}{\mu_1 + \mu_2} \right|_0^\infty . \end{aligned} \tag{10.9}$$

Evaluating the last line of (10.9) yields (10.7).

To prove the last claim, note that

$$\begin{aligned} P(Y > x) &= P(X_1 > x, X_2 > x) \\ &= P(X_1 > x)P(X_2 > x) \\ &= e^{-\mu_1 x} e^{-\mu_2 x} \\ &= e^{-(\mu_1 + \mu_2)x}. \end{aligned} \tag{10.10}$$

The second line follows from independence and (10.1) yields the result. □

In this chapter we treat certain continuous time models in which the service time of a customer follows an exponential distribution. Let us now address the customer arrival process.

The *Poisson process* is the most important stochastic process for modeling customer arrivals, and it has several equivalent definitions. For our purposes the following description will suffice. A Poisson process with rate λ, denoted PoisP (λ), is a process that counts the number of customers arriving in any interval of time $[0, t]$. The system is assumed empty at time 0. The time that elapses until the arrival of the first customer as well as the successive times between customer arrivals (*interarrival* times) are all independent exp(λ) random variables.

To generate a PoisP (λ), we sample from an exp(λ) distribution to obtain the time of arrival of the first customer. We then sample independently from the same distribution to obtain the elapsed time between the arrival of the first and second customers. This process is continued to obtain the time of arrival of the third and subsequent customers.

The Poisson process is the correct model for completely random customer arrivals. This follows from the memoryless property of the exponential distribution that yields the interarrival times. Assume that it has been at least y units of time since the previous customer arrived and that there has been no new arrival. In the purely random situation this information should not make it either more or less probable that an arrival would occur in a certain time interval from that point on. The exponential distribution has this valuable property.

The parameter λ is the *rate* of the Poisson process. If $\lambda = 3$ customers per minute, then on average three customers will arrive in any one-minute period. Note that if these customers are being served by a single exp($\mu = 2$) server, then we have trouble on our hands! On average, 3 customers are arriving to the system every minute, but only 2 are being served, and this system is unstable.

10.2 CONTINUOUS TIME MARKOV DECISION CHAINS

In this section we discuss a mathematical structure, called a *continuous time Markov decision chain* (CTMDC), that is useful for modeling the control of certain systems occurring in continuous time. As the explanation proceeds, we carry along an illustrative example.

The CTMDC, denoted Ψ, has a state space S that is a countable set. Associated with each $i \in S$ is a nonempty finite set A_i of actions available in i. Assume that action $a \in A_i$ is chosen. Then a cost is incurred. This may consist of both an instantaneous cost $G(i, a)$ incurred immediately and a cost rate $g(i, a)$ in effect until the next transition. The time until the next transition is exponentially distributed with parameter $\upsilon(i, a)$. The new state is chosen according to a probability distribution $(P_{ij}(a))_{j \in S}$. The theory can be developed allowing $P_{ii}(a) > 0$. However, for most systems it is the case that $P_{ii}(a) \equiv 0$, and we assume this. That is, when a transition occurs, it is a "real" transition that can be observed (an actual change of state) rather than a "dummy" transition from a state to itself.

Example 10.2.1. Service Rate Control of the M/M/1 Queue. This is a continuous time analogue of Example 2.1.2. Customers arrive to a single server according to a Poisson process with rate λ. If the queue is nonempty, then the action set is $A = \{a_1, a_2, \ldots, a_K\}$, where $0 < a_1 < \ldots < a_K$. If action a is chosen, then the service time is exponentially distributed with parameter a.

There is a nonnegative holding cost rate $H(i)$ for holding i customers in the queue and a nonnegative cost rate $c(a)$ in effect when serving at rate a. In this model there are no instantaneous costs.

We set $S = \{i | i \geq 0\}$. State 0 means that a service has just been completed, leaving the queue empty. In state 0 there are no actions (null action). State $i \geq 1$ means that either a service has just been completed, and the served customer has departed (leaving a nonempty queue with i customers), or a new customer has just arrived (boosting the number in the queue to i). The action set is A. Note that if a new customer arrives and a service is ongoing, we allow a new service rate to be chosen. Because of the memoryless property of the exponential distribution, we do not have to take elapsed service into account and may assume that a fresh service starts at that point.

The cost rates are given by

$$
\begin{aligned}
g(0) &= 0, \\
g(i, a) &= c(a) + H(i), \qquad i \geq 1.
\end{aligned} \tag{10.11}
$$

When the system is in state 0, then the waiting time until a customer arrives (and the system enters state 1) is exponentially distributed with parameter λ. Hence $\upsilon(0) = \lambda$ and $P_{01} = 1$.

Now assume that the system is in state $i \geq 1$ and action a is chosen. Then two "exponential clocks" are started. One measures the time until the next arrival, which occurs with rate λ. The other measures the time until the service is finished, which occurs with rate a. Hence the time until a transition (to either $i+1$ or $i-1$) is governed by the minimum of these two clocks. By Proposition 10.1.3 the transition time is exponentially distributed with parameter $\lambda + a$. Thus $\upsilon(i, a) = \lambda + a$. The probabilities are found using (10.7). (Recall that there is a probability of zero that these two exponential clocks register exactly the same time. Hence we cannot have both an arrival and a service completion at the same time.)

To summarize the transition rates and probabilities, we have

$$
\begin{aligned}
v(0) &= \lambda, \quad P_{01} = 1, \\
v(i,a) &= \lambda + a, \\
P_{ii+1}(a) &= \frac{\lambda}{\lambda + a}, \\
P_{ii-1}(a) &= \frac{a}{\lambda + a}, \qquad i \geq 1.
\end{aligned}
\tag{10.12}
$$

□

This completes the specification of the CTMDC Ψ and the development of an example to illustrate this concept.

10.3 AVERAGE COST OPTIMIZATION OF A CTMDC

In this section we give a set of assumptions that enable us to compute an average cost optimal stationary policy for Ψ. The first point that needs to be addressed is the following: What constitutes a policy for Ψ, and what does it mean for a policy to be average cost optimal?

Informally we define a policy θ to be a nonanticipatory rule for choosing actions. It may depend on the history of the process through the present state and may randomize among actions. The history includes the past states of the process, the actions chosen in those states, and the times spent in those states. A stationary policy is defined as in Chapter 2.

There are two common definitions of the average cost under an arbitrary policy. The first definition, and perhaps the most natural, considers the expected cost incurred under the policy during the interval $[0, t)$, divided by t, and then takes the limit supremum of this quantity as $t \to \infty$. However, we employ the second definition, which considers the expected cost incurred during n transitions, divided by the expected time for those transitions, and then takes the limit supremum as $n \to \infty$. Formally let $E_\theta[C_n]$ be the total expected cost incurred under θ during the first n transition periods. Let $E_\theta[T_n]$ be the total expected time taken up under θ for the first n transition periods. Then we define

$$
\begin{aligned}
J_\theta^\Psi(i) &= \limsup_{n \to \infty} \frac{E_\theta[C_n | X_0 = i]}{E_\theta[T_n | X_0 = i]}, \\
J^\Psi(i) &= \inf_\theta J_\theta^\Psi(i), \qquad i \in S.
\end{aligned}
\tag{10.13}
$$

To be fully rigorous, we need to do some work as in Section 2.3 to convince ourselves that the quantities in the first line of (10.13) are well defined. For the sake of brevity, this argument is omitted.

We are interested in conditions under which $J^\Psi(i)$ is identically equal to a (finite) constant J^{Ψ} and there exists a computable average cost optimal stationary policy.

Note that if the process is in state i and action a is chosen, then the expected time until a change of state is given by $\tau(i,a) =: 1/v(i,a)$. Here is a basic assumption.

Assumption (CTB). There exist constants B and τ such that

$$0 < \tau < \inf_{i,a} \tau(i,a) \le \sup_{i,a} \tau(i,a) \le B < \infty. \tag{10.14}$$

□

Note that CTB stands for *continuous time bounded*. If the expected transition times were unbounded, then the time to make a transition could stretch out as time progressed, leading to "bad behavior." Similarly, if the expected transition times could be arbitrarily small, then a potentially infinite number of transitions could occur in a finite interval, which again is "bad behavior."

The following result is the analogue of Lemma 7.2.1:

Lemma 10.3.1. Let Ψ be a CTMDC satisfying Assumption (CTB), and let e be a stationary policy for Ψ. Assume that there exist a (finite) constant Z and a (finite) function z that is bounded below in i such that

$$Z\tau(i,e) + z(i) \ge G(i,e) + g(i,e)\tau(i,e) + \sum_j P_{ij}(e)z(j), \qquad i \in S. \tag{10.15}$$

then $J_e^{\Psi}(i) \le Z$ for $i \in S$.

Proof: The proof involves some modifications of the proof of Lemma 7.2.1. Only the necessary changes in that proof are indicated. To avoid confusion with continuous time, let us employ k for the discrete time index. It is proved by induction that $E_e[z(X_k)] \le ZBk + z(i)$ for $k \ge 0$.

Equation (7.6) becomes

$$\begin{aligned} &E_e[G(X_k,e) + g(X_k,e)\tau(X_k,e)] \\ &\quad \le ZE_e[\tau(X_k,e)] + E_e[z(X_k)] - E_e[z(X_{k+1})], \qquad k \ge 0. \end{aligned} \tag{10.16}$$

Add the terms in (10.16) for $k = 0$ to $n - 1$, and divide by the sum of the expected transition times to obtain

$$\begin{aligned}
\frac{\sum_{k=0}^{n-1} E_e[G(X_k,e)+g(X_k,e)\tau(X_k,e)]}{\sum_{k=0}^{n-1} E_e[\tau(X_k,e)]}
&\le Z + \frac{z(i) - E_e[z(X_n)]}{\sum_{k=0}^{n-1} E_e[\tau(X_k,e)]} \\
&\le Z + \frac{z(i)+L}{n\tau},
\end{aligned} \tag{10.17}$$

where $-L$ is a (finite) lower bound for z and τ is from Assumption (CTB).

We now take the limit supremum of both sides of (10.17). The limit supremum of the left side of (10.17) is $J_e^{\Psi}(i)$, and this yields the result. □

The plan is to introduce a (discrete time) *auxillary* Markov decision chain Δ that is closely connected to the continuous time process Ψ. We may then form an approximating sequence (Δ_N) for Δ and use the computational method introduced in Chapter 8 to compute an average cost optimal stationary policy for Δ. Under a certain assumption this policy is also optimal for Ψ. The development may be represented schematically as

$$\Psi \Rightarrow \Delta \Rightarrow (\Delta_N), \tag{10.18}$$

where Ψ is the original continuous time infinite state process, Δ is the discrete time infinite state auxillary process, and (Δ_N) is the approximating sequence for Δ consisting of finite state processes for which computation can be carried out.

Let us now define the (discrete time) MDC Δ. Its states and actions are the same as those of Ψ. The costs and transition probabilities are given by

$$\begin{aligned}
C(i,a) &= G(i,a)v(i,a) + g(i,a), \\
P_{ij}^*(a) &= \begin{cases} \tau v(i,a)P_{ij}(a), & j \neq i, \\ 1 - \tau v(i,a), & j = i. \end{cases}
\end{aligned} \tag{10.19}$$

Note from (10.14) that $\tau v(i,a) = \tau/\tau(i,a) < 1$. If the process is in state i, then in each slot the probability of transitioning to $j \neq i$ is proportional to the probability in Ψ. There is also a nonzero probability of remaining in state i. This may be contrasted to Ψ for which $P_{ii}(a) \equiv 0$.

Observe that the sets of policies for Δ and for Ψ are not identical. A policy for Ψ can only choose a new action when a state transition takes place. A policy for Δ may choose a new action in each time slot, even if the state remains the same. However, it is easy to see that the sets of stationary policies are identical.

Here is the crucial lemma that makes Δ useful.

Lemma 10.3.2. Let Ψ be a CTMDC satisfying Assumption (CTB), let Δ be an auxillary MDC, and let e be a stationary policy.

(i) Assume that there exist a (finite) constant Z and a (finite) function w such that

$$Z + w(i) \geq C(i, e) + \sum_j P^*_{ij}(e)w(j), \qquad i \in S. \tag{10.20}$$

Then Z and $z = \tau w$ satisfy (10.15).

(ii) Assume that there exist a (finite) constant Z and a (finite) function z such that (10.15) holds. Then Z and $w = z/\tau$ satisfy (10.20).

Proof: To prove (i), assume that (10.20) holds. Substituting z/τ for w into (10.20) and using (10.19) yields (10.15) after some algebraic manipulation. The proof of (ii) is similar. Problem 10.10 asks you to fill in the details. □

We now make the following assumption linking the minimum average costs in Ψ and in Δ. It is assumed that Assumption (CTB) holds and that an auxillary MDC Δ has been formed.

Assumption (CTAC). We have $J^{\Delta}(.) \leq J^{\Psi}(.)$, where $J^{\Delta}(.)$ is the minimum average cost in Δ. □

Now assume that we have an approximating sequence $(\Delta_N)_{N \geq N_0}$ for Δ. The following result is the analogue of Theorem 8.1.1 and allows us to compute an average cost optimal stationary policy for Ψ.

Theorem 10.3.3. Let Ψ be a CTMDC satisfying Assumption (CTB), let Δ be an auxillary MDC such that Assumption (CTAC) holds, and let (Δ_N) be an approximating sequence for Δ satisfying the (AC) assumptions. Note that (8.1) becomes

$$J^N + r^N(i) = \min_a \left\{ C(i, a) + \sum_{j \in S_n} P^*_{ij}(a; N) r^N(j) \right\},$$
$$i \in S_N, N \geq N_0. \tag{10.21}$$

Then:

(i) The quantity $J^* = \lim_{N \to \infty} J^N$ is the minimum average cost in Δ and Ψ.

(ii) Any limit point e^* of a sequence e^N of stationary policies realizing the minimum in (10.21) is average cost optimal for Δ and Ψ.

Proof: Since the (AC) assumptions hold for (Δ_N) and Δ, the proof and conclusions of Theorem 8.1.1 hold for (Δ_N) and Δ. Hence from that result we conclude that the quantity J^* is the minimum average cost in Δ, e^* is average cost optimal in Δ, and (8.3) becomes

$$J^* + w(i) \geq C(i, e^*) + \sum_j P^*_{ij}(e^*)w(j), \qquad i \in S. \tag{10.22}$$

It follows from Lemma 10.3.2 that $Z = J^*$ and $z = \tau w$ satisfy (10.15).

By (AC3) it is the case that w (and hence z) is bounded below in i. Lemma 10.3.1 implies that $J^{\Psi}_{e*}(.) \leq J^*$. Then $J^{\Psi}_{e*}(.) \leq J^* \equiv J^{\Delta}(.) \leq J^{\Psi}(.) \leq J^{\Psi}_{e*}(.)$, where the second to the last inequality follows from (CTAC). Hence these terms are all equal. This proves that e^* is average cost optimal for Ψ with constant average cost J^*. □

The development in this section allows us to replace Ψ by the auxillary MDC Δ, obtain an approximating sequence for Δ, compute average cost optimal stationary policies in the AS, and know that any limit point of these is optimal for Ψ. In carrying out this program, we already know how to verify the (AC) assumptions. The problem, of course, is the verification of Assumption (CTAC). We will not be able to fully explicate its verification here. Rather we now indicate how it can be shown.

Remark 10.3.4. Assume that we have been able to come up with a (finite) constant Z that is a *lower bound* for the average costs in Ψ, a stationary policy f, and a (finite and bounded below) function z satisfying (10.15). This may be accomplished by emulating, for Ψ, the development of the (SEN) assumptions in Chapter 7. It will then follow from Lemma 10.3.1 that f is average cost optimal for Ψ with constant average cost Z.

Now assume that Assumption (CTB) holds and that an auxillary MDC Δ is given. It follows from Lemma 10.3.2(ii) that Z, $w = z/\tau$, and f satisfy (10.20). But it then follows from Lemma 7.2.1 that $J^{\Delta}_f(i) \leq Z$. Hence $J^{\Delta}(.) \leq J^{\Delta}_f(i) \leq Z \equiv J^{\Psi}(.)$, and Assumption (CTAC) holds. □

Remark 10.3.5. Assume that Assumption (CTB) holds, and let e be a stationary policy. Then e induces a MC in Δ. We call this MC. Similarly e induces what is known as a continuous time Markov chain (CTMC) in Ψ. We call this CTMC. This method uses some results from the theory of average costs for continuous time Markov chains. We will give the idea but omit the background material.

The communicating classes of MC and CTMC are the same. Let R be a class. Then R is positive recurrent in MC if and only if it is positive recurrent in CTMC. In fact it may be shown that

$$\pi_i^{\Delta}(e) = \frac{\tau(i,e)\pi_i^{\Psi}(e)}{\sum_{j \in R} \tau(j,e)\pi_j^{\Psi}(e)}, \qquad i \in R. \tag{10.23}$$

The denominator on the right of (10.23) is a normalizing constant, which we denote by γ.

Using this, we may show that the average cost on R is the same for MC and CTMC. This follows since

$$\begin{aligned} J_{e,R}^{\Delta} &= \sum_{i \in R} C(i,e)\pi_i^{\Delta}(e) \\ &= \sum_{i \in R} [G(i,e)v(i,e) + g(i,e)]\pi_i^{\Delta}(e) \\ &= \frac{1}{\gamma} \sum_{i \in R} [G(i,e) + g(i,e)\tau(i,e)]\pi_i^{\Psi}(e) \\ &= J_{e,R}^{\Psi}. \end{aligned} \tag{10.24}$$

The second line follows from (10.19), and the third line from (10.23). The last line follows from CTMC theory.

Now let us assume that $J^{\Delta}(.)$ is a constant J^{Δ} and $J^{\Psi}(.)$ is a constant J^{Ψ}. Moreover assume that there exists an average cost optimal stationary policy f for Ψ inducing a CTMC with a positive recurrent class R. Note that

$$J^{\Delta} \le J_{f,R}^{\Delta} = J_{f,R}^{\Psi} = J^{\Psi}. \tag{10.25}$$

Hence Assumption (CTAC) holds. □

10.4 SERVICE RATE CONTROL OF THE M/M/1 QUEUE

It is time to do some computation! In this section we compute an average cost optimal stationary policy for Example 10.2.1.

Assumption (CTB) is satisfied with $B = 1/\lambda$ and $\tau = 1/[2(\lambda + a_K)]$. We may then define the auxillary MDC Δ. Its costs and transition probabilities are given by

$$\begin{aligned} C(0) &= 0, \\ C(i,a) &= c(a) + H(i), \qquad i \ge 1, \\ P_{01}^{*} &= \tau\lambda, \qquad P_{00}^{*} = 1 - \tau\lambda, \\ P_{ii+1}^{*}(a) &= \tau\lambda, \quad P_{ii-1}^{*}(a) = \tau a, \quad P_{ii}^{*}(a) = 1 - \tau(\lambda + a), \qquad i \ge 1. \end{aligned} \tag{10.26}$$

Equation (10.26) follows from (10.11–12) and (10.19). Note from (10.3) that the transition probabilities may be interpreted as arising from a discretization with step size τ, in which terms that are $o(\tau)$ are ignored.

We operate under the following basic assumptions (BA):

(BA1). We have $\lambda < a_K$.

(BA2). The holding cost rate $H(i)$ is increasing in i, and there exists a (finite) constant D and a nonnegative integer n such that $H(i) \le Di^n$ for $i \ge 0$.

The approximating sequence (Δ_N) is defined by letting $S_N = \{0, 1, \ldots, N\}$. Note that there is excess probability only in state N, and we send this probability to N.

It is possible to use previously derived results to verify that the (AC) assumptions hold for (Δ_N). In particular, we will fit Δ into the structure treated in Example 7.6.4 and apply the results developed there. Consider Example 7.6.4 under the assumption of a Bernoulli arrival process with P(a single customer arrives) $= \tau\lambda$. Under action a, the probability of a service finishing in a slot is τa. Then the mean customer arrival rate $\tau\lambda$ is less than the maximum customer service rate τa_K. Hence the basic assumptions in Example 7.6.4 hold. It then follows from Example 8.3.2 that the (AC) assumptions hold for Example 10.2.1 and that an average cost optimal stationary policy for Δ may be computed using value iteration.

Assumption (CTAC) may be shown to hold using Remark 10.3.4, and we will assume that this has been done.

We will compute an optimal policy under the assumption that $H(i) = Hi$, for a positive constant H. It is likely (unless there are ties) that the optimal policy computed using (AC) will be increasing in i and eventually choose a_k. The optimal policy may be given as a sequence of $K - 1$ intervals, with the interpretation as in Section 8.5.

The expressions for the VIA 6.6.4 are given by

$$\begin{aligned} w_n(0) &= (1 - \tau\lambda)u_n(0) + \tau\lambda u_n(1), \\ w_n(i) &= Hi + \min_a\{c(a) + \tau a u_n(i-1) + [1 - \tau(\lambda + a)]u_n(i) \\ &\quad + \tau\lambda u_n(i+1)\}, \qquad 1 \le i \le N-1, \\ w_n(N) &= HN + \min_a\{c(a) + \tau a u_n(N-1) + (1 - \tau a)u_n(N)\}, \\ u_{n+1}(i) &= w_n(i) - w_n(0), \qquad 0 \le i \le N. \end{aligned} \tag{10.27}$$

The second and third equations in (10.27) may be evaluated in the same loop by introducing an auxillary variable that equals $i+1$ for $1 \le i \le N-1$ and equals N for $i = N$. Note that the equations in (10.27) are almost the same as those in (8.11) with changes in the costs and transition probabilities. ProgramSeven gives this computation.

We would like to have a *benchmark policy* to compare with the optimal policy. Assume that rate a satisfies $\lambda < a$. Then the policy $d(a)$ that always serves at rate a has finite average cost and can be implemented with open-loop control. Our benchmark policy d serves at the rate a that minimizes $J^{\Psi}_{d(a)}$. That is, under d we serve at the constant rate that yields $J^{\Psi}_{d} = \min_{a>\lambda}\{J^{\Psi}_{d(a)}\}$.

Proposition 10.4.1. Assume that a satisfies $\lambda < a$, and let $d(a)$ be the policy that always serves at rate a. Then letting $\rho_a = \lambda/a$, we have

$$J^{\Psi}_{d(a)} = \rho_a c(a) + \frac{H\rho_a}{1-\rho_a}. \tag{10.28}$$

Proof: It is well-known from the theory of M/M/1 queues that $\pi^{\Psi}_i(d(a)) = (1-\rho_a)\rho^i_a$, where $\rho_a < 1$ is the *utilization factor*; in other words, the probability the server is busy (e.g., Gross and Harris 1998). The cost rate is in effect whenever the server is busy giving the first term. The second term is the expected holding cost and is easy to derive from $\sum i\pi^{\Psi}_i(d(a))$. □

Remark 10.4.2. It is intuitively clear and may be proved by induction on (10.27) that if H and $c(a)$ are multiplied by a positive constant, then the optimal average cost is multiplied by that constant, and the optimal policy is unchanged. For this reason we assume that $H = 1$ in all our scenarios. In all scenarios we used the weaker convergence criterion (Version 1) of the VIA. □

Checking Scenarios 10.4.3. For the first check, we let $\lambda = 3.0$ and $H = 0.0$. The service rates are 4.0, 5.0, and 5.5 with respective cost rates 2.0, 5.0, and 6.0. Because there is no holding cost, it is optimal to always use the smallest rate, and (10.28) yields $J^{\Psi} = 1.5$. This is born out by the program.

For the second check, we let $\lambda = 5.0$ and $H = 1.0$. The service rates are 6.0, 8.0, and 10.0 with cost rates identically equal to 4.0. Because the cost rates are the same, it is optimal to always use the largest rate, and (10.28) yields $J^{\Psi} = 3.0$. This is born out by the program. □

Scenarios 10.4.4. Table 10.1 gives the results. Recall that the optimal policy is given as two intervals, with the interpretation that for queue levels above the maximum shown, it is optimal to serve at the fastest rate. A dash in the table means that entry is identical to the corresponding entry in the previous column.

In Scenario 1 note that $c(a) = 2a + 5$. It is optimal to serve at the fastest rate. This is similar to an effect discussed in Proposition 7.6.7(ii). We might try to prove a similar result for this model.

In Scenario 2 we have an unstable slowest rate and a very expensive fastest rate. It is optimal to serve at the slowest rate when there are one or two customers in the queue. Then it is optimal to switch to the middle rate and serve at this rate until a queue level of at least 80. We suspect that for sufficiently

Table 10.1 Results for Scenarios 10.4.4

Scenario	1	2	3	4	5	6	7	8
λ	3.0	2.0	2.0	2.0	5.0	5.0	10.0	20.0
Service rates	2.0	1.0	—	—	5.0	5.1	10.2	24.0
	4.0	4.0			5.5	5.3	10.6	27.0
	8.0	7.0			5.8	6.0	12.0	30.0
Costs	9.0	1.0	1.0	1.0	0.0	0.0	0.0	1.0
	13.0	50.0	50.0	50.0	10.0	10.0	10.0	1.5
	21.0	500.0	150.0	100.0	100.0	25.0	25.0	5.0
N	48	84	—	—	—	—	—	—
J_d^Ψ	$a = 8.0$	$a = 4.0$	—	—	$a = 5.5$	$a = 6.0$	—	$a = 27.0$
	8.475	26.0			19.091	25.833		3.968
J^Ψ	8.475	21.091	—	20.971	17.043	15.193	—	3.902
Savings	0.0	4.909	—	5.029	2.048	10.640	—	0.066
Optimal policy	∅∅	[1, 2] [3, >80]	[1, 2] [3, 38]	[1, 2] [3]	[1, 7] [8, >80]	[1, 12] ∅	—	∅ [1, 8]

large queue level, it is optimal to serve at the fastest rate, but this level was not located for an approximation level of 84.

Scenarios 3 and 4 explore the effect on Scenario 2 when the cost of the fastest rate is backed off. In Scenario 3 it is reduced to 150. Here the break point to switch to the fastest rate is 39. In Scenario 4 it is reduced further to 100. The break point is reduced to 4. These are interesting results that you are asked to explore further in Problem 10.11.

In Scenario 5 there is a free rate equal to the arrival rate, and a quite expensive fastest rate. The break point to switch to the fastest rate was not located for an approximating level of 84.

In Scenario 6 all the rates are stable with modest increases in cost. In this interesting example, the optimal policy switches from the slowest rate to the fastest rate at a queue level of 13. Note that in Scenario 7 the arrival rate and service rates have been doubled, while the cost rates are the same. The optimal policy and minimum average cost are identical to those in Scenario 6. You are asked to explore this in Problem 10.12.

In Scenario 8 we have fairly large arrival and service rates. Note that it is never optimal to serve at the slowest rate, even though the queue is stable under this rate. □

10.6 M/M/K QUEUE WITH DYNAMIC SERVICE POOL

In this model customers arrive to a single queue according to a Poisson process with rate λ. There is a (finite) pool of K independent servers, each capable of serving a single customer, and the service time of that customer is exponentially distributed with rate μ.

The action set $A = \{0, 1, \dots, K\}$. If action k is chosen, the interpretation is that k servers are available for service (i.e., turned on), while $K - k$ servers are turned off. If there are currently i customers in the queue with $1 \le i \le k$, then all these will be serviced. If $k < i$, then only k customers will be serviced. Note that $k = 0$ means that all the servers are turned off.

There is a holding cost rate $H(i)$ charged for every unit of time that the queue contains i customers (where $H(0) = 0$). There is a cost rate $c(k)$ operative for each unit of time that k servers are turned on. It is natural to assume that $c(k)$ is an increasing function with $c(0) = 0$, but this is not required.

Now assume that k^* servers are presently turned on and a new action k is chosen. There is a matrix $D(k^*, k)$ of instantaneous charges, where $D(k, k) = 0$. If $k > k^*$, then $D(k^*, k)$ is a one-time activation charge for turning on some servers; if $k < k^*$, then it is a one-time deactivation charge for turning off some servers. The holding and service cost rates, as well as the instantaneous charges are all nonnegative.

Let's model this as a CTMDC. Let $S = \{(i, k) | i \ge 0,\ k \in A\}$. The state $(i,\ k)$ means that there are currently i customers in the queue, k servers are turned on, and either a service has just been completed or a new customer has arrived. [See Fig. 10.1 for an M/M/5 system in state (7,4). Note that all 7 customers are considered to be in the queue.] The action set is A in every state. The costs are

$$G((i, k^*), k) = D(k^*, k), \quad g((i, k^*), k) = H(i) + c(k). \tag{10.29}$$

Note that the cost rate is charged on the number of available servers, whether or not they are actually serving.

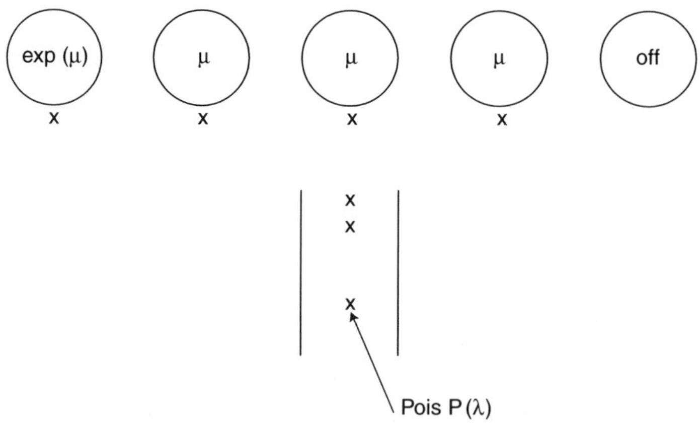

Figure 10.1 M/M/5 dynamic service pool system in state (7, 4).

The transition rates are given by

$$v((i,k^*),k) = \lambda + \min\{i,k\}\mu, \qquad i,k \geq 0. \tag{10.30}$$

The transition probabilities are given by

$$P_{(i,k^*)(i+1,k)}(k) = \frac{\lambda}{\lambda + \min\{i,k\}\mu},$$

$$P_{(i,k^*)(i-1,k)}(k) = \frac{\min\{i,k\}\mu}{\lambda + \min\{i,k\}\mu}, \qquad i,k \geq 0. \tag{10.31}$$

This completes the specification of this example as a CTMDC Ψ.

We now develop the auxillary MDC Δ. Clearly Assumption (CTB) holds, and we may set $\tau =: 1/[2(\lambda + K\mu)]$.

The costs in Δ are given by

$$C((i,k^*),k) = D(k^*,k)(\lambda + \min\{i,k\}\mu) + H(i) + c(k), \qquad i,k \geq 0. \tag{10.32}$$

This follows from (10.19) and (10.29). The transition probabilities are

$$P^*_{(i,k^*)(i+1,k)}(k) = \tau\lambda,$$

$$P^*_{(i,k^*)(i-1,k)}(k) = \tau \min\{i,k\}\mu,$$

$$P^*_{(i,k^*)(i,k^*)}(k) = 1 - \tau(\lambda + \min\{i,k\}\mu), \qquad i,k \geq 0. \tag{10.33}$$

This follows from (10.19) and (10.31) and completes the specification of Δ.

We operate under the following basic assumptions (BA):

(BA1). We have $\lambda < K\mu$.

(BA2). The holding cost rate $H(i)$ is increasing in i, and there exists a (finite) constant D and a nonnegative integer n such that $H(i) \leq Di^n$ for $i \geq 0$.

An approximating sequence for Δ is obtained by not allowing more than N customers in the buffer. Hence $S_N = \{(i,k) | 0 \leq i \leq N,\ 0 \leq k \leq K\}$. In Δ the only possible transition from a state in S_N to the outside occurs if the system is in state (N, .), Suppose that action k is chosen. If an arrival occurs, then the system would transition to (N + 1, k). The probability of this event is given to state (N, k). This defines the ATAS (Δ_N).

It may be argued that the (AC) assumptions hold for (Δ_N) and that the VIA is valid. In the interest of brevity and because this argument is similar to ones presented previously, it is omitted. Assumption (CTAC) may be shown to hold using Remark 10.3.4, and we will assume that this has been done.

We will compute an optimal policy under the assumption that $H(i) = Hi$ and $D(k^*, k) = D|k^* - k|$ for positive constants H and D. The base point for the calculations is $(0, K)$. The expressions for the VIA 6.6.4 are given by

$$\begin{aligned} w_n(i,k^*) &= Hi + \min_k\{c(k) + D|k^* - k|(\lambda + \min\{i,k\}\mu) + \tau\lambda u_n(z^*,k) \\ &\quad + \tau\min\{i,k\}\mu u_n(z_*,k) + [1 - \tau(\lambda + \min\{i,k\}\mu)]u_n(i,k^*)\}, \\ &\quad 0 \le i \le N, \\ u_{n+1}(.) &= w_n(.) - w_n(0,K). \end{aligned} \tag{10.34}$$

The auxillary variable z^* equals $i+1$ for $i < N$ and equals N for $i = N$. The auxillary variable z_* equals $i-1$ for $i > 0$ and equals 0 for $i = 0$. ProgramEight gives the computation.

Remark 10.5.1. It is easy to prove, by induction on n, that if H, $c(k)$, and D are each multiplied by a positive constant, then the optimal average cost is multiplied by that constant and the optimal policy is unchanged. For this reason we may assume that $H = 1$ in all our scenarios. □

Let $d(k)$ be the stationary policy that always has k servers turned on. For $\lambda/\mu < k \le K$ the queue is stable under $d(k)$. Note that if the process starts in a transient state (i, k^*), $k^* \ne k$, it will reach the positive recurrent class under $d(k)$ in one step and with finite cost. From the theory of continuous time queueing systems, we may obtain a formula for the average cost under $d(k)$. We then use as our benchmark the policy $d(k)$ with the smallest average cost. The next result gives the details.

Proposition 10.5.2. Let k be a positive integer satisfying $\lambda/\mu < k \le K$, and let $d(k)$ be the stationary policy that always has k servers turned on. Let

$$\gamma = \frac{\lambda}{\mu} \quad \text{and} \quad \eta(k) = \sum_{n=0}^{k-1} \frac{\gamma^n}{n!}.$$

Then

$$J_{d(k)}^{\Psi} = c(k) + H\gamma\left[1 + \frac{\gamma^k}{(k-\gamma)\{\gamma^k + (k-1)!(k-\gamma)\eta(k)\}}\right]. \tag{10.35}$$

Proof: Because k servers are constantly turned on, the average cost includes a term of $c(k)$ per unit time. The second term is H times the average steady state number in the system. An expression for the average number in the system is given in Gross and Harris (1998), and the second term in (10.35) follows from that after some algebraic manipulation. For the interested reader, Problem *10.14 asks you to derive (10.35) from the expression in Gross and Harris (1998). As a check on (10.35) note that the second term reduces to the second term in (10.28) for $k = 1$. □

It is difficult to optimize (10.35) analytically. Hand calculations may be done for the appropriate values of k and the one yielding the minimum value then defines the benchmark. Alternatively, a short program to perform the optimization can be written.

Checking Scenarios 10.5.3. For the first check, we let $K = 5$, $\lambda = 3.0$, $\mu = 1.0$, $H = 1.0$, $D = 0.0$, and $c(k) \equiv 2.0$. Because the service cost rate is constant, it is clear that the policy $d(5)$ is optimal. We calculate $J^{\Psi}_{d(5)} = 5.354$ from (10.35), and this is born out by the program. The computed optimal policy turns servers off as customers depart and turns them on as customers enter. Because $D = 0.0$, there is no penalty for doing this, and the program is set up in such a way that it will be done. Note that (10.35) still applies to calculate the minimum average cost under the computed optimal policy. (Why?)

For the second check, we let $K = 4$, $\lambda = 5.0$, $\mu = 2.0$, $H = 1.0$, $D = 5.0$, and $c(k) \equiv 1.0$. Because the service cost rate is constant, it is clear that the policy $d(4)$ is optimal. We calculate $J^{\Psi}_{d(4)} = 4.033$ from (10.35), and this is born out by the program. □

Scenarios 10.5.4. Table 10.2 gives the results. In all scenarios we have $H = \mu = 1.0$. Scenarios 1 through 4 explore the situation with $K = 5$, $c(k) = 2k$, $\lambda = 3.0$, and D increasing. Note that in these scenarios the benchmark is constant. As D increases, we expect the minimum average cost to approach the benchmark, and that is what happens.

The determination of an optimal policy requires some explanation. Let us begin with Scenario 2, since it is most representative of the method. The program output gives the optimal pool size k for any number in the system and current value k^*. Thus, if the current state is $i = 0$ and $k^* = 1$, then the optimal choice is $k = 1$; that is, do nothing. We claim that state (0, 1) is transient under the Markov chain induced by the optimal policy e.

To verify that (0, 1) is transient, let us identify the positive recurrent class in the MC induced by e. Here is the method. Begin with a larger state in which all servers are turned on, say (6, 5). Then the printout yields $e(6, 5) = 5$. Assume that a service completion occurs so that the new state is (5, 5). Then $e(5, 5) = 5$. If a service completion occurs, the new state is (4, 5), and we see that $e(4, 5) = 4$. This means that if the queue length decreases to 4, then it is optimal to turn off one server. If a service completion occurs in (4, 4), then the new state is (3, 4) and $e(3, 4) = 4$. We continue to work our way down in this fashion. If a service completion occurs in (3, 4), then the new state is (2, 4) and

Table 10.2 Results for Scenarios 10.5.4

Scenario	1	2	3	4	5	6	7
Parameters	$K = 5$ $\lambda = 3.0$ $\mu = 1.0$ $D = 0.0$	— $D = 1.0$	— $D = 3.0$	— $D = 10.0$	$K = 7$ $\lambda = 3.0$ $\mu = 1.0$ $D = 1.0$	$K = 7$ $\lambda = 5.0$ $\mu = 1.0$ $D = 1.0$	$K = 10$ $\lambda = 7.0$ $\mu = 1.0$ $D = 1.0$
Cost rates	$c(k) = 2k$	—	—	—	—	$c(k) = k/2$ $0 \leq k \leq 5$ $c(6) = 5.0$ $c(7) = 10.0$	$c(k) = 3k$
J_d^{Ψ}	$k = 4$ 12.528	—	—	—	—	$k = 6$ 12.938	$k = 9$ 35.347
J^{Ψ}	9.354	11.317	12.024	12.388	11.246	11.652	31.937
Savings	3.174	1.211	0.504	0.14	1.282	1.286	3.41
Optimal policy	[0] (1, 2, 3, 4, 5, 4, 3, 2, 1)	[2] (4, 6, 8, 4, 2, 1)	[3] (7, 10, 3, 1)	[4] (15, 3)	[2] (4, 6, 8, 10, 12, 6, 5, 4, 2, 1)	[5] (8, 13, 7, 4)	[3] (5, 6, 8, 10, 11, 13, 15, 9, 7, 6, 5, 3, 2, 0)

$e(2, 4) = 3$. So, if the queue length decreases to 2, it is optimal to turn off another server. If a service completion occurs in (2, 3), then the new state is (1, 3) and $e(1, 3) = 2$. So, if the queue length decreases to 1, it is optimal to turn off another server. If a service completion occurs in (1, 2), then the new state is (0, 2) and $e(0, 2) = 2$. This means that no change is optimal.

We now begin to work our way up. If the process is in state (0, 2) and an arrival occurs, then the new state is (1, 2) and $e(1, 2) = 2$. We continue until it is seen that $e(4, 2) = 3$. This means that if the queue length increases to 4, then another server should be turned on. Continuing, we see that yet another will be turned on when the queue length reaches 6. Finally all 5 servers will be turned on when the length reaches 8. Then the process repeats itself.

Hence it is clear that it is never optimal to have less than two servers turned on and that states such as (0, 1) are transient. The positive recurrent class under the optimal policy may be given as [2] (4, 6, 8, 4, 2, 1). The first term indicates the minimum number of servers to be always turned on. If the queue is empty, then 2 should be turned on. When the queue length increases to 4, then a third server should be turned on. When it increases to 6, then a fourth server should be turned on. Finally, when it increases to 8, then all the servers should be on. Similarly, when it decreases to 4, then one server should be turned off. When it decreases to 2, then another should be turned off. When it decreases to 1, then a third server should be turned off, leaving 2 on.

Notice that as the queue length increases, the number of servers turned on lags behind the queue length. Similarly, when all the servers are turned on and the queue length begins to decrease, there is a lag in turning them off. This phenomenon is known as *hysteresis*, and it occurs since $D > 0$. Note that there is no hysteresis in the optimal policy for Scenario 1. The effect of increasing D in Scenarios 1 through 4 is to increase the minimum number turned on and to increase the hysteretic effect.

In specifying an optimal policy, it is only necessary to specify its positive recurrent class. The reason is that if the process reaches this class in finite expected time and with finite expected cost, then its average cost will equal the minimum average cost. If the process starts in a transient state, the controller can immediately turn all servers on, run the system until it empties, and then turn off all but the minimum number indicated. From this point on, the system will operate within the positive recurrent class.

The process of identifying the positive recurrent class seems complicated, but with a little practice it will be easy (and fun!) to scan the output and identify the positive recurrent class induced by an optimal policy. (Notice how easily the optimal policy might be programmed into a control mechanism.) Problem 10.15 asks you to run ProgramEight for Scenarios 3 through 7 and identify the optimal policies as given in Table 10.2.

Scenario 5 is the same as Scenario 2 except that the pool of servers has been increased to 7. It is conceivable that the benchmark policy could change, but in this case it doesn't. Because we are not charging for having a certain pool size, it is clear that the minimum average cost is a decreasing function of K and hence has a limit. Thus the minimum average cost for Scenario 5 must be less than that for Scenario 2. Now compare the optimal policies. Notice that they are the same on the "ends" but differ in the middle because of the availability of servers 6 and 7.

Scenario 6 explores an example in which the cost rate is nonlinear in the number of servers turned on. Scenario 7 explores a situation with a fairly large pool of 10 servers. □

10.6 CONTROL OF A POLLING SYSTEM

Consider a polling system as in Fig. 1.7. Stations 1, 2, ..., K are arranged in a ring. Each station has an infinite buffer, and customers arrive to station k according to a Poisson process with rate λ_k. The service time of a customer at station k follows an exponential distribution with rate μ_k. The server travels around the ring counterclockwise from station 1 to station 2, and so on, and finally back to station 1. The *walking time* for the server to get from station k to station $k + 1$ is exponentially distributed with rate ω_k. Note that station $K + 1$ is station 1. The arrival processes, service times, and walking times are all independent.

We say that the server is *walking* if it is presently undergoing a walk. If it is presently at a station, we say it is *stationary*. Let $\mathbf{i} = (i_1, \ldots, i_K)$ be the vector of buffer occupanices. The state space for the CTMDC Ψ for this model consists of all tuples of the form ($\mathbf{i}$, k, z), where $z = 0$ or 1 and $\mathbf{i}$ is the vector of current buffer occupancies. The state is ($\mathbf{i}$, k, 0) if the server is walking from station k and a new customer has just arrived to the system (it is counted in $\mathbf{i}$). There are no actions in this state. The state is ($\mathbf{i}$, k, 1) if exactly one of the following holds: (1) a walk terminating at k has just been completed, (2) the server has just finished a service at k (the customer has departed and is not counted in $\mathbf{i}$), or (3) the server is stationary at k and a customer has just arrived to the system (it is counted in $\mathbf{i}$). The action set in these states is $\{a, b\}$, where a = remain at the current station and b = initiate a walk. Note that a server is allowed to initiate a walk whenever it arrives to a new station, completes a service, or is stationary and observes a new customer arriving. Also note that the server has the option of choosing to stay at a station even if the buffer of that station is empty.

One might also assume that if the server has just arrived to a station and chooses action a, then an additional *setup* time is incurred. Our model does not treat this elaboration.

A holding cost rate $H(\mathbf{i})$ is charged on the current buffer contents. A walking cost rate W_k is charged for each unit of time spent walking from station k to station $k + 1$. It is possible to develop more elaborate cost models, but for illustrative purposes this will suffice. Note that if $H(\mathbf{i}) = \sum i_k$ and $W_k \equiv 0$, then an average cost optimal policy minimizes the expected long run average number of customers in the system.

The costs in the CTMDC model are

$$\begin{aligned} g((\mathbf{i},k,0)) = g((\mathbf{i},k,1),b) &= H(\mathbf{i}) + W_k, \\ g((\mathbf{i},k,1),a) &= H(\mathbf{i}). \end{aligned} \tag{10.36}$$

To develop the transition rates and probabilities, let e_j be a K-dimensional unit vector with a 1 in the jth position and 0's elsewhere. Let $\lambda = \sum \lambda_k$ be the total arrival rate. The transition rates are given by

$$\begin{aligned} v((\mathbf{i},k,0)) &= v((\mathbf{i},k,1),b) = \lambda + \omega_k, \\ v((\mathbf{i},k,1),a) &= \begin{cases} \lambda, & i_k = 0, \\ \lambda + \mu_k, & i_k \geq 1. \end{cases} \end{aligned} \tag{10.37}$$

The transition probabilities are given by

$$\begin{aligned} P_{(\mathbf{i},k,0)(\mathbf{i}+\mathbf{e}_j,k,0)} &= P_{(\mathbf{i},k,1)(\mathbf{i}+\mathbf{e}_j,k,0)}(b) = \frac{\lambda_j}{\lambda+\omega_k}, \qquad 1 \leq j \leq K, \\ P_{(\mathbf{i},k,0)(\mathbf{i},k+1,1)} &= P_{(\mathbf{i},k,1)(\mathbf{i},k+1,1)}(b) = \frac{\omega_k}{\lambda+\omega_k}, \\ P_{(\mathbf{i},k,1)(\mathbf{i}+\mathbf{e}_j,k,1)}(a) &= \frac{\lambda_j}{\lambda + I(i_k \neq 0)\mu_k}, \qquad 1 \leq j \leq K, \\ P_{(\mathbf{i},k,1)(\mathbf{i}-\mathbf{e}_k,k,1)}(a) &= \frac{I(i_k \neq 0)\mu_k}{\lambda + I(i_k \neq 0)\mu_k}. \end{aligned} \tag{10.38}$$

Here I is an indicator function, enabling us to handle the cases of an empty buffer or a nonempty buffer with the same expression. This completes the specification of Ψ.

We now develop the discrete time auxillary MDC Δ. Let $\varphi =: \max_k\{\mu_k, \omega_k\}$. Clearly Assumption (CTB) holds, and we may set $\tau =: 1/[2(\lambda+\varphi)]$.

The costs in Δ are given by

$$\begin{aligned} C((\mathbf{i},k,0)) = C((\mathbf{i},k,1),b) &= H(\mathbf{i}) + W_k, \\ C((\mathbf{i},k,1),a) &= H(\mathbf{i}). \end{aligned} \tag{10.39}$$

This follows from (10.19) and (10.36).

The transition probabilities are

$$
\begin{aligned}
P^*_{(\mathbf{i},k,0)(\mathbf{i}+\mathbf{e}_j,k,0)} = P^*_{(\mathbf{i},k,1)(\mathbf{i}+\mathbf{e}_j,k,0)}(b) &= \tau\lambda_j, \qquad 1 \le j \le K,\\
P^*_{(\mathbf{i},k,0)(\mathbf{i},k+1,1)} = P^*_{(\mathbf{i},k,1)(\mathbf{i},k+1,1)}(b) &= \tau\omega_k,\\
P^*_{(\mathbf{i},k,0)(\mathbf{i},k,0)} = P^*_{(\mathbf{i},k,1)(\mathbf{i},k,1)}(b) &= 1 - \tau(\lambda + \omega_k),\\
P^*_{(\mathbf{i},k,1)(\mathbf{i}+\mathbf{e}_j,k,1)}(a) &= \tau\lambda_j, \qquad 1 \le j \le K,\\
P^*_{(\mathbf{i},k,1)(\mathbf{i}-\mathbf{e}_k,k,1)}(a) &= \tau I(i_k \neq 0)\mu_k\\
P^*_{(\mathbf{i},k,1)(\mathbf{i},k,1)}(a) &= 1 - \tau(\lambda + I(i_k \neq 0)\mu_k).
\end{aligned}
\tag{10.40}
$$

This follows from (10.19) and (10.37–38).

Let $\rho_k = \lambda_k/\mu_k$, and let $\rho = \sum \rho_k$. We will compute an average cost optimal stationary policy under the following basic assumptions (BA):

(BA1). We have $\rho < 1$.

(BA2). We have $H(\mathbf{i}) = \sum H_k i_k$ for positive constants H_k.

The approximating sequence (Δ_N) is defined as follows: No buffer is allowed to contain more than N customers. If a customer arrival would cause a particular buffer to overflow, then the probability of that event is given to the appropriate state with buffer content N. Assume, for example, that $N = 10$, $\mathbf{i} = (8, 10, 3, 5)$, and the server is serving at station 3. Thus the current state is ((8, 10, 3, 5), 3, 1). If there is an arrival to station 2, then the system would transition to ((8, 11, 3, 5), 3, 1). The probability of this event is given to ((8, 10, 3, 5), 3, 1). Other cases are handled similarly.

Note that (BA1) is the condition for stability of the polling system under a stationary policy known as *exhaustive service*, denoted d. This operates as follows: Whenever the server arrives to a station, it serves customers at that station until the buffer completely empties, and it then walks to the next station and repeats the process. Upon arrival to a station with an empty buffer, it immediately initiates a walk. Note that when the system is empty, the server will continually cycle until a customer enters the system.

One may show that the (AC) assumptions hold for Δ and (Δ_N), and that Assumption (CTAC) holds. We omit this lengthy argument.

The expressions for the VIA 6.6.4 are given by

$$
w_n(\mathbf{i},k,0) = H(\mathbf{i}) + W_k + \tau \sum_j \lambda_j u_n(\mathbf{i}+\mathbf{e}_j,k,0)
$$

$$
+ \tau\omega_k u_n(\mathbf{i},k+1,1) + [1 - \tau(\lambda + \omega_k)]u_n(\mathbf{i},k,0),
$$

$$w_n(\mathbf{i}, k, 1) = H(\mathbf{i}) + \min\left\{ \tau \sum_j \lambda_j u_n(\mathbf{i} + \mathbf{e}_j, k, 1) + \tau \mu_k I(i_k \neq 0) u_n(\mathbf{i} - \mathbf{e}_k, k, 1) \right.$$

$$+ [1 - \tau(\lambda + \mu_k I(i_k \neq 0))] u_n(\mathbf{i}, k, 1),\ W_k + \tau \sum_j \lambda_j u_n(\mathbf{i} + \mathbf{e}_j, k, 0)$$

$$\left. + \tau \omega_k u_n(\mathbf{i}, k+1, 1) + [1 - \tau(\lambda + \omega_k)] u_n(\mathbf{i}, k, 1) \right\},$$

$$u_{n+1}(.) = w_n(.) - w_n(\mathbf{0}, 1, 1). \tag{10.41}$$

ProgramNine gives the computation.

We will develop a benchmark for the special case in which the holding cost coefficients equal 1, the walking costs equal 0, and the mean service rates are all equal. The benchmark is the average cost under the exhaustive service policy d. In this situation the average cost is precisely the expected number of customers in the system in steady state. Note that operating the system under d is essentially open-loop control, since implementation of this policy only requires the server to know when the currently served queue empties out.

Proposition 10.6.1. Let d be the policy of exhaustive service, and assume that $H_k \equiv 1$, $W_k \equiv 0$, and $\mu_k \equiv \mu$. Note that $\rho_k = \lambda_k/\mu$ and $\rho = \lambda/\mu$. Let

$$r =: \sum_k \frac{1}{\omega_k}, \qquad r^* =: \sum_k \frac{1}{\omega_k^2},$$

be the mean (respectively, the variance) of the total walking time of one *polling cycle* (one trip around the ring). Then

$$J_d^\Psi = \frac{\rho}{1-\rho} + \frac{\lambda r^*}{2r} + \frac{r[\lambda - (\sum \lambda_k^2/\mu)]}{2(1-\rho)}. \tag{10.42}$$

**Proof:* We will apply a pseudoconservation law derived in Boxma and Groenendijk (1987). This applies to a polling system with Poisson arrivals and general service and walking times (first and second moments of these quantities must be finite). Some notation is required, which later will be specialized to the case in the proposition.

Let us assume that the system operating under d is in steady state and introduce the following random variables, which apply to a station k: Let Y_k be the service time, and note that $\rho_k = \lambda_k E[Y_k]$. Let Q_k be the waiting time (i.e., the time until service begins) and T_k the system time. Note that $T_k = Q_k + Y_k$, and hence $E[T_k] = E[Q_k] + E[Y_k]$. Let L_k be the number of customers. Finally let $L = \sum_k L_k$ be the total number of customers in the system.

It follows from *Little's formula* (a well-known result in queueing theory) that $E[L_k] = \lambda_k E[T_k]$. This expresses the intuitively appealing idea that the average number of customers in a queueing system in steady state equals the average arrival rate of customers to the system times the average time a customer spends in the system. Summing over k and using the above relationships yields

$$E[L] = \sum_k \lambda_k E[Q_k] + \rho. \tag{10.43}$$

Let Z be a random variable representing the total walking time in a polling cycle. The form of the pseudoconservation law given in Takagi (1990, p. 278) is

$$\sum_k \rho_k E[Q_k] = \frac{\rho \sum \lambda_k E[Y_k^2]}{2(1-\rho)} + \frac{\rho \operatorname{var}(Z)}{2E[Z]} + \frac{E[Z](\rho - \sum \rho_k^2)}{2(1-\rho)}. \tag{10.44}$$

Observe that if the mean service times are constant, say $E[Y_k] \equiv b$, then b may be factored out of the left side of (10.44). Then using (10.43) and a bit of algebraic manipulation, we obtain

$$E[L] = \frac{\lambda \sum \lambda_k E[Y_k^2]}{2(1-\rho)} + \frac{\lambda \operatorname{var}(Z)}{2E[Z]} + \frac{E[Z](\lambda - b\sum \lambda_k^2)}{2(1-\rho)} + \rho. \tag{10.45}$$

In the situation of the proposition, we have $E[L] = J_d^\Psi$, $E[Z] = r$, $\operatorname{var}[Z] = r^*$, $b = 1/\mu$, and $E[Y_k^2] \equiv 2/\mu^2$. Substituting these quantities into (10.45) and simplifying yields (10.42). □

Checking Scenario 10.6.2. The program was run with $\lambda_1 = 0.25$, $\lambda_2 = \lambda_3 = 0.5$, $\mu_k \equiv 2.0$, $\omega_k \equiv 1.0$, $W_k \equiv 2.0$, $H_1 = H_3 = 0$, and $H_2 = 1.0$. In this case there is no incentive to serve customers as stations 1 or 3, and hence the server should remain stationary at station 2. The value of J^Ψ should be the average number of customers in an M/M/1 queue with utilization factor $\lambda_2/\mu_2 = 0.5/2.0 = 0.25$. This yields $J^\Psi = 0.25/0.75 = 1/3$ from the second term in (10.28). This is born out by the program which yields an optimal policy identically equal to 0 1 0. Here 0 means walk at stations 1 and 3, and 1 means remain stationary at station 2. □

Scenarios 10.6.3. Table 10.3 gives the results. In all scenarios we set $H_k \equiv 1$. The value of ρ is a measure of system loading. The convergence to

Table 10.3 Results for Scenarios 10.6.3

Scenario	1	2	3	4	5	6
Parameters	$\lambda_k \equiv 0.25$ $\mu_k \equiv 1.25$ $\omega_k \equiv 1.00$ $W_k \equiv 0.0$	— $W_k \equiv 2.0$	$\lambda_k \equiv 0.25$ $\mu_k \equiv 1.25$ $\omega_k \equiv 0.5$ $W_k \equiv 0.0$	$\lambda_1 = 1.00$ $\lambda_2 = 0.25$ $\lambda_3 = 0.25$ $\mu_k \equiv 4.0$ $\omega_k \equiv 1.0$ $W_k \equiv 0.0$	$\lambda_1 = 0.15$ $\lambda_2 = 0.1$ $\lambda_3 = 0.2$ $\mu_1 = 1.00$ $\mu_2 = 0.5$ $\mu_3 = 0.9$ $\omega_1 = 1.00$ $\omega_2 = 0.5$ $\omega_3 = 2.0$ $W_k \equiv 1.0$	$\lambda_k \equiv 0.25$ $\mu_1 = 1.00$ $\mu_2 = 2.0$ $\mu_3 = 3.0$ $\omega_k \equiv 1.00$ $W_k \equiv 0.0$
ρ	0.6	—	—	0.375	0.572	0.458
N	20	20	20	20	25	20
J_d^Ψ	4.125	NA	6.750	4.275	NA	NA
J^Ψ	3.95	4.56	6.64	3.76	3.27	2.88
Optimal policy	(0, 0, 0) 1 1 1	(0, 0, 0) 1 1 1 (0, 0, 1) 1 0 1 (0, 1, 0) 0 1 1 (1, 0, 0) 1 1 0	(0, 0, 0) 1 1 1 (0, 0, 1) 1 1 1 (0, 1, 0) 1 1 1 (1, 0, 0) 1 1 1	(0, 0, 0) 1 0 1 (0, 0, 1) 1 0 1 (0, 1, 0) 1 1 0	(0, 0, 0) 1 1 1 See text	(0, 0, 0) 1 1 1 See text

the optimal policy is much more rapid than the convergence to J^Ψ. It should be noted that larger values of N might yield a slightly more accurate value of J^Ψ. The optimal policy is indicated only for those states where it deviates from the exhaustive policy d for at least one station. The deviations are indicated by giving the state followed by a triple of numbers, with 0 indicating that it is optimal to walk and 1 that it is optimal to remain at the corresponding station.

Scenario 1 is a symmetric situation with no walking cost rate. The minimum long run average number of customers in the system differs from the average number under d by a modest 4.2%. When the system is empty, it is optimal to remain stationary. This is the only deviation from d, which would have the server cycle until reaching a station with a customer. It might be conjectured that there are deviations from d when the system is extremely imbalanced. However, a moment's thought will convince the reader that this is not so. The reason is that "a bird in the hand is worth two in the bush." That is, there is no incentive for the server to forsake serving a customer in its present location and begin a walk to reach another station where there may be many more customers. Hence the optimal policy for Scenario 1 differs from exhaustive only when the system is empty.

Scenario 2 is identical to Scenario 1 except that there is a cost for walking of 2 per unit time. The value of J^{Ψ} suffers an increase of 13.4%. Interestingly the optimal policy differs slightly from that in Scenario 1. The optimal policy again remains stationary when the system is empty. Note that the three states with 1 customer in the system are symmetric images of each other, as is the indicated optimal policy. The only deviation from d occurs when the server is at an empty station that is two walks away from the station with the customer. In this case it is optimal to remain at that station.

Scenario 3 is identical to Scenario 1 except that the walking rate has been cut in half. This means that the expected time to complete each walk is doubled. In this case the savings over exhaustive is slight. The optimal policy remains stationary both when the system is empty and when it contains exactly 1 customer.

Scenario 4 is a system with identical service rates but imbalanced arrival rates, with station 1 receiving customers at a rate 4 times that of stations 2 or 3. The savings in the minimum average number in the system over d is 12%. The service at station 2 is exhaustive. When there is exactly one customer in the system and that customer is at station 3, then the optimal policy behaves exhaustively at 3 but is stationary at 1. See Fig. 10.2. This is because it anticipates the next customer arriving there rather than at station 2. A similar explanation holds for the remaining exception to d.

Scenarios 5 and 6 consider situations in which the service rates are unequal. For unequal service rates there will typically be massive deviations from d, and

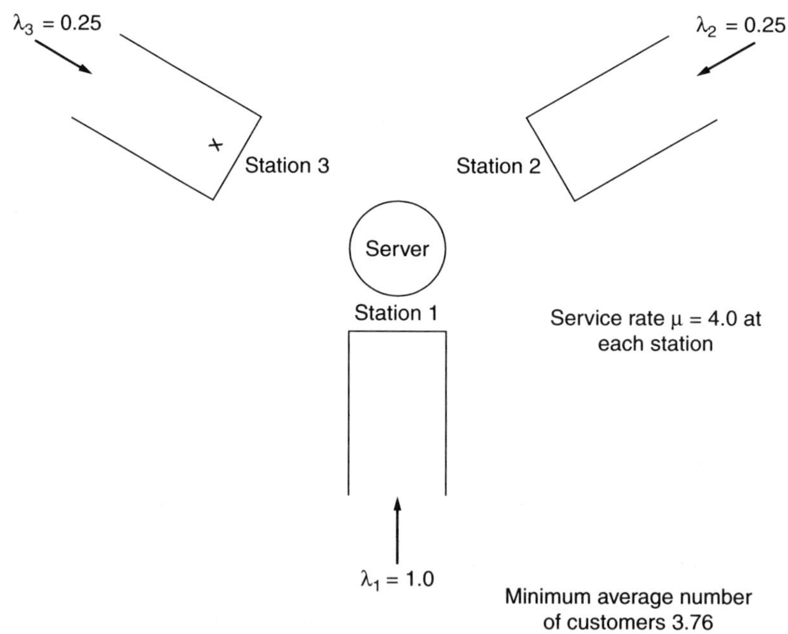

Figure 10.2 Scenario 4 from Table 10.3.

one should not attempt to print the output. Instead, it can be readily scanned to identify the pattern of deviation. For these cases more work should be done to obtain a "user friendly" form for the output.

In Scenario 5 the arrival rates, service rates, and walking rates are all unequal. It costs 1 per unit time when walking. It is optimal to remain stationary when the system is empty. We will indicate the other deviations from d for buffer occupancies up to 10. There are deviations with optimal actions 0 0 1. These occur in states (0, 1 or 2, $\geq$ 9), (0, 3 or 4, $\geq$ 10), (0, 5 or 6, $\geq$ 11), (0, 7 or 8, $\geq$ 12), (0, 9 or 10, $\geq$ 13). Notice that the deviation occurs only at station 2. Its buffer is nonempty, but it is optimal to walk to station 3 when the imbalance reaches a certain level. The reason is that the service rate at 3 exceeds that at 2. So does the arrival rate, but we suspect that this is a smaller factor.

All other deviations (except one) have optimal actions 1 0 1, and these occur in states for which station 1 is nonempty and the imbalance between 2 and 3 exceeds a certain amount. The deviation occurs only at station 2. These states are (1, 1 or 2, $\geq$ 8), (1, 3 or 4, $\geq$ 9), (1, 5 or 6, $\geq$ 10), (1, 7 or 8, $\geq$ 11), (1, 9 or 10, $\geq$ 12), and continuing in a similar fashion as the occupancy of buffer 1 increases. For example, when its occupancy is 4, we have (4, 1 or 2, $\geq$ 5), (4, 3 or 4, $\geq$ 6), (4, 5 or 6, $\geq$ 7), (4, 7 or 8, $\geq$ 8), and (4, 9 or 10, $\geq$ 9). When the state is (10, 1, 0), the actions are 1 0 0, which again is a deviation at station 2. Last we have action 1 0 1 in state (10, 1, 1).

Scenario 6 has equal arrival rates and unequal service rates, and again we see substantial deviations from d. Actions 0 1 1 are optimal in states (1, 0, $\geq$ 9), (1, 1, $\geq$ 8), (1, 2, $\geq$ 7), (1, 3, $\geq$ 6), (1, 4, $\geq$ 5), (1, 5 to 7, $\geq$ 4), (1, 8, $\geq$ 2), (1, 9, $\geq$ 1), and (1, 10, $\geq$ 0). In state (2, 0, $\geq$ 9) action 0 0 1 is optimal. In state (2, 1, $\geq$ 8) action 0 1 1 is optimal. It continues in this fashion until state (9, 0, $\geq$ 12) in which 0 0 1 is optimal and state (9, 1, $\geq$ 11) in which 0 1 1 is optimal.

The interesting conclusion is that we may see substantial deviations from d under unequal service rates, but less deviation when the service rates are equal and the arrival rates are unequal. The intuitive reason is that unequal arrival rates induce only "potential differences" between the stations and cause the optimal policy to exhibit a mild anticipatory effect. However, unequal service rates are "real differences" between the stations and cause much more of an effect.

Much additional work remains to be done to understand the optimal control of polling systems. □

BIBLIOGRAPHIC NOTES

The subject of uncontrolled continuous time queueing systems is a vast one. Kleinrock (1975), Gross and Harris (1998), Cooper (1981), and Wolff (1989) are some standard references.

Jewell (1963) contains foundational material on the control of continuous time systems. The approach we have followed of introducing an auxillary MDC for the CTMDC is due to Schweitzer (1971) who developed it for the finite state space case. Our approach of applying the (AC) assumptions to the auxillary MDC and assuming Assumption (CTAC) is new. Sennott (1989b) develops an existence theory for CTMDCs that may be used to verify Assumption (CTAC) as in Remark 10.3.4.

See also Ross (1970), Lippman (1975b), Serfozo (1979), Puterman (1994), Bertsekas (1987) and (1995, Vol 2), Spieksma (1990), and Kitaev and Rykov (1995). Stidham and Weber (1993) contains a summary of recent results as well as a valuable bibliography. The focus of most of this work is on obtaining structural results for optimal policies rather than on computing optimal policies.

Tijms (1994) presents some material on stochastic dynamic programming, including some computational results. In particular, the model in Section 10.5 is treated. The Schweitzer transformation is applied to obtain an auxillary MDC. The computation is performed by truncating the state space and assuming that if 20 or more customers are present in the system, then all the servers will be turned on. A computation is done with $K = 10, \lambda = 7.0, \mu = 1.0, H = 10.0, D = 10.0$, and $c(k) = 30k$. The computation produces an upper bound on the minimum average cost of 319.5 and a lower bound of 319.3. According to our theory, H, D, and $c(k)$ may be divided by 10 without affecting the optimal policy. Note that this produces our Scenario 7 with an optimal average cost of 31.937. Agreement is sweet!

The literature on polling models is voluminous, and we mention only a few references. A seminal work is Takagi (1986), where the stability criterion under exhaustive service was derived heuristically. It is shown rigorously in Altman et al. (1992) and Georgiadis and Szpankowski (1992). See also Fricker and Jaibi (1994). Takagi (1990, 1997) are useful survey articles containing many references.

The pseudoconservation law employed in Proposition 10.6.1 is due to Boxma and Groenendijk (1987). An equivalent form of this result is in Takagi (1990, p. 278). It is possible to derive the average number in the system under general service rates. This may be done using Little's formula and results giving the expected waiting time at each station. These quantities may be calculated recursively. See Takagi (1997), Cooper et al. (1996), and Srinivasan et al. (1995).

Some results on the control of polling systems are beginning to appear, and a few papers are discussed in Takagi (1997). We mention Browne and Yechiali (1989) and Kim et al. (1996). In the former paper, the control problem is formulated as a semi-Markov decision process and some heuristic rules for minimizing the cycle time are given. In the latter paper, various algorithms are compared for the optimization of a polling system identical to ours except that the buffers are truncated. The control of polling systems is a subject wide open for further research and discovery.

PROBLEMS

10.1. Prove that the exponential distribution is the only continuous distribution with the memoryless property. *Hint:* Use the fact that the only real-valued monotonic function r satisfying $r(x+y) = r(x)r(y)$ for $x, y \geq 0$, and with $r(0) = 1$, is $r(x) = e^{\alpha x}$ for some constant α.

10.2. The random variable X has a $\Gamma(n, \lambda)$ distribution, where $\lambda > 0$ and n is a positive integer, if its density (the derivative of F_X) is given by

$$f_X(x) = \frac{\lambda}{(n-1)!} e^{-\lambda x}(\lambda x)^{n-1}, \qquad x \geq 0.$$

Let $X_1, X_2, \ldots, X_n$ be independent exp(λ) random variables. Prove that $X = X_1 + X_2 + \ldots + X_n$ has a $\Gamma(n, \lambda)$ distribution. *Hint:* Prove this by induction on n. Use a conditioning technique similar to that in (10.9) to obtain an expression for $P(X > y)$. Then differentiate $P(X \leq y)$ to obtain the density.

10.3. Let W_n be the waiting time until the nth arrival in a PoisP(λ). Show that W_n has a $\Gamma(n, \lambda)$ distribution.

10.4. Assume that customer arrivals to a system follow a PoisP(λ). Show that the number of customers arriving in $[0, t]$ has a Poisson distribution with parameter λt. *Hint:* Calculate the probability of n customers arriving by conditioning on the value of W_n from Problem 10.3. It is also the case that the number of customers arriving in any interval of length t has the same distribution. Argue informally why this should be true.

10.5. Develop a CTMDC model for Example 10.2.1 if there is an instantaneous cost for initiating a service at rate a. Assume that if a new customer arrives, then another rate may be chosen and another instantaneous cost incurred (even if the same rate is selected).

10.6. Develop a CTMDC model for Example 10.2.1 if there is an instantaneous cost for *changing* the service rate. Assume that the action set A also applies to state 0 so that a rate may be chosen (or remain in effect) in anticipation of the next arrival.

10.7. Consider an M/M/1 queue with service rate μ and controllable arrival rate. All customers are admitted. Just after a new customer arrives or just after a service completion, the controller chooses from action set $\{1, 2, \ldots, K\}$, where action k means that the time until the next customer arrival follows a PoisP(λ_k) process. Assume that there is a nonnegative instantaneous cost $C(k)$ associated with action k as well as a cost rate $c(k)$ and a holding cost rate $H(i)$ as in Example 10.2.1. Model this system as a CTMDC.

10.8. Model the continuous time version of the routing to parallel queues example treated in Section 8.6 as a CTMDC. Assume that a holding cost rate of $H_1(i) + H_2(j)$ is charged when there are i customers in the first queue and j customers in the second queue.

10.9. Consider a CTMDC on $\{1, 2, 3, \ldots\}$ with one action in each state and $P_{i,i+1} \equiv 1$. Assume that

$$g(i) = v(i) = \begin{cases} 1, & i \text{ odd}, \\ 2, & i \text{ even}. \end{cases}$$

Calculate $J^{\Psi}(1)$.

10.10. Complete the proof of Lemma 10.3.2.

10.11. Run ProgramSeven for the following scenarios. Each one is as in Scenario 3 of Table 10.1 except that the cost of fastest service is changed to the value indicated. Discuss your results.

(a) 140.0

(b) 125.0

(c) 110.0

10.12. In ProgramSeven show that if λ and each service rate a are multiplied by the same positive constant, then the optimal policy and minimum average cost are unchanged. *Hint:* Prove this by induction on n using (10.27). What is the relation of the new value of τ to the old value?

10.13. Run ProgramSeven for the scenarios below and discuss the results. Each scenario has the value of λ followed by the three service rates and their respective costs:

(a) $\lambda = 1.0$; $a = 0.9, 1.2, 1.5$; $c(a) = 1.0, 3.0, 6.0$.

(b) $\lambda = 0.5$; $a = 0.5, 0.75, 1.0$, $c(a) = 0.0, 5.0, 10.0$.

(c) $\lambda = 8.0$; $a = 7.0, 8.0, 9.0$, $c(a) = 0.0, 20.0, 40.0$.

***10.14.** Derive the expression in (10.35).

10.15. Run ProgramEight for Scenarios 3 through 7 in Scenarios 10.5.4, and verify the positive recurrent class for the optimal policy in each case. Make additional runs of your choice, and discuss the results.

10.16. Consider the situation in Proposition 10.6.1, and assume that we have a second system with the same parameters except that each walking time parameter is cut in half. Let $J_d^{\Psi+}$ be the average number of customers in the second system. Prove that $J_d^{\Psi+} = 2J_d^{\Psi} - \rho/(1-\rho)$.

10.17. Consider the situation in Proposition 10.6.1, and assume that we have a second system with the same parameters except that the arrival rates, as well as the common service rate, are doubled (so that ρ remains constant). Let $J_d^{\Psi+}$ be the average number of customers in the second system. Prove that the expression in Problem 10.16 also holds in this case.

10.18. Run ProgramNine for the following scenarios and discuss your results. In each scenario except (d) set $H_k \equiv 1.0$.

(a) $\lambda_k \equiv 0.5$, $\mu_k \equiv 3$, $\omega_k \equiv 0.75$, $W_k \equiv 1.0$.

(b) $\lambda_1 = 5.0$, $\lambda_2 = 0.5$, $\lambda_3 = 1.0$, $\mu_k \equiv 10$, $\omega_k \equiv 5.0$, $W_k \equiv 0.0$. Let N be 25 or 30.

(c) This system is as in Scenario 5 of Table 10.3 except that $W_k \equiv 0.0$.

(d) This system is as in (a) except that $H_2 = 3.0$.

APPENDIX A

Results from Analysis

Certain results from analysis are used repeatedly throughout the book and are collected here for the convenience of the reader. Standard statements and proofs of some of these results involve measure theory. However, the material in this book does not require that level of generality. For this reason all the proofs provided here are tailored to our special case. The proofs are not requisite to an understanding of the text and may be omitted.

Sections A.1 and A.2 contain the most frequently used theorems from analysis. Section A.3 contains basic material on power series. Section A.4 contains an important Tauberian theorem that provides a link between the infinite horizon discounted cost criterion and the average cost criterion. Section A.5 contains an example illustrating this theorem.

A.1 USEFUL THEOREMS

In this section a collection of useful results is presented.

Proposition A.1.1. Let $(q(a))_{a \in A}$ be a probability distribution on the finite (nonempty) set A. Let $u: A \to (-\infty, \infty]$ be a function. Then $\sum_{a \in A} q(a)u(a) \geq \min_{a \in A} \{u(a)\}$, and equality occurs if and only if the probability distribution is concentrated on the subset $B = \{b \in A | u(b) = \min_{a \in A} \{u(a)\}\}$.

Proof: (Recall the convention that $0 \cdot \infty = 0$. So any terms with $q(a) = 0$ may be discarded. The distribution is concentrated on B if $q(a) = 0$ for $a \notin B$. To simplify notation, the subscripts on the minimum and summation are omitted.) If $u \equiv \infty$, then it is easily seen that the claims hold.

Now assume that $\min \{u(a)\} = w < \infty$. Then $u(a) \geq w$, and hence $\sum q(a)u(a) \geq w \sum q(a) = w$. This proves the first statement.

Since the minimization is over a finite set, the set B of minimizing actions must be nonempty. If q is concentrated on B, then it is clear that we have equality. Now let us assume that there exists $a^* \in A - B$ such that $q(a^*) > 0$.

We show that equality cannot hold. Let $u(a^*) = w + \delta$, where $\delta > 0$. Then $\sum q(a)u(a) \geq w \sum_{a \neq a^*} q(a) + (w + \delta)q(a^*) = w + q(a^*)\delta > w$. □

Let us informally review what is meant by the *limit infimum* (respectively, *limit supremum*) of a sequence of extended real-valued numbers. The limit infimum (respectively, supremum) is the smallest (respectively, largest) limit point of the sequence. The *limit* exists if and only if the limit infimum equals the limit supremum, and the limit is then this quantity.

Consider the sequence $0, \frac{1}{2}, 0, \frac{2}{3}, 0, \frac{3}{4}, 0, \frac{4}{5}, \ldots$. The limit infimum equals 0, and the limit supremum is the limit of the subsequence $\frac{1}{2}, \frac{2}{3}, \frac{3}{4}, \ldots$, which equals 1. Since $0 < 1$, the limit does not exist. For the sequence $5, \infty, 5, 5, -1, 5, 5, 5, -2, 5, 5, 5, 5, -3, \ldots$, the limit supremum equals 5 and the limit infimum equals $-\infty$.

An alternative definition of the limit supremum of the sequence u_n is $\limsup_{n \to \infty} \{u_n\} = \lim_{M \to \infty} \sup_{n \geq M} \{u_n\}$, with a similar definition for the limit infimum. It can be seen that the two definitions agree.

Remark A.1.2. Section A.1 and A.2 deal with various functions $u(., N)$. These functions are always assumed to be defined for integers $N \geq N_0$, where N_0 is some nonnegative integer. We sometimes deal with sequences S_N of sets, and likewise these are assumed to be defined for $N \geq N_0$. □

The next result shows that a limit infimum may be passed through a minimization over a finite set.

Proposition A.1.3. Let A be a finite (nonempty) set and $u(a, N)$ an extended real-valued function of $a \in A$ and N.

(i) Then $\liminf_{N \to \infty} \min_{a \in A} \{u(a, N)\} = \min_{a \in A} \{\liminf_{N \to \infty} u(a, N)\}$.

(ii) If $\lim_{N \to \infty} u(a, N)$ exists for every a, then $\lim_{N \to \infty} \min_{a \in A} \{u(a, N)\} = \min_{a \in A} \{\lim_{N \to \infty} u(a, N)\}$.

Proof: (To simplify notation, drop the subscript on min and let $\to \infty$ be understood.) To prove (i), observe that $\min\{u(a, N)\} \leq u(a, N)$. Hence $\liminf_N \min\{u(a, N)\} \leq \liminf_N u(a, N)$. This implies that

$$\liminf_{N} \min\{u(a, N)\} \leq \min\{\liminf_{N} u(a, N)\}. \tag{A.1}$$

We need to show that (A.1) is an equality.

Consider two cases. First suppose that $\liminf_N \min\{u(a, N)\} = -\infty$. This means that there exists a subsequence N_r such that $\lim_r \min\{u(a, N_r)\} = -\infty$. Since A is a finite set, there must exist a^* and a subsequence of N_r (call it N_s for notational convenience) such that $u(a^*, N_s) = \min\{u(a, N_s)\}$ for all s. This implies that $\lim_s u(a^*, N_s) = -\infty$. This clearly implies that equality holds in (A.1).

Now assume that $\liminf_N \min\{u(a,N)\} > -\infty$ and that equality fails. This implies that there exist a sequence N_r and $\epsilon > 0$ such that $\min\{u(a,N_r)\} + \epsilon \leq \min\{\liminf_N u(a,N)\}$. Since A is a finite set, there must exist a^* and a subsequence N_s such that $u(a^*,N_s) = \min\{u(a,N_s)\}$. This is easily seen to yield a contradiction. Hence equality must hold in (A.1).

The proof of (ii) is omitted. □

Example A.1.4. This example shows that A.1.3(i) does not hold with min replaced by max. Let $A = \{a_1, a_2\}$, and define $u(a_1,N)$ to equal 0 for N even and 1 for N odd (whereas $u(a_2,N)$ equals 1 for N even and 0 for N odd). Then $\liminf_N u(.,N) = 0$. Hence $\max\{\liminf_N u(.,N)\} = 0$. Now $\max\{u(.,N)\} \equiv 1$, and hence $\liminf_N \max\{u(.,N)\} = 1$. □

The next result shows that the limit infimum of a finite sum of terms equals or exceeds the sum of the limit infimum of each term.

Proposition A.1.5. Let G be a finite (nonempty) set and $u(j,N)$ a function of $j \in G$ and N with values in $(-\infty, \infty]$. Then

$$\liminf_{N\to\infty} \sum_{j\in G} u(j,N) \geq \sum_{j\in G} (\liminf_{N\to\infty} u(j,N)) \tag{A.2}$$

under the condition that there is no indeterminate form in the summation on the right of (A.2).

Proof: (An indeterminate form occurs in a summation if one summand equals ∞ and another equals $-\infty$. The notation is simplified by omitting the index of summation and $\to \infty$.) The condition on u implies that some values may be ∞ but none can be $-\infty$. Hence an indeterminate form cannot occur in the summation on the left of (A.2). Let $\liminf_N u(.,N) = u(.)$.

Consider three cases. First assume that $u(j^*) = -\infty$ for some j^*. Avoiding an indeterminate form on the right means that $u(j) < \infty$ for $j \neq j^*$. Then $\sum u(j) = -\infty$, and the result holds.

Next assume that $u(j^*) = \infty$ for some j^*. Avoiding an indeterminate form on the right means that $u(j) > -\infty$ for $j \neq j^*$. Let $H = \{j | u(j) = \infty\}$. There exists N^* such that $u(j,N) \geq u(j) - 1$ for $N \geq N^*$ and $j \in G - H$. Recall that $|G - H|$ denotes the cardinality of the set. Then $\sum u(j,N) = \sum_{j\in H} u(j,N) + \sum_{j\notin H} u(j,N) \geq \sum_{j\in H} u(j,N) + \sum_{j\notin H} u(j) - |G - H|$. Taking the limit infimum of both sides yields $\liminf_N \sum u(j,N) = \infty$, and the result holds.

Finally assume that u is finite-valued, and let $\sum u(j) = U$ and $\epsilon > 0$. There exists N^* such that $u(j,N) \geq u(j) - \epsilon/|G|$ for $N \geq N^*$. Then $\sum u(j,N) \geq U - \epsilon$ for $N \geq N^*$. Taking the limit infimum of both sides and using the fact that ϵ is arbitrary yields the result. □

Example A.1.6. This shows that the inequality in (A.2) may be strict. Let $G = \{j, j^*\}$. Define $u(j, N)$ to be 0 for N even and 1 for N odd, while $u(j^*, N)$ is 1 for N even and 0 for N odd. Then it is easily seen that the right side of (A.2) is 0, whereas the left side is 1. □

The next result generalizes Proposition A.1.5 for the case of a nonnegative function.

Proposition A.1.7. Let S be a countable set and $u(j, N)$ a function of $j \in S$ and N with values in $[0, \infty]$. Then

$$\liminf_{N \to \infty} \sum_{j \in S} u(j, N) \geq \sum_{j \in S} (\liminf_{N \to \infty} u(j, N)). \tag{A.3}$$

Proof: Recall that the sum of an infinite series is defined as the limit of its sequence of partial sums if that limit exists. Since the terms of the series on the right of (A.3) are nonnegative, the sequence of partial sums is increasing. Hence the limit exists (it may be ∞). To prove (A.3), it is sufficient to show that

$$\liminf_{N} \sum_{j \in S} u(j, N) \geq \sum_{j \in G} (\liminf_{N} u(j, N)), \tag{A.4}$$

where G is an arbitrary finite subset of S.

Now

$$\begin{aligned} \liminf_{N} \sum_{j \in S} u(j, N) &\geq \liminf_{N} \sum_{j \in G} u(j, N) \\ &\geq \sum_{j \in G} (\liminf_{N} u(j, N)). \end{aligned} \tag{A.5}$$

The first line follows from the nonnegativity of u and the second line from Proposition A.1.5. This completes the proof. □

The next result is a variant of Proposition A.1.7.

Proposition A.1.8. Let S be a countable set and (S_N) an increasing sequence of subsets of S such that $\cup S_N = S$. Let $u(j, N)$ be a function of $j \in S_N$ (or of $j \in S$) and N taking values in $[0, \infty]$. Then

$$\liminf_{N\to\infty} \sum_{j\in S_N} u(j,N) \geq \sum_{j\in S} (\liminf_{N\to\infty} u(j,N)). \tag{A.6}$$

Proof: Define the function u^* by

$$u^*(j,N) = \begin{cases} u(j,N), & j \in S_N, \\ 0, & j \in S - S_N, \end{cases} \tag{A.7}$$

and note that $\liminf_N u^*(.,N) = \liminf_N u(.,N)$. Then

$$\sum_{j\in S_N} u(j,N) = \sum_{j\in S} u^*(j,N), \tag{A.8}$$

and the result follows from Proposition A.1.7. □

Example A.1.9. This shows that Proposition A.1.8 may fail if u can take on negative values. Let $S = \{1, 2, \ldots\}$ and $S_N = \{1, 2, \ldots, 2N\}$. Let $u(j,N)$ equal 0 for $1 \leq j \leq N$, and equal -1 for $N < j \leq 2N$. Then $\liminf_N u(.,N) \equiv 0$, and hence the right side of (A.6) is 0. However, $\sum_{S_N} u(j,N) = -N$, and hence the left side of (A.6) is $-\infty$. □

Proposition A.1.10. Let $(u_n)_{n\geq 0}$ be a sequence of real numbers, and let $w_n = \sum_0^{n-1} u_k$, for $n \geq 1$. Then

$$\liminf_{n\to\infty} u_n \leq \liminf_{n\to\infty} \frac{w_n}{n} \leq \limsup_{n\to\infty} \frac{w_n}{n} \leq \limsup_{n\to\infty} u_n. \tag{A.9}$$

Proof: We prove the leftmost inequality. Fix a positive integer M. Then for $n > M$ we have

$$\begin{aligned} w_n &= \sum_{k=0}^{M-1} u_k + \sum_{k=M}^{n-1} u_k \\ &\geq \sum_{k=0}^{M-1} u_k + (n-M) \inf_{M\leq k\leq n-1} \{u_k\} \\ &\geq \sum_{k=0}^{M-1} u_k + (n-M) \inf_{k\geq M} \{u_k\}. \end{aligned} \tag{A.10}$$

Let us divide both sides of (A.10) by n and then take the limit infimum as $n \to \infty$. This yields $\liminf_n w_n/n \geq \inf_{k \geq M} \{u_k\}$. Then let $M \to \infty$ to obtain the result.

The proof of the rightmost inequality is similar. □

A.2 FATOU'S LEMMA AND THE DOMINATED CONVERGENCE THEOREM

We use the results in Section A.1 to prove several famous results.

Proposition A.2.1 (Fatou's Lemma). Let S be a countable set and $(P_j)_{j \in S}$ a probability distribution. Let $u(j, N)$ be a function of $j \in S$ and N, taking values in $[-L, \infty]$ for some nonnegative (finite) constant L. Then

$$\liminf_{N \to \infty} \sum_{j \in S} P_j u(j, N) \geq \sum_{j \in S} P_j (\liminf_{N \to \infty} u(j, N)). \tag{A.11}$$

Proof: Recall that $0 \cdot \infty = 0$. Hence, if any $P_j = 0$, then that term may be discarded. So assume that $P_j > 0$ for all j.

Let $r(j, N) = u(j, N) + L$, and note that $r \geq 0$. We have

$$\sum_{j \in S} P_j u(j, N) = \sum_{j \in S} P_j r(j, N) - L. \tag{A.12}$$

Then from Proposition A.1.7 it follows that

$$\begin{aligned} \liminf_N \sum_{j \in S} P_j u(j, N) &= \liminf_N \sum_{j \in S} P_j r(j, N) - L \\ &\geq \sum_{j \in S} (\liminf_N r(j, N)) - L \\ &= \sum_{j \in S} P_j [\liminf_N u(j, N) + L] - L \\ &= \sum_{j \in S} P_j (\liminf_N u(j, N)). \end{aligned} \tag{A.13}$$

This completes the proof. □

The next example shows that (A.11) may fail if the function is unbounded below.

Example A.2.2. Let $S = \{1, 2, \dots\}$, and let $P_j = 1/2^j$. In a moment we will need the fact that $\sum_{j=N+1}^{\infty} 1/2^j = 1/2^N$. Define $u(j,N)$ to equal 0 for $j \le N$ and to equal -2^{2N} for $j > N$. For j fixed, observe that $\liminf_N u(j,N) = 0$ and hence that the right side of (A.11) is 0. But $\sum P_j u(j,N) = -2^{2N} (\sum_{j=N+1}^{\infty} 1/2^j) = -2^N$. Hence the left side of (A.11) is $-\infty$. □

The following result gives a sufficient condition for passing a limit through an infinite summation:

Theorem A.2.3 (Dominated Convergence Theorem). Assume that the following hold:

(i) S is a countable set with probability distribution $(P_j)_{j \in S}$.
(ii) $u(j,N)$ and $w(j,N)$ are finite functions of $j \in S$ and N such that $|u| \le w$.
(iii) $\text{Lim}_{N\to\infty} u(.,N) = u(.)$ and $\lim_{N\to\infty} w(.,N) = w(.)$ exist.
(iv) $\text{Lim}_{N\to\infty} \sum_{j\in S} P_j w(j,N)$ exists and equals $\sum_{j\in S} P_j w(j) < \infty$.

Then $\lim_{N\to\infty} \sum_{j\in S} P_j u(j,N)$ exists and equals $\sum_{j\in S} P_j u(j)$.

Proof: Fatou's lemma may be employed to give a simple proof of this result. Let $\sum_{j\in S} P_j w(j) = W$. Since $|u| \le w$, it follows that $\sum_{j\in S} P_j u(j) = U$ exists and $|U| \le W$.

Note that $w + u \ge 0$. Applying Proposition A.2.1 to this function yields

$$\liminf_N \sum_j P_j(w(j,N) + u(j,N)) \ge \sum_j P_j(w(j) + u(j))$$
$$= W + U. \tag{A.14}$$

But note that

$$\liminf_N \sum_j P_j(w(j,N) + u(j,N))$$
$$= \liminf_N \left(\sum_j P_j w(j,N) + \sum_j P_j u(j,N) \right)$$
$$= W + \liminf_N \sum_j P_j u(j,N), \tag{A.15}$$

since the limit of the first term on the right exists. Then (A.14–15) imply that $\liminf_N \sum_j P_j u(j,N) \ge U$.

We also have $w - u \geq 0$. Applying Proposition A.2.1 to this function yields

$$\liminf_N \sum_j P_j(w(j,N) - u(j,N)) \geq \sum_j P_j(w(j) - u(j)) = W - U. \tag{A.16}$$

But note that

$$\begin{aligned}\liminf_N \sum_j P_j(w(j,N) - u(j,N)) &= \liminf_N \left(\sum_j P_j w(j,N) - \sum_j P_j u(j,N) \right) \\ &= W - \limsup_N \sum_j P_j u(j,N),\end{aligned} \tag{A.17}$$

Then (A.16–17) imply that $\limsup_N \sum_j P_j u(j,N) \leq U$. This proves the result. □

An important special case of the dominated convergence theorem occurs when the function $w(j,N)$ is independent of N.

Corollary A.2.4. Assume that the following hold:

(i) S is a countable set with probability distribution $(P_j)_{j \in S}$.
(ii) $u(j,N)$ is a function of $j \in S$ and N such that $\lim_{N \to \infty} u(.,N) = u(.)$ exists.
(iii) w is a finite function on S such that $|u| \leq w$ and $\sum_{j \in S} P_j w(j) < \infty$.

Then $\lim_{N \to \infty} \sum_{j \in S} P_j u(j,N)$ exists and equals $\sum_{j \in S} P_j u(j)$.

We now treat the counterparts of Fatou's lemma and the dominated convergence theorem for the case in which the probability distribution may also be a function of N.

Proposition A.2.5 (Generalized Fatou's Lemma). Assume that the following hold:

(i) S is a countable set with probability distribution $(P_j)_{j \in S}$.
(ii) (S_N) is an increasing sequence of subsets of S such that $\cup S_N = S$.

(iii) $(P_j(N))_{j \in S_N}$ is a probability distribution on S_N satisfying $\lim_{N \to \infty} P_j(N) = P_j$ for $j \in S$.

(iv) $u(j, N)$ is a function of $j \in S_N$ and N taking values in $[-L, \infty]$, for some nonnegative (finite) constant L.

Then

$$\liminf_{N \to \infty} \sum_{j \in S_N} P_j(N)u(j,N) \geq \sum_{j \in S} P_j(\liminf_{N \to \infty} u(j,N)). \tag{A.18}$$

Proof: Let $r(j,N) = u(j,N) + L$. Then

$$\sum_{j \in S_N} P_j(N)u(j,N) = \sum_{j \in S_N} P_j(N)r(j,N) - L. \tag{A.19}$$

The proof follows in a manner similar to (A.13), using Proposition A.1.8. □

Theorem A.2.6 (Generalized Dominated Convergence Theorem). Assume that (i–iii) from Proposition A.2.5 hold, and in addition assume the following:

(iv) There exist finite functions $u(j,N)$ and $w(j,N)$ of $j \in S_N$ and N such that $|u| \leq w$.

(v) $\text{Lim}_{N \to \infty} u(.,N) = u(.)$ and $\lim_{N \to \infty} w(.,N) = w(.)$ exist.

(vi) $\text{Lim}_{N \to \infty} \sum_{j \in S_N} P_j(N)w(j,N)$ exists and equals $\sum_{j \in S} P_j w(j) < \infty$.

Then $\lim_{N \to \infty} \sum_{j \in S_N} P_j(N)u(j,N)$ exists and equals $\sum_{j \in S} P_j u(j)$.

Proof: A proof can be given using the generalized Fatou's lemma and following the ideas in the proof of the dominated convergence theorem. □

An important special case occurs when the function $w(j,N)$ is a constant.

Corollary A.2.7. Assume that (i–iii) from Proposition A.2.5 hold and in addition that:

(iv) There exists a function $u(j,N)$ of $j \in S_N$ and N such that $\lim_{N \to \infty} u(.,N) = u(.)$ exists, and

(v) there exists a (finite) constant w such that $|u| \leq w$.

Then $\lim_{N \to \infty} \sum_{j \in S_N} P_j(N)u(j,N)$ exists and equals $\sum_{j \in S} P_j u(j)$.

A.3 POWER SERIES

This section presents some elementary facts about power series. For the proofs of these results, the reader should consult a book on analysis such as Apostol (1974).

Let $\alpha \in [0, \infty)$, and let u_n be a sequence of nonnegative terms with $u_0 < \infty$. The series

$$U(\alpha) = \sum_{n=0}^{\infty} \alpha^n u_n \tag{A.20}$$

is a *power series* (about the origin). Note that we consider only power series with nonnegative terms. Since the terms are nonnegative, it is the case that the sequence of partial sums is increasing and hence the sum $U(\alpha)$ always exists (it may be ∞). (Note that $U(\alpha)$ denotes both the series itself and its sum. This is a regrettable notational confusion that is enshrined in mathematical history.)

We are interested in determining those values of α for which the sum $U(\alpha) < \infty$; in this case we say that the series *converges*. It is the case that (A.20) converges (to $u_0 < \infty$) for $\alpha = 0$. The number

$$R = \left(\limsup_{n \to \infty} \sqrt[n]{u_n} \right)^{-1} \in [0, \infty] \tag{A.21}$$

is the *radius of convergence* of the power series. If $R = 0$, then (A.20) converges only for $\alpha = 0$. If $R = \infty$, then (A.20) converges for $\alpha \in [0, \infty)$. If $0 < R < \infty$, then (A.20) converges for $\alpha \in [0, R)$ and diverges to ∞ for $\alpha \in (R, \infty)$. Its status for $\alpha = R$ must be checked.

Remark A.3.1. Let us assume that $R > 0$. Then $U(\alpha)$ is a differentiable function of $\alpha \in (0, R)$. Its derivative is the power series

$$\frac{dU(\alpha)}{d\alpha} = \sum_{n=1}^{\infty} n\alpha^{n-1} u_n \tag{A.22}$$

which is obtained by differentiating (A.20) term by term. The amazing result is that the radius of convergence of (A.22) is also R. This can be seen from (A.21). Hence this procedure can be repeated on (A.22) as many times as one wishes to find higher derivatives. The radius of convergence never changes. □

Remark A.3.2. Let $U(\alpha)$ (respectively, $W(\alpha)$) be a power series with radius of convergence $R_1 > 0$ (respectively, $R_2 > 0$). Their product is the power series

$$\begin{aligned} U(\alpha)W(\alpha) &= (u_0 + \alpha u_1 + \alpha^2 u_2 + \ldots)(w_0 + \alpha w_1 + \alpha^2 w_2 + \ldots) \\ &= u_0 w_0 + \alpha(u_0 w_1 + u_1 w_0) + \alpha^2(u_0 w_2 + u_1 w_1 + u_2 w_0) + \ldots \end{aligned} \quad (A.23)$$

which has radius of convergence $R = \min\{R_1, R_2\}$ and converges to the product of the individual sums on $[0, R)$. □

Remark A.3.3. The best-known and most important power series is the *geometric series*, obtained when $u_n \equiv B$, for some (finite) positive B. In this case the radius of convergence is $R = 1$, and we have

$$\begin{aligned} U(\alpha) &= B(1 + \alpha + \alpha^2 + \alpha^3 + \ldots) \\ &= \frac{B}{1 - \alpha}, \qquad \alpha \in [0, 1). \end{aligned} \quad (A.24)$$

A useful related formula is

$$B(\alpha + 2\alpha^2 + 3\alpha^3 + \ldots) = \frac{B\alpha}{(1 - \alpha)^2}, \qquad \alpha \in [0, 1). \quad (A.25)$$

This is obtained from (A.24) by differentiating the power series and then multiplying through by α. □

A.4 A TAUBERIAN THEOREM

In this section we prove an important result for power series. In the theory of Markov decision chains, this result provides a crucial link between the infinite horizon discounted cost and average cost optimization criteria. The reader need only understand the statement of Theorem A.4.2. The rest of the material in this section is starred.

The following lemma is used in the proof of Theorem A.4.2. It involves the function $r(\alpha)$ whose graph appears in Fig. A.1. This function has a jump discontinuity at e^{-1}. Note that

$$\int_0^1 r(x)\, dx = \int_{e^{-1}}^1 \frac{dx}{x} = 1. \quad (A.26)$$

***Lemma A.4.1.** Given $\epsilon > 0$, there exist continuous functions $s(\alpha)$ and $s^*(\alpha)$ for $\alpha \in (0, 1)$ such that $s^* \le r \le s$ and

$$1 - \epsilon \le \int_0^1 s^*(x)\, dx \le \int_0^1 s(x)\, dx \le 1 + \epsilon. \quad (A.27)$$

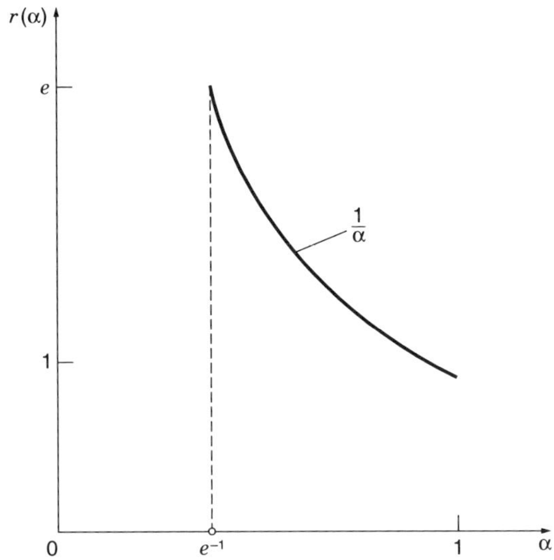

Figure A.1 Graph of $r(\alpha)$.

Proof: The function s is indicated in Fig. A.2, and the function s^* in Fig. A.3. Clearly we have $s^* \leq r \leq s$. It is easy to see that (A.27) will hold for appropriate choices of δ and γ. The details are omitted. □

Here is the fundamental result. It is called a Tauberian theorem after the mathematician A. Tauber (1866–1947), who studied results of this type.

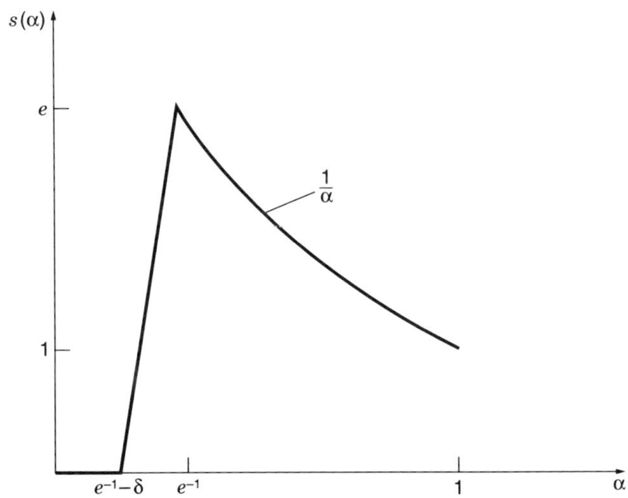

Figure A.2 Graph of $s(\alpha)$.

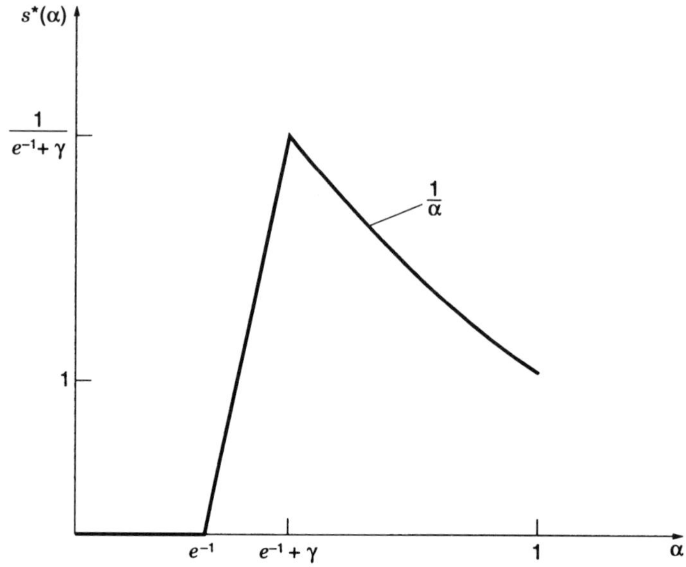

Figure A.3 Graph of $s^*(\alpha)$.

Theorem A.4.2. Let $U(\alpha)$ be a power series as defined in (A.20). Let $w_n = \sum_0^{n-1} u_k$ for $n \geq 1$. Then

$$\liminf_{n\to\infty} \frac{w_n}{n} \leq \liminf_{\alpha\to 1^-} (1-\alpha)U(\alpha) \leq \limsup_{\alpha\to 1^-} (1-\alpha)U(\alpha) \leq \limsup_{n\to\infty} \frac{w_n}{n}. \tag{A.28}$$

The following statements are equivalent:

(i) All the terms in (A.28) are equal and finite.
(ii) $\text{Lim}_{n\to\infty} w_n/n$ exists and is finite.
(iii) $\text{Lim}_{\alpha\to 1^-} (1-\alpha)U(\alpha)$ exists and is finite.

**Proof:* We first take care of a special case. Assume that $u_{n_0} = \infty$ for some n_0. Then $U(\alpha) = \infty$, and so the middle terms of (A.28) are both ∞. Moreover $w_n = \infty$ for $n \geq n_0 + 1$. This implies that the outer terms are both ∞. Thus (A.28) holds in this case, with all terms equal to ∞.

Now assume that $u_n < \infty$ for all n. Let R be the radius of convergence of $U(\alpha)$. We consider two cases.

First assume that $R < 1$. Then $U(\alpha) = \infty$ for $\alpha \in (R, 1)$. This implies that the middle terms of (A.28) are both ∞. It follows from (A.21) that $\limsup_{n\to\infty} \sqrt[n]{u_n} > 1$. This implies that there exist $\epsilon > 0$ and a subsequence n_k such that

$$(u_{n_k})^{1/n_k} \geq 1 + \epsilon, \qquad \text{all } n_k. \tag{A.29}$$

Then $u_{n_k} \geq (1 + \epsilon)^{n_k}$. And hence

$$\frac{(1 + \epsilon)^{n_k}}{n_k + 1} \leq \frac{u_{n_k}}{n_k + 1} \leq \frac{w_{n_k + 1}}{n_k + 1}. \tag{A.30}$$

As we let $k \to \infty$, the term on the left of (A.30) approaches ∞. This implies that the rightmost term in (A.28) equals ∞, and this proves that (A.28) holds.

Now assume that $R \geq 1$. It follows from Remarks A.3.2–3 that

$$\begin{aligned}\left(\sum_{n=0}^{\infty} \alpha^n\right)\left(\sum_{n=0}^{\infty} \alpha^n u_n\right) &= (1 + \alpha + \alpha^2 + \ldots)(u_0 + \alpha u_1 + \alpha^2 u_2 + \ldots) \\ &= u_0 + \alpha(u_0 + u_1) + \alpha^2(u_0 + u_1 + u_2) + \ldots \\ &= \sum_{n=0}^{\infty} \alpha^n w_{n+1},\end{aligned} \tag{A.31}$$

and this power series converges to $U(\alpha)/(1 - \alpha)$ for $\alpha \in [0, 1)$.

Therefore for $\alpha \in [0, 1)$, and for any positive integer M we have

$$\begin{aligned}(1 - \alpha)U(\alpha) &= (1 - \alpha)^2 \sum_{n=0}^{\infty} \alpha^n w_{n+1} \\ &= (1 - \alpha)^2 \sum_{n=0}^{M-1} \alpha^n w_{n+1} + (1 - \alpha)^2 \sum_{n=M}^{\infty} (n + 1)\alpha^n \left(\frac{w_{n+1}}{n + 1}\right) \\ &\leq (1 - \alpha)^2 \sum_{n=0}^{M-1} \alpha^n w_{n+1} + \sup_{n \geq M} \left(\frac{w_{n+1}}{n + 1}\right) (1 - \alpha)^2 \sum_{n=0}^{\infty} (n + 1)\alpha^n \\ &= (1 - \alpha)^2 \sum_{n=0}^{M-1} \alpha^n w_{n+1} + \sup_{n \geq M} \left(\frac{w_{n+1}}{n + 1}\right).\end{aligned} \tag{A.32}$$

Eq. (A.32) yields $\limsup_{\alpha \to 1^-} (1 - \alpha)U(\alpha) \leq \sup_{n \geq M} w_{n+1}/(n + 1)$. Then letting $M \to \infty$ yields the rightmost inequality in (A.28).

It remains to show that the leftmost inequality in (A.28) holds. From the second equality in (A.32), it follows that

$$\begin{aligned}
(1-\alpha)U(\alpha) &\geq (1-\alpha)^2 \sum_{n=M}^{\infty} (n+1)\alpha^n \left(\frac{w_{n+1}}{n+1} \right) \\
&\geq \inf_{n \geq M} \left(\frac{w_{n+1}}{n+1} \right) (1-\alpha)^2 \left[\sum_{n=0}^{\infty} (n+1)\alpha^n - \sum_{n=0}^{M-1} (n+1)\alpha^n \right] \\
&= \inf_{n \geq M} \left(\frac{w_{n+1}}{n+1} \right) \left[1 - (1-\alpha)^2 \sum_{n=0}^{M-1} (n+1)\alpha^n \right]. \qquad \text{(A.33)}
\end{aligned}$$

Eq. (A.33) yields $\liminf_{\alpha \to 1^-} (1-\alpha)U(\alpha) \geq \inf_{n \geq M} w_{n+1}/(n+1)$. Then letting $M \to \infty$ yields the leftmost inequality in (A.28). This completes the proof of (A.28).

It is clearly the case that (i) $\Leftrightarrow$ (ii) $\Rightarrow$ (iii). So to complete the proof, it remains to show that (iii) implies (ii). We give an elegant but nonelementary proof due to Karamata (see Titchmarsh, 1939).

Let $f(\alpha)$ be an integrable function of $\alpha \in (0,1)$, and let $U_f(\alpha)$ be the series

$$U_f(\alpha) := \sum_{n=0}^{\infty} \alpha^n u_n f(\alpha^n). \qquad \text{(A.34)}$$

Note that this is not necessarily a power series, but for each $\alpha \in (0,1)$ it is a series of real numbers. Let $\lim_{\alpha \to 1^-} (1-\alpha)U(\alpha) := L < \infty$, and consider the statement

$$\lim_{\alpha \to 1^-} (1-\alpha)U_f(\alpha) = L \int_0^1 f(x)\, dx. \qquad \text{(A.35)}$$

We will prove that (A.35) holds for polynomial functions, then for continuous functions, and then finally for the function r from Fig. A.1.

Let $p(\alpha)$ be a polynomial function. Clearly it is sufficient to show that (A.35) holds for terms of the form $p(\alpha) = \alpha^k$ for k a positive integer. Then

$$\begin{aligned}
(1-\alpha)U_p(\alpha) &= (1-\alpha) \sum_{n=0}^{\infty} u_n (\alpha^{k+1})^n \\
&= \left[\frac{1-\alpha}{1-\alpha^{k+1}} \right] \left\{ (1-\alpha^{k+1}) \sum_{n=0}^{\infty} u_n (\alpha^{k+1})^n \right\}. \qquad \text{(A.36)}
\end{aligned}$$

Now let $\alpha \to 1^-$. The term in square brackets approaches $1/(k+1) = \int_0^1 x^k\, dx$.

The term in curly brackets approaches L by assumption. Hence (A.35) holds for polynomials.

Now let $s(\alpha)$ be continuous, and fix $\epsilon > 0$. By a theorem of Weierstrauss (see Apostol, 1974), there exists a polynomial p such that

$$p(\alpha) - \epsilon \leq s(\alpha) \leq p(\alpha) + \epsilon, \qquad 0 < \alpha < 1. \tag{A.37}$$

This implies that

$$\int_0^1 p(x)\,dx - \epsilon \leq \int_0^1 s(x)\,dx \leq \int_0^1 p(x)\,dx + \epsilon. \tag{A.38}$$

It follows from (A.37) that $(1-\alpha)U_s(\alpha) \leq (1-\alpha)\ \{U_p(\alpha) + \epsilon U(\alpha)\}$. We have

$$\begin{aligned}\limsup_{\alpha \to 1^-} (1-\alpha)U_s(\alpha) &\leq L\left(\int_0^1 p(x)\,dx + \epsilon\right)\\ &\leq L\left(\int_0^1 s(x)\,dx + 2\epsilon\right).\end{aligned} \tag{A.39}$$

The first inequality in (A.39) follows from (A.35) for p, and the second inequality follows from the leftmost inequality in (A.38). Using similar reasoning, we find a lower bound for the limit infimum. Since $\epsilon > 0$ is arbitrary, this proves (A.35) for s.

The proof of (A.35) for the function r uses Lemma A.4.1 and what has just been proved for continuous functions. Because it is quite similar to the reasoning we have just gone through, we omit the argument.

Let us see how (A.35) for the function r may be used to complete the proof. Note that $\alpha^n \geq e^{-1}$ if and only if $n \leq -(\ln \alpha)^{-1}$. So we have

$$\begin{aligned}(1-\alpha)U_r(\alpha) &= (1-\alpha) \sum_{n=0}^{[-(\ln \alpha)^{-1}]} u_n\\ &= (1-\alpha)w_{[-(\ln \alpha)^{-1}]+1}.\end{aligned} \tag{A.40}$$

Here [] denotes the greatest integer function, so [5.3] = 5, [8.9] = 8, and so on.

The limit of the quantity on the left side of (A.40) exists and equals L for any sequence of discount factors approaching 1. Suppose that we let $\alpha = e^{-1/n}$. The right side of (A.40) becomes

$$\{(1 - e^{-1/n})(n + 1)\}\left(\frac{w_{n+1}}{n+1}\right). \tag{A.41}$$

As $n \to \infty$, the term in curly brackets in (A.41) approaches 1. Because the left side of (A.40) approaches L, it is the case that the limit of the term in round brackets in (A.41) must exist and equal L. This proves that (ii) holds. □

A.5 AN EXAMPLE

In this section we give an example illustrating Theorem A.4.2. Under "most common circumstances" all of the terms in (A.28) are equal, and hence the limits exist. Here is how to construct an example for which some of the inequalities are strict.

Example A.5.1. The example is a sequence of 0s and 1s. Let $(q_n)_{n \geq 1}$ be a sequence of positive integers, to be specified later. Figure A.4 shows the sequence, which consists of blocks, with first q_1 1s, then q_1 0s, and so on. Let u_n be the nth member of the sequence. If we begin the indexing with 0, then we have $u_n = 1$ for $0 \leq n \leq q_1 - 1$, and so on.

Then w_n/n = (# of 1s in first n terms)/n. It is readily seen that this proportion is minimized by taking the subsequence $n = 2q_1, 2(q_1 + q_2), \ldots$, and the minimum proportion is $\frac{1}{2}$. This implies that $\liminf_{n \to \infty} w_n/n = \frac{1}{2}$.

The proportion is maximized by taking the sequence $n = q_1, 2q_1 + q_2, 2(q_1 + q_2) + q_3$, and so on, and for this subsequence we have the following values for w_n/n:

$$1, \quad \frac{q_1 + q_2}{2q_1 + q_2}, \quad \frac{q_1 + q_2 + q_3}{2(q_1 + q_2) + q_3}, \ldots. \tag{A.42}$$

Let $s_n = \sum_{k=1}^{n} q_k$. Then for $n \geq 2$ the sequence in (A.42) becomes $(1 + s_{n-1}/s_n)^{-1}$.

Our task now is to find values of q_k that make this quantity approach a number greater than $\frac{1}{2}$. Neither of the simple choices of $q_k \equiv q$ or $q_k = k$ will work. In each case the resulting sequence has a limit equal to $\frac{1}{2}$.

Let *Choice One* be $q_1 = 1$ and inductively $q_{k+1} = s_k$. This implies that $s_n = 2s_{n-1}$ and yields a limit supremum of 2/3. Let *Choice Two* be $q_1 = 1$ and inductively $q_{k+1} = (k+1)s_k$. This implies that $s_n = (n+1)s_{n-1}$ and yields a limit supremum of 1.

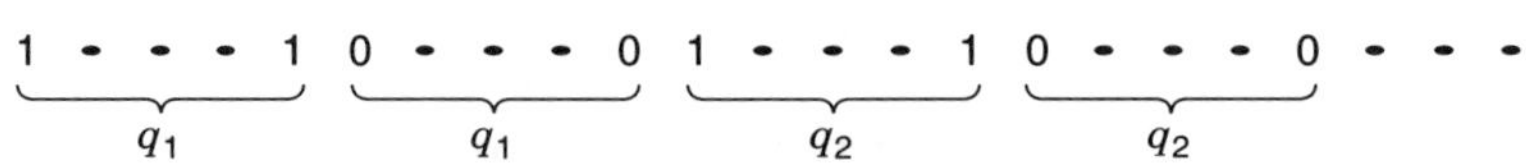

Figure A.4 Example A.5.1.

In either case the equivalence of (ii) and (iii) in Theorem A.4.2 implies that the middle inequality in (A.28) is strict. It is much more difficult to construct an example for which the leftmost inequality, say, is strict. Such a construction is given in Liggett and Lippman (1969), and we do not present it here. □

BIBLIOGRAPHIC NOTES

Some of these results appear in any good book on analysis, for example, Apostol (1974). Some of the results are modifications of known results, and some of the proofs have been developed for this text.

The proof of Theorem A.2.3 is an elaboration of a cryptic proof in Royden (1968, p. 232).

The continuous time version of Theorem A.4.2 appears in Widder (1941). A proof for the discrete case was given in Sennott (1986b). The proof of (iii) $\Rightarrow$ (ii) is due to Karamata and appears in Titchmarsh (1939, pp. 227–229).

Langen (1991) gives some results in a more theoretical setting similar to those in Section A.2.

APPENDIX B

Sequences of Stationary Policies

Throughout the book we deal with sequences of stationary policies. The concept of a stationary policy that is a *limit point* of such a sequence is fundamental. This idea has two variants, one for the MDC Δ and one for an AS (Δ_N). The proof of Proposition B.3 is optional, since it utilizes certain concepts from topology. The reader who desires to pursue this proof should consult a general topology text such as Pervin (1964) for the relevant background. The proof of Proposition B.5 depends only on the statement of Proposition B.3. Finally Proposition B.6 is a related result for functions. Its proof is also optional.

A *sequence* of stationary policies for Δ is a map from the natural numbers $\{1,2,3,\ldots\}$ to the set of stationary policies for Δ. Thus $f_1, f_2, f_3, \ldots$ is a sequence of stationary policies, where 1 is mapped to f_1, 2 is mapped to f_2, and so on. These policies do not have to be distinct. We could have $f_n \equiv f$.

We could also have the sequence $e_2, e_4, e_6, \ldots$, where 1 is mapped to e_2, 2 is mapped to e_4, and so on. Or we could have the sequence $d_{1/2}, d_{2/3}, d_{3/4}, \ldots$, where 1 is mapped to $d_{1/2}$, 2 to $d_{2/3}$, and so on. Informally, a sequence of stationary policies is just a list of them, with the proviso that there be infinitely many policies in the list (although the policies do not have to be distinct).

Now suppose that we have a sequence f_r of stationary policies. Then a *subsequence* of this sequence is a selection, in order, of policies from the list that also forms a sequence. For instance, if $f_1, f_2, f_3, f_4, f_5, f_6, f_7, f_8, f_9, \ldots$ is the original sequence, then $f_2, f_4, f_6, f_8, \ldots$ is a subsequence. Moreover there can be subsequences of subsequences. Note that $f_4, f_8, \ldots$ is a subsequence of the subsequence. Every subsequence of a subsequence is a subsequence of the original sequence. And every subsequence is a sequence in its own right.

What about notation? If f_r is the original sequence, then a subsequence is denoted f_{r_k}, where r_k denotes an appropriate selection from the original indexes. If we need to consider a subsequence of a subsequence, we denote it by f_{r_s} or some other appropriate notation; triple subscripts are not employed.

Here is the definition of a limit point of a sequence of stationary policies.

Definition B.1. Let f_r be a sequence of stationary policies for Δ. The stationary policy f is a *limit point* of the sequence if there exists a subsequence f_{r_k} such that given $i \in S$, it is the case that $f_{r_k}(i) = f(i)$ for sufficiently large index r_k (how large may depend on i). We denote this by $\lim_k f_{r_k} = f$ or by $f_{r_k} \to f$. □

This says that for a given state i, the policies in the subsequence choose the same action at i as the policy f, as long as we have gone "far enough out" in the subsequence. The amount necessary to go out may vary with i. Here is an example to clarify this concept.

Example B.2. Let $S = \{0, 1, 2, \ldots\}$, and assume that there are actions a and b available in each state. The transition probabilities and costs are irrelevant and are omitted.

Let f be the policy that always chooses a. Let e_n be the policy that chooses a in state $0 \le i \le n$, and b in states $i \ge n + 1$. Then $e_n \to f$. To prove this, fix i and choose n so large that $i \le n$. Then $e_n(i) = a = f(i)$ for $n \ge i$.

Let d be the policy that always chooses b, and consider the sequence e_1, d, e_2, d, e_3, d, $\ldots$. Then e_1, e_2, e_3, $\ldots$ is a subsequence converging to f, and d, d, d, $\ldots$ is a subsequence converging to d (a "trivial" subsequence). Notice that this sequence has two limit points. Can you construct a sequence with two nontrivial converging subsequences? □

Here is the first result.

Proposition B.3. Every sequence of stationary policies for Δ has at least one limit point.

**Proof:* For each i the finite action set A_i may be considered a compact metric space in its discrete topology. Consider the topological product space $A^* = \Pi_{i \in S} A_i$. There is a one-to-one correspondence between the points of A^* and the stationary policies for Δ. This comes about through the identification of a stationary policy d with the element $(d(i))_{i \in S}$ in A^*.

Since the topological product of compact topological spaces is compact, it follows that A^* is a compact topological space. It is known that a countable product of metric spaces is metrizable. That is, it has a metric compatible with the product topology. By means of this result it follows that A^* is a compact metric space.

In a compact metric space it is the case that every sequence of points has a convergent subsequence. So, if f_r is a sequence of stationary policies, then there exist a stationary policy f and a subsequence f_{r_k} converging to f in the product topology. This means the following: Given i, we have $f_{r_k}(i)$ converging to $f(i)$ in the topological space A_i. But since this is a finite discrete space, convergence implies that $f_{r_k}(i) = f(i)$ for sufficiently large index r_k. But this is precisely the notion of convergence in Definition B.1, and hence $f_{r_k} \to f$. □

The fact that S is countable is used crucially in this proof. If S is uncountable, then A^* is still a compact topological space, but it is not metrizable. It is the case that every net in a compact topological space has a convergent subnet, where the notion of net generalizes that of a sequence. However, it is not necessarily the case that every sequence has a convergent subsequence. The finiteness of the action sets is also crucial to the proof. Do you see why?

It is necessary to have a similar result involving an AS for Δ.

Definition B.4. Let (Δ_N) be an AS for Δ. For each N let e^N be a stationary policy for Δ_N. The stationary policy e for Δ is a *limit point* of the sequence e^N if there exists a subsequence e^{N_r} such that given $i \in S$, it is the case that $e^{N_r}(i) = e(i)$ for sufficiently large index N_r. □

Note the difference between Definitions B.1 and B.4. In the first case the stationary policies f_r are defined on S, while in the second case the stationary policy e^N is defined only on S_N. Here is the second result.

Proposition B.5. Let (Δ_N) be an AS for Δ. Every sequence e^N of stationary policies for (Δ_N) has a limit point.

Proof: For each $i \in S$ choose and fix an arbitrary $a_i \in A_i$. Define the stationary policy f_N for Δ by

$$f_N(i) = \begin{cases} e^N(i), & i \in S_N, \\ a_i, & i \in S - S_N. \end{cases}$$

Then f_N is a sequence of stationary policies for Δ, and by Proposition B.3 it has a limit point. Hence there exist a stationary policy e for Δ and a subsequence N_r such that $f_{N_r} \to e$. Then, given $i \in S$, we have $f_{N_r}(i) = e(i)$ for $N_r \geq s$. Here s is an index dependent on i. Now choose and fix N^* such that $i \in S_N$ for $N \geq N^*$. Then for $N_r \geq \max\{s, N^*\}$ we have $f_{N_r}(i) = e^{N_r}(i) = e(i)$. This proves the result. □

Here is a related result for functions using the same proof technique as that of Proposition B.3.

Proposition B.6. Let $L(i)$ and $M(i)$ be nonnegative (finite) functions on S. Assume that $u_r(i)$ is a sequence of functions on S with $-L \leq u_r \leq M$ for all r. Then there exist a subsequence r_k and a function w, with $-L \leq w \leq M$, satisfying $\lim_{k \to \infty} u_{r_k}(i) = w(i)$ for all $i \in S$.

**Proof:* Note that $[-L(i), M(i)]$ is a closed interval of the real line and hence is a compact metric space. The product space $\Pi_{i \in S}[-L(i), M(i)]$ is a compact metric space. Moreover there is a one-to-one correspondence between points of the product space and functions y on S with $-L \le y < M$. Namely y is identified with the point $(y(i))_{i \in S}$.

Hence u_r is a sequence in the product space. Since every sequence in a compact metric space has a convergent subsequence, there exist a subsequence r_k and a function w such that $u_{r_k} \to w$ in the product topology. But this means that pointwise convergence holds. Hence $\lim_{k \to \infty} u_{r_k}(i) = w(i)$ for all $i \in S$. □

BIBLIOGRAPHIC NOTES

The background in topology appears in any good text such as Pervin (1964). Proposition B.3 appears in Sennott (1989a), and Proposition B.5 in Sennott (1997a).

APPENDIX C

Markov Chains

This appendix deals with Markov chains on a countable state space. Section C.1 summarizes background material on Markov chains and Section C.2 treats Markov chains with an associated cost structure. Section C.3 deals with the special results that apply when the state space is finite.

Some of the results in these sections are attributed, some are proved, and some are stated without proof. The proofs of the latter may be found in any book containing a good treatment of countable state Markov chains. For example, the reader may consult Karlin and Taylor (1975), Taylor and Karlin (1984), Ross (1996), or Cinlar (1975). The most advanced treatment is Chung (1967).

In Sections C.4 and C.5 we present results involving approximating sequences for Markov chains.

The proofs that are given are for the convenience of the interested reader. It is not necessary to read these proofs to understand how the results in this appendix are applied in the text.

C.1 BASIC THEORY

A *Markov chain* (MC) Γ is a discrete time process defined on a countable state space S for $t = 0, 1, 2, \ldots$. Associated with $i \in S$ is a probability distribution $(P_{ij})_{j \in S}$, where P_{ij} is the probability that Γ will transition to state j during the next slot, given that it is currently in state i. We assume that $\sum_j P_{ij} = 1$ for all i, and hence the process cannot leave S. The characteristic property of a MC is the *memoryless property*. If the chain is currently in state i, then its future evolution depends only on i and not on the history of the chain prior to that time. A more rigorous definition of a MC is found in the references.

Let $\mathbf{P}$ be the matrix of transition probabilities. Let $P_{ij}^{(2)}$ be the probability of transitioning from i to j in two slots. Then $P_{ij}^{(2)} = \sum_k P_{ik}P_{kj}$, and these probabilities are the entries of the product matrix $\mathbf{P}^2$. In general, $P_{ij}^{(t)}$ is the probability of transitioning from i to j in t slots and is given by the ijth entry of the product matrix $\mathbf{P}^t$. We let $P_{ij}^{(0)} = \delta_{ij}$.

As Γ moves from state to state forever, we are interested in classifying the types of behavior that a state may exhibit. For states i and j, if there exists $t \geq 0$ such that $P_{ij}^{(t)} > 0$, then we say that i *leads to* j. If i leads to j and j leads to i, then we say that i and j *communicate*. Communication is an equivalence relation on S (that is, every state communicates with itself; if i communicates with j, then j communicates with i; and finally, if i communicates with j and j communicates with k, then i communicates with k). This implies that S decomposes into disjoint equivalence classes of communicating states. If S is a single communicating class, then Γ is *irreducible*.

Let X_t be the state of the MC at time t. Given initial state $X_0 = i$, let T be a random variable denoting the time to return to i. There are two possibilities, either $P(T < \infty) < 1$ or $P(T < \infty) = 1$.

If $P(T < \infty) < 1$, then we have $P(T = \infty) > 0$. This means that there is a positive probability of never returning to i, and we say that i is *transient*. Note that the chain may well visit i several times. However, after each visit there is a fixed positive probability of never returning. Hence the visits form a sequence of repeated independent Bernoulli trials that eventually result in never returning to i. Therefore a transient state is visited only finitely many times during any evolution of the MC.

If $P(T < \infty) = 1$, then state i is visited infinitely many times, and we say that i is *recurrent*. There are two types of recurrent states. Note that $E[T]$ denotes the expected time of a first return (first passage) to i. If $E[T] = \infty$, then i is said to be *null recurrent*. In this case the chain returns to i infinitely many times but the mean time for any return is infinite. If $E[T] < \infty$, then i is said to be *positive recurrent*. In this case the chain returns to i infinitely many times, and the mean time between any two visits is finite. As notation we set $E[T] = m_{ii}$.

These properties are class properties; that is to say, every state in a communicating class is either transient, null recurrent, or positive recurrent. A null recurrent class must be infinite. Positive and null recurrent classes are *closed*, since no state in such a class can lead to a state outside the class.

Example C.1.1. To facilitate understanding of these concepts consider the MC whose structure is shown in Fig. C.1. It is seen that S consists of three copies of the nonnegative integers. We choose the distribution $(p_j > 0)_{j \geq 1}$ such that $\lambda = \sum j p_j < \infty$ and the distribution $(q_j > 0)_{j \geq 1}$ such that $\sum j q_j = \infty$.

Each row forms a communicating class. From any middle state $i^* \geq 1$, there is a probability of $\frac{1}{2}$ of transitioning to 0^*. From 0^* there is a probability of $\frac{1}{2}$ of never returning to the middle row. Hence the middle row is a transient class.

It is clear that the other two classes are recurrent. Conditioning on the first state visited shows that $m_{00} = 1 + \sum j p_j = 1 + \lambda < \infty$, and hence the top row is a positive recurrent class. Similarly we have $m_{0\#0\#} = 1 + \sum j q_j = \infty$, and hence the bottom row is a null recurrent class. □

Let

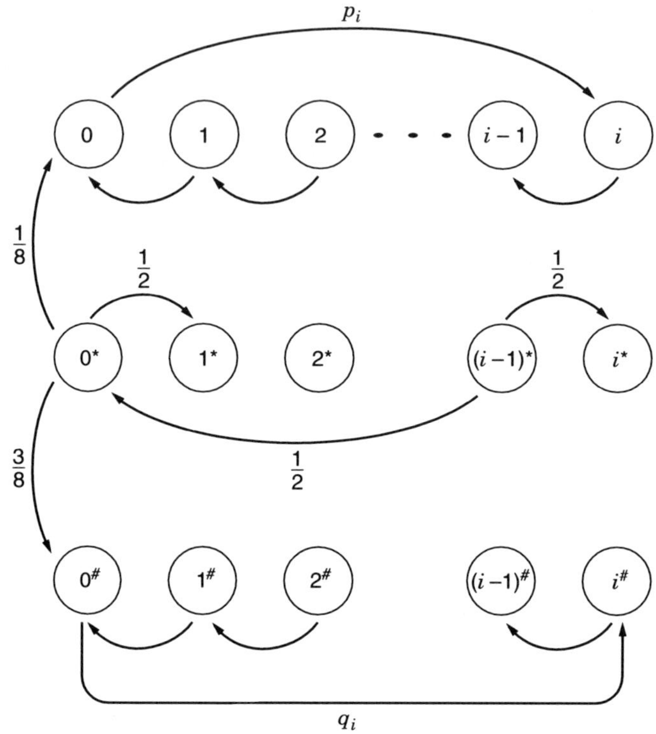

Fig. C.1 Example C.1.1.

$$Q_{ij}^{(n)} =: \frac{1}{n} \sum_{t=0}^{n-1} P_{ij}^{(t)}, \qquad i,j \in S \tag{C.1}$$

and note that this is the expected number of visits to state j per unit time in $[0, n-1]$ when starting in state i.

It is the case that $\pi_j = \lim_{n\to\infty} Q_{jj}^{(n)}$ exists. The quantity π_j is the *steady state probability* of being in state j. Assuming that $X_0 = j$, it may be thought of as the limiting average number of visits to j per unit time, or alternatively as the probability that a random observer finds the chain in state j after a long time has elapsed.

If j is transient or null recurrent then $\pi_j = 0$. If j is positive recurrent, then $\pi_j > 0$. It is the case that $\pi_j = (m_{jj})^{-1}$, where for j transient or null recurrent we have $m_{jj} = \infty$ and the quotient is interpreted as 0.

Now let T_{ij} be a random variable denoting the first passage time to go from i to j, namely the number of transitions required to first reach j from i. Then Chung (1967) proves that $\lim_{n\to\infty} Q_{ij}^{(n)} = P(T_{ij} < \infty)\pi_j$. This will be 0 unless both j is positive recurrent and i leads to j. If R is a positive recurrent class, then $\lim_{n\to\infty} Q_{ij}^{(n)} = \pi_j$, for all $i, j \in R$.

Proposition C.1.2. Let R be a positive recurrent class.

(i) We have $\pi_j = \sum_{i \in R} P_{ij}\pi_i$ for $j \in R$ and $\sum_{j \in R} \pi_j = 1$. The nonnegative solution to these equations is unique.

(ii) For $i, j \in R$ let e_{ij} be the expected number of visits to j during a first passage from i to i. Then $\pi_j = e_{ij}/m_{ii} = \pi_i e_{ij}$. (See Chung, 1967.)

Example C.1.3. In Example C.1.1 recall that the top row is a positive recurrent class. We have $\pi_0 = (m_{00})^{-1} = (1 + \lambda)^{-1}$. For $i \geq 1$ it follows that $\pi_i = \pi_0 e_{0i} = (1 + \lambda)^{-1} \sum_{j=i}^{\infty} p_j$. The steady state probabilities for states in the second and third rows are 0. We have $P(T_{i^*j} < \infty) = \frac{1}{4}$ and $P(T_{i^*j\#} < \infty) = \frac{3}{4}$ for all $i^*, j, j^\#$. □

Now assume that R is a positive recurrent class. Then R is *aperiodic* if $\pi_j = \lim_{n\to\infty} P_{ij}^{(n)}$ for $i, j \in R$. Note that this is a stronger convergence requirement than the one introduced above, which involves averaging. A sufficient condition for R to be aperiodic is that $P_{ii} > 0$ for some $i \in R$. A necessary and sufficient condition is the following: There exist an element $i \in R$ and positive integers n and m, with greatest common divisor equal to 1, such that $P_{ii}^{(n)}$ and $P_{ii}^{(m)}$ are both positive. The requirement of aperiodicity of a positive recurrent class rules out "periodic" behavior.

Now fix a state i and a nonempty set $G \subset S$. We introduce some important concepts. The *taboo* probability ${}_GP_{ik}^{(t)}$ is the probability of transitioning from i to k in t slots while avoiding the taboo set G. The initial state i and the terminal state k may lie in G, but none of the intermediate states are allowed to be in G. Note that ${}_GP_{ik}^{(1)} = P_{ik}$ and ${}_GP_{ik}^{(0)} = \delta_{ik}$.

Let T_{iG} be the first passage time from i to G, namely the number of transitions required to first reach G from i. If $i \in G$, then the chain must make at least one transition before returning to G. Thus it is always the case that $T_{iG} \geq 1$. Let ${}_Gu_{ik}$ be the expected number of visits to k in a first passage from i to G. For $k \in G$ we have ${}_Gu_{ik} = 0$ if $i \notin G$, and ${}_Gu_{ik} = \delta_{ik}$ if $i \in G$. Note that ${}_iu_{ij}$ generalizes the quantity e_{ij}, which was introduced earlier for i and j elements of a positive recurrent class.

Let $m_{iG} = E[T_{iG}]$ be the expected first passage time. If $P(T_{iG} < \infty) = 1$, then the chain eventually reaches G and m_{iG} may be finite or infinite. If $P(T_{iG} < \infty) < 1$, then $m_{iG} = \infty$. If $G = \{j\}$, then the expected first passage time is denoted m_{ij}.

Proposition C.1.4. Let G be a nonempty subset of S.

(i) For $k \notin G$ we have ${}_Gu_{ik} = \delta_{ik} + \sum_{t=1}^{\infty} {}_GP_{ik}^{(t)}$.

(ii) We have $m_{iG} = \sum_{k \in S} {}_G u_{ik}$.

(iii) It is the case that

$$ {}_G P_{ik}^{(t+1)} = \sum_{j \notin G} P_{ij}\, {}_G P_{jk}^{(t)}, \qquad i,k \in S, t \geq 1. \tag{C.2} $$

$$ {}_G u_{ik} = \delta_{ik} + \sum_{j \notin G} P_{ij}\, {}_G u_{jk}, \qquad i,k \in S \tag{C.3} $$

$$ m_{iG} = 1 + \sum_{j \notin G} P_{ij} m_{jG}, \qquad i \in S. \tag{C.4} $$

(iv) If G is contained in a positive recurrent class R, then $\pi_j = \sum_{i \in G} \pi_i\, {}_G u_{ij}$ for $j \in R$ and $\sum_{i \in G} \pi_i m_{iG} = 1$.

(v) If R is a positive recurrent class, then $m_{ij} < \infty$ for all $i, j \in R$.

Proof: It is easy to see that the expressions in (i–ii) hold. Equation (C.2) follows by conditioning on the first state visited. For $k \in G$ it is easily seen that (C.3) holds. For $k \notin G$ we sum both sides of (C.2), for $t = 1$ to ∞, add and subtract appropriate terms, and employ (i) to obtain (C.3). Equation (C.4) follows by summing (C.3) over k and employing (ii). This verifies (iii).

The first equation in (iv) follows from Grassman et al. (1985), and we omit the proof. The reader may note that if $G = \{i\}$, then this equation reduces to the one in Proposition C.1.2(ii). The second equation follows by summing both sides of the first equation over $j \in R$.

Let us prove (v). Equation (C.4) yields

$$ m_{jj} = 1 + \sum_{k \neq j} P_{jk} m_{kj}. \tag{C.5} $$

We know that $m_{jj} < \infty$. Fix $i \neq j$. Since j leads to i, there exists $t > 0$ such that $P_{ji}^{(t)} > 0$. By choosing the smallest such t, we have ${}_j P_{ji}^{(t)} > 0$. Employing the expression in (C.4) allows us to iterate (C.5) $t - 1$ times to obtain

$$ m_{jj} \geq \sum_{k \neq j} {}_j P_{jk}^{(t)} m_{kj}. \tag{C.6} $$

Here some nonnegative terms have been discarded from the right side. It follows that we must have $m_{ij} < \infty$. □

Proposition C.1.5. Let G be a nonempty subset of S. Assume that there exist a (finite) nonnegative function y on S and $\epsilon > 0$ such that

$$\sum_j P_{ij}[y(j) - y(i)] \le -\epsilon, \qquad i \notin G. \tag{C.7}$$

Then for $i \notin G$ we have $P(T_{iG} < \infty) = 1$ and $m_{iG} \le y(i)/\epsilon$.

Proof: Using the fact that y is nonnegative, (C.7) implies that

$$y(i) \ge \epsilon + \sum_{j \notin G} P_{ij} y(j), \qquad i \notin G. \tag{C.8}$$

Iterating this n times yields

$$y(i) \ge \epsilon + \epsilon\left(\sum_{t=1}^{n} \sum_{j \notin G} {}_G P_{ij}^{(t)}\right) + \sum_{j \notin G} {}_G P_{ij}^{(n+1)} y(j). \tag{C.9}$$

As a shorthand let $T = T_{iG}$. Since $T \ge 1$, we have $P(T > 0) = 1$. Then we see from (C.9) that $y(i) \ge \epsilon \sum_{t=0}^{n} P(T > t)$, where the last term on the right of (C.9) has been dropped, since y is nonnegative. Now $P(T > t) \ge P(T = \infty)$, and hence $y(i) \ge \epsilon(n+1)P(T = \infty)$. Letting $n \to \infty$ yields a contradiction unless $P(T = \infty) = 0$. Thus $P(T < \infty) = 1$. Using a familiar property of nonnegative random variables, the inequality yields $E[T] = \sum_{t=0}^{\infty} P(T > t) \le y(i)/\epsilon$. □

Corollary C.1.6. Assume that there exist a distinguished state z, a (finite) nonnegative function y on S, and $\epsilon > 0$ such that

$$\sum_j P_{zj} y(j) < \infty,$$

$$\sum_j P_{ij}[y(j) - y(i)] \le -\epsilon, \qquad i \ne z. \tag{C.10}$$

Then $P(T_{iz} < \infty) = 1$ for all i. Moreover $m_{iz} \le y(i)/\epsilon$ for $i \ne z$. Finally $m_{zz} < \infty$, and hence z is positive recurrent.

Proof: Choosing $G = \{z\}$ in Proposition C.1.5 proves the claims concerning $i \ne z$.

Now $P(T_{zz} < \infty) = \sum_j P_{zj} P(T_{zz} < \infty | X_1 = j) = P_{zz} + \sum_{j \ne z} P_{zj} P(T_{jz} < \infty) = 1$. Then $m_{zz} = 1 + \sum_{j \ne z} P_{zj} m_{jz} \le 1 + (\sum_{j \ne z} P_{zj} y(j))/\epsilon < \infty$. □

The function y in the two results above is called a *Lyapunov function* and the use of such functions is crucial to the results in the book. If $S = \{0, 1, 2, \ldots\}$, then a particularly useful choice is $y(i) = i$. In this case we say that $\gamma_i =: \sum_j P_{ij}(j - i)$ is the *drift* at i. It measures the expected movement of Γ in one transition. The following result is proved in Sennott et al. (1983):

Proposition C.1.7. Assume that Γ is positive recurrent on $S = \{0, 1, 2, \ldots\}$ and that $P_{ij} = 0$ for $i \geq 2$ and $j < i-1$. That is, the chain can transition downward only one state at a time. Then $\sum_i \pi_i \gamma_i = 0$.

C.2 MARKOV CHAINS WITH COSTS

Assume that to each state i is attached a (finite) nonnegative cost $C(i)$. In this section we discuss the important ideas related to a Markov chain with costs, which we continue to refer to as the MC Γ.

Let

$$J_i^{(n)} =: \frac{1}{n} E\left[\sum_{t=0}^{n-1} C(X_t)|X_0 = i\right]$$

$$= \sum_j C(j) Q_{ij}^{(n)}, \qquad i \in S, \tag{C.11}$$

be the expected cost incurred per unit time in $[0, n-1]$ when starting in state i.

Given that $m_{iG} < \infty$, we let c_{iG} be the expected cost of a first passage from i to G. Since costs may be 0, it doesn't make sense to talk about the expected cost of a first passage without knowing that the expected first passage time is finite.

Proposition C.2.1. Let R be a positive recurrent class.

(i) For $i \in R$, $\lim_{n \to \infty} J_i^{(n)}$ exists and equals the (finite or infinite) constant $J_R =: \sum_{j \in R} \pi_j C(j)$.
(ii) For $i \in R$ we have $J_R = c_{ii}/m_{ii}$.
(iii) $J_R = \sum_{j \in R} \pi_j E[C(X_n)|X_0 = j]$ for $n \geq 0$.

Proof: We first prove (ii). Fix $i \in R$. It follows from Proposition C.1.2(ii) that $J_R = \sum_{j \in R} C(j) e_{ij}/m_{ii} = c_{ii}/m_{ii}$. Observe that $m_{ii} < \infty$ but we may have $c_{ii} = \infty$.

To prove (i), note that $Q_{ij}^{(n)} \to \pi_j$ for $i \in R$. From (C.11) and Proposition A.1.7 it then follows that $\lim\inf_{n \to \infty} J_i^{(n)} \geq J_R$, Thus, if $J_R = \infty$, the limit exists and equals ∞ for every i.

Now assume that $J_R = c_{ii}/m_{ii} < \infty$. Then (i) follows from the renewal reward theorem. For example, see Ross (1996). A MC proof is given by Chung (1967, p. 93). The lengths of successive first returns to a state i in a positive recurrent class are independent and identically distributed, and hence these successive first-passage times constitute a renewal process. In renewal theory the length of such a first passage is called a *cycle*. The renewal reward theorem says that the average cost, namely the limit of $J_i^{(n)}$, is given by the expected cost incurred during a cycle (which is c_{ii}) divided by the expected length of a cycle (which is m_{ii}).

We prove (iii) by induction on n. It holds for $n = 0$ by definition. Now assume that it holds for n. Then

$$\begin{aligned}\sum_{j \in R} \pi_j E[C(X_{n+1})|X_0 = j] &= \sum_{k \in R} C(k) \sum_{j \in R} \pi_j P_{jk}^{(n+1)} \\ &= \sum_{k \in R} C(k) \sum_{s \in R} \left(\sum_{j \in R} \pi_j P_{js} \right) P_{sk}^{(n)} \\ &= \sum_{k \in R} C(k) \sum_{s \in R} \pi_s P_{sk}^{(n)} \\ &= J_R. \end{aligned} \tag{C.12}$$

The first line follows by definition of the expectation. The interchange of the order of summation is justified, since all terms are nonnegative. The second line follows from the basic discussion in Section C.1. The third line follows from Proposition C.1.2(i). The fourth line follows by a rearrangement of the terms and an application of the induction hypothesis. □

The next three results are the cost counterparts to the expected first passage time results in Section C.1.

Proposition C.2.2. Let G be a nonempty subset of S.

(i) Assume that $m_{iG} < \infty$ for some i. Then $c_{iG} = \sum_k C(k)\ {}_G u_{ik}$.

(ii) Under the hypothesis of (i), we have

$$c_{iG} = C(i) + \sum_{j \notin G} P_{ij} c_{jG}. \tag{C.13}$$

(iii) If G is contained in a positive recurrent class R, then $J_R = \sum_{i \in G} \pi_i c_{iG}$.

(iv) If R is a positive recurrent class with $J_R < \infty$, then $c_{ij} < \infty$ for all $i, j \in R$.

Proof: It is clear that (i) holds. Part (ii) follows by multiplying both sides of (C.3) by $C(k)$, summing over k, and applying (i).

Part (iii) follows by multiplying the first equation in Proposition C.1.4(iv) by $C(j)$ and summing over $j \in R$.

To prove (iv), observe that from Proposition C.1.4(v) it follows that $m_{ij} < \infty$. Moreover $c_{jj} < \infty$. The proof is now similar to the proof of Proposition C.1.4(v). □

Proposition C.2.3. Let G be a nonempty subset of S such that $m_{iG} < \infty$ for all $i \notin G$. Assume that there exist a (finite) nonnegative function r on S and a finite subset $H \subset S - G$ such that

$$\sum_j P_{ij}[r(j) - r(i)] \leq -C(i), \qquad i \notin G \cup H,$$

$$\sum_j P_{ij} r(j) < \infty, \qquad i \in H. \tag{C.14}$$

Then there exists a (finite) nonnegative constant F such that $c_{iG} \leq r(i) + F m_{iG}$ for $i \notin G$. If $H = \emptyset$, then $c_{iG} \leq r(i)$ for $i \notin G$.

Proof: Let $C = \max_{i \in H} C(i)$ and $D = \max_{i \in H} \sum_j P_{ij} r(j)$. These are both finite constants. Let $F = C + D$.

Let $X_0 = i, X_1, \ldots, X_n \in G$ be a first passage with $X_t \notin G$ for $0 \leq t < n$. If $X_t \notin H$, then from the first inequality in (C.14) it follows that $C(X_t) + E[r(X_{t+1})|X_t] \leq r(X_t)$. If $X_t \in H$, then $C(X_t) \leq C$, and the second inequality in (C.14) yields $E[r(X_{t+1})|X_t] \leq D$. Hence in either case we have $C(X_t) + E[r(X_{t+1})|X_t] \leq C + D + r(X_t) = F + r(X_t)$. Taking the expectation of both sides of this inequality yields

$$E[C(X_t)] + E[r(X_{t+1})] \leq F + E[r(X_t)], \qquad 0 \leq t < n. \tag{C.15}$$

Note that it follows from (C.15) by induction that $E[r(X_t)] < \infty$ for $0 \leq t \leq n$. We now add the terms in (C.15) to obtain $E[\sum_{t=0}^{n-1} C(X_t)] \leq r(i) + Fn$. If this is multiplied by the probability that the first passage is of length n and summed over n, then we obtain the first result. The proof for $H = \emptyset$ is an obvious modification of this proof. □

Corollary C.2.4. Assume that $m_{iz} < \infty$ for some distinguished state z and all i. Assume that there exist a (finite) nonnegative function r on S and a finite subset H^* containing z such that

$$\sum_j P_{ij} r(j) < \infty, \qquad i \in H^*,$$

$$\sum_j P_{ij}[r(j) - r(i)] \leq -C(i), \qquad i \notin H^*. \tag{C.16}$$

Then there exists a (finite) nonnegative constant F such that $c_{iz} \leq r(i) + Fm_{iz}$ for $i \neq z$. If $H^* = \{z\}$, then $c_{iz} \leq r(i)$ for $i \neq z$. Finally we have $c_{zz} < \infty$.

Proof: We apply Proposition C.2.3 with $G = \{z\}$ and $H = H^* - \{z\}$ to obtain the first two claims.

We have assumed that $m_{zz} < \infty$, and hence it makes sense to talk about c_{zz}. We have $c_{zz} = C(z) + \sum_{j \neq z} P_{zj} c_{jz} \leq C(z) + \sum_{j \neq z} P_{zj}[r(j) + Fm_{jz}] = C(z) - F + Fm_{zz} + \sum_{j \neq z} P_{zj} r(j) < \infty$. □

The following type of MC is frequently employed in the text:

Definition C.2.5. Assume that there exists a distinguished state z such that $m_{iz} < \infty$ and $c_{iz} < \infty$ for all $i \in S$. Then the MC is z *standard.* □

This definition entails the following powerful implications.

Proposition C.2.6. Assume that Γ is z standard.

(i) The state space S decomposes into a positive recurrent class R containing z and a set U of transient states.
(ii) The average cost J_R on R is finite.
(iii) $\text{Lim}_{n \to \infty} J_i^{(n)}$ exists and equals J_R for all i.

Proof: By assumption $m_{zz} < \infty$, and hence the communicating class R containing z is positive recurrent. Clearly any state in $U = S - R$ must be transient, since it leads to z. Since $c_{zz} < \infty$ it follows from Proposition C.2.1(ii) that $J_R < \infty$. This proves (i–ii).

By Proposition C.2.1(i) it is only necessary to prove (iii) for transient states. If the process starts in $i \in U$, then in a finite expected amount of time and with finite expected cost, it will be in state z, and the average cost associated with z is J_R. The delayed renewal reward theorem then gives (iii). See Heyman and Sobel (1982, p. 184). Intuitively the result follows because there is an initial renewal interval with a different distribution, namely the first passage to z, and from then on the renewal process behaves as discussed in the proof of Proposition C.2.1. □

Remark C.2.7. (i) If Corollaries C.1.6 and C.2.4 hold, then Γ is z standard.

(ii) It follows from Propositions C.1.4(v) and C.2.2(iv) that if Γ is irreducible and positive recurrent with finite average cost, then it is z standard for any state z. □

C.3 MARKOV CHAINS WITH FINITE STATE SPACE

The results in Sections C.1 and C.2 apply when the MC has a countable state space, that is, a finite or denumerably infinite state space. However, when S is finite, additional results of a special nature hold.

Throughout this section we assume that Γ is a Markov chain defined on a finite state space S. Then Γ has at least one positive recurrent class. Let R_1, R_2, ..., R_K be a list of the positive recurrent classes. Let U be the set of states not in a positive recurrent class. Since a null recurrent class must be infinite, Γ has no null recurrent classes and the states in U must be transient. Moreover it is the case that from $i \in U$ some positive recurrent class is reached in finite expected time and with finite expected cost. If $p_k(i)$ denotes the probability that class R_k is reached first, then we have $\sum_k p_k(i) = 1$.

We know from Proposition C.2.1(i) that the average cost on R_k is a constant J_k. Since the costs are bounded, it follows that $J_k < \infty$. It may be seen that $J(i) = \sum_k p_k(i)J_k$. That is, the average cost at an arbitrary state i is a convex combination of the average costs on the positive recurrent classes. It is clear that the average cost function is a constant J if and only if $J_k \equiv J$.

For S finite we say that the MC is *unichain* if there is just one positive recurrent class R. In this case the average cost function must be constant. If the distinguished state z is an arbitrary element of R, then the chain is z standard. It is the case that a MC with a finite state space is z standard if and only if it is unichain with positive recurrent class containing z.

C.4 APPROXIMATING SEQUENCES FOR MARKOV CHAINS

In this section we have a Markov chain Γ with costs on a denumerable state space S. We are interested in constructing an approximating sequence of finite state Markov chains. The following definition is the MC counterpart of Definition 2.5.1:

Definition C.4.1. The sequence $(\Gamma_N)_{N \geq N_0}$ is an *approximating sequence* (AS) for Γ if there exists an increasing sequence $(S_N)_{N \geq N_0}$ of nonempty finite subsets of S such that $\cup S_N = S$. Each Γ_N is a MC with costs on S_N. Given $i \in S_N$ the cost at i equals $C(i)$, and there is a transition probability distribution $(P_{ij}(N))_{j \in S_N}$ satisfying $\lim_{N \to \infty} P_{ij}(N) = P_{ij}$ for $i, j \in S$. □

Quantities such as first passage times in the AS will be denoted by $m_{iG}(N)$, and so on. The next result provides some general relationships.

Proposition C.4.2. Let the AS be given, and let G be a finite nonempty subset of S.

(i) $\text{Lim}_{N \to \infty}\, {}_GP_{ik}^{(t)}(N) = {}_GP_{ik}^{(t)}$ for $i, k \in S$, $t \geq 1$.

(ii) ${}_Gu_{ik}(N) = {}_Gu_{ik} = \delta_{ik}$ for $k \in G$, $i \in S$, and N sufficiently large. In general, we have $\liminf_{N \to \infty} {}_Gu_{ik}(N) \geq {}_Gu_{ik}$ for $i, k \in S$.

(iii) $\text{Lim inf}_{N \to \infty}\, m_{iG}(N) \geq m_{iG}$ for $i \in S$.

Proof: Since G is finite, we may assume that N is so large that $G \subset S_N$. We prove (i) by induction. Now ${}_GP_{ik}^{(1)}(N) = P_{ik}(N) \to P_{ik} = {}_GP_{ik}^{(1)}$, and hence the statement is true for $t = 1$. Now assume that it is true for t. Then (C.2) for S_N yields

$$ {}_GP_{ik}^{(t+1)}(N) = \sum_{j \in S_N - G} P_{ij}(N)\,{}_GP_{jk}^{(t)}(N). \tag{C.17} $$

Taking the limit of both sides of (C.17), employing Corollary A.2.7 with bounding function 1 and the induction hypothesis, yields the result for $t + 1$ and proves (i).

The first statement in (ii) is clear. To prove the second statement, consider the equation in Proposition C.1.4(i) for S_N. Take the limit infimum of both sides of this equation and employ Proposition A.1.7, what has just been proved, and Proposition C.1.4(i) to obtain the result. This proves (ii).

To prove (iii), consider the equation in Proposition C.1.4(ii) for S_N. Take the limit infimum of both sides of this equation and employ Proposition A.1.8, what has just been proved, and Proposition C.1.4(ii) to obtain the result. □

Proposition C.4.3. Let an AS be given. Then the following hold:

(i) If $i \in S$ is transient or null recurrent, then $\lim_{N \to \infty} \pi_i(N) = \pi_i = 0$.

(ii) Let R be a positive recurrent class in Γ. Then given a sequence N_r, there exist a subsequence N_s of N_r and a constant b, with $0 \leq b \leq 1$, such that $\lim_{s \to \infty} \pi_i(N_s) = b\pi_i$ for $i \in R$.

**Proof:* Assume that i is transient or null recurrent. Then $m_{ii} = \infty$, and it follows from Proposition C.4.2(iii) that $m_{ii}(N) \to \infty$. Then $\pi_i(N) = 1/m_{ii}(N) \to 0$, which proves (i).

To prove (ii), fix the sequence N_r. We may assume that the sequence of functions $\pi_i(N_r)$ is defined on all of S by setting $\pi_i(N) = 0$ for $i \notin S_N$. Note that $0 \leq \pi_i(N_r) \leq 1$. By Proposition B.6 there exist a subsequence N_s of N_r and a function q_i, with $0 \leq q_i \leq 1$, such that $\pi_i(N_s) \to q_i$ for all i.

If i is transient or null recurrent, then from (i) it follows that $q_i = 0$. If

i is positive recurrent, then from Proposition C.4.2 we have $1/q_i = \lim_{s \to \infty} m_{ii}(N_s) \geq \liminf_{N \to \infty} m_{ii}(N) \geq m_{ii} = 1/\pi_i$. This implies that $q_i \leq \pi_i$.

Consider the equation

$$\pi_j(N) = \sum_{i \in S_N} P_{ij}(N)\pi_i(N), \qquad j \in S_N. \tag{C.18}$$

In Proposition C.1.2(i) we gave this equation as valid for a positive recurrent class in Γ_N. However, it is easily seen that (C.18) holds in the general case, in which there may be transient states and multiple positive recurrent classes.

We now take the limit infimum through values of N_s of both sides of (C.18) and employ Proposition A.2.5 to obtain

$$q_j \geq \sum_i P_{ij} q_i, \qquad j \in S. \tag{C.19}$$

Now assume that $j \in R$. On the right side of (C.19), notice that P_{ij} can be positive in only two cases, namely $i \in R$ or i transient. If i is transient, then $q_i = 0$, and we may omit that term. Hence (C.19) yields (*): $q_j \geq \sum_{i \in R} P_{ij} q_i$ for $j \in R$. Since $q_j \leq \pi_j$, it follows that $\sum_{j \in R} q_j \leq 1$. If we assume that the inequality in (*) is strict for some j^* and sum both sides over $j \in R$, then we obtain a contradiction. Hence equality holds in (*). We then iterate (*) n times and average to obtain

$$q_j = \sum_{i \in R} Q_{ij}^{(n)} q_i, \qquad j \in R. \tag{C.20}$$

We wish to take the limit of the right side of (C.20) and pass the limit through the summation. To justify this, we may use Corollary A.2.4 with bounding function 1. Note that $(q_j)_{j \in R}$ may not be a probability distribution, but since $\sum_{j \in R} q_j =: b \leq 1$, a term with the extra probability (multiplied by 0) can be added to the right side of (C.20). This easily yields $q_j = b\pi_j$ for $j \in R$. □

The constant b in Proposition C.4.3 depends on the positive recurrent class and on the sequence N_r. Note that $b \equiv 1$ for all positive recurrent classes and sequences, is equivalent to $\pi_i(N) \to \pi_i$ for $i \in S$, and we will use the two expressions interchangeably. The next example shows that we may have $b < 1$.

Example C.4.4. Let $S = \{0, 1, 2, \ldots\}$ with $P_{00} = 1$ and $P_{ii-1} = 1$ for $i \geq 1$. Then $\pi_0 = 1$ and $\pi_i = 0$ for $i \geq 1$. We construct two approximating sequences with $S_N = \{0, 1, \ldots, N\}$ for $N \geq 2$.

To define AS_1, let $P_{00}(N) = 1 - N^{-1}$, $P_{0N}(N) = N^{-1}$, $P_{ii-1}(N) = 1$ for

$1 \leq i \leq N-1$, and $P_{NN}(N) = 1$. This satisfies $\pi_i(N) = 0$, $0 \leq i \leq N-1$, and $\pi_N(N) = 1$. In this case $b = 0$.

To define AS$_2$, let the transition probabilities be as in AS$_1$ except set $P_{NN-1}(N) = 1$. This makes Γ_N into an irreducible MC, and it is easy to see, using reasoning similar to that in Example C.1.3, that $\pi_0(N) = \frac{1}{2}$ and $\pi_i = \frac{1}{2N}$ for $1 \leq i \leq N$. In this case $b = \frac{1}{2}$. □

Here is a result related to the cost structure.

Proposition C.4.5. Let an AS be given, and let G be a finite nonempty subset of S. Assume that $m_{iG} < \infty$ for some i and that $m_{iG}(N) < \infty$ for sufficiently large N. Then $\liminf_{N\to\infty} c_{iG}(N) \geq c_{iG}$.

Proof: From Proposition C.2.2(i) it follows that

$$c_{iG}(N) = \sum_{k \in S_N} C(k)_G u_{ik}(N). \tag{C.21}$$

We then take the limit infimum of both sides of (C.21) and employ Proposition A.1.8 and Propositions C.4.2(ii) and C.2.2(i) to obtain the result. □

Proposition C.4.6. Let an AS be given, and let R be a positive recurrent class in Γ. Then the following are equivalent:

(i) $\pi_i(N) \to \pi_i$ for $i \in R$.
(ii) $m_{zz}(N) \to m_{zz}$ for some $z \in R$.
(iii) $m_{iG}(N) \to m_{iG}$ for any nonempty finite subset G of R and $i \in R$.

Now assume that any (and hence all) of the above conditions hold. Then the following are equivalent:

(iv) $J(i)(N) \to J_R$ for $i \in R$.
(v) $c_{zz}(N) \to c_{zz}$ for some $z \in R$.
(vi) $c_{iG}(N) \to c_{iG}$ for any nonempty finite subset G of R and $i \in R$.

Proof: Observe that (ii) is equivalent to $\pi_z(N) \to \pi_z$, and if this holds, then we must have $b = 1$ for R. Thus (i) and (ii) are equivalent. Clearly (iii) implies (ii), so it remains to prove that (i) implies (iii).

So let G be a nonempty finite subset of R, and fix a subsequence N_r. Let $\liminf_{r\to\infty} m_{iG}(N_r) =: w(i)$ for $i \in S$.

Part (i) implies that for sufficiently large N, the finite set G is contained in a positive recurrent class $R(N)$ of Γ_N. Then Proposition C.1.4(iv) yields

$$1 = \sum_{i \in G} \pi_i(N) m_{iG}(N). \tag{C.22}$$

We first show that $m_{iG}(N) \to m_{iG}$ for $i \in G$. Take the limit infimum of both sides of (C.22) through values N_r to obtain

$$\begin{aligned} 1 &\geq \sum_{i \in G} \pi_i w(i) \\ &\geq \sum_{i \in G} \pi_i \left(\liminf_{N \to \infty} m_{iG}(N) \right) \\ &\geq \sum_{i \in G} \pi_i m_{iG} \\ &= 1. \end{aligned} \tag{C.23}$$

The second line is clear and the third line follows from Proposition C.4.2(iii). The fourth line follows from Proposition C.1.4(iv). Hence all the terms in (C.23) are equal. This readily implies that $w(i) = m_{iG}$, which yields $m_{iG}(N) \to m_{iG}$, for $i \in G$.

Now consider (C.4) for Γ_N. Taking the limit infimum of both sides through values N_r yields

$$w(i) \geq 1 + \sum_{j \notin G} P_{ij} w(j), \qquad i \in S. \tag{C.24}$$

Now fix $k \in R - G$. It is easy to see that there must exist $i^* \in G$ and $t \geq 1$ such that ${}_G P_{i^*k}^{(n)} > 0$. Iterating (C.24) $n - 1$ times yields

$$\begin{aligned} w(i^*) &\geq 1 + \sum_{t=1}^{n-1} \sum_{j \notin G} {}_G P_{i^*j}^{(t)} + \sum_{j \notin G} {}_G P_{i^*j}^{(n)} w(j) \\ &\geq 1 + \sum_{t=1}^{n-1} \sum_{j \notin G} {}_G P_{i^*j}^{(t)} + \sum_{j \notin G} {}_G P_{i^*j}^{(n)} m_{jG} \\ &= m_{i^*G}. \end{aligned} \tag{C.25}$$

The second line follows from Proposition C.4.2(iii). The third line follows from (C.4) for Γ iterated $n-1$ times. Since $w(i^*) = m_{i^*G}$, it follows that all the terms in (C.25) are equal. This readily implies that $w(k) = m_{kG}$ and proves (iii).

Now assume that (i–iii) hold, and fix $i \in R$. It is easily seen that i and z are elements of a positive recurrent class $R(N)$ for N sufficiently large. It follows from Proposition C.2.1(ii) that $J(i)(N) = J(z)(N) = c_{zz}(N)/m_{zz}(N)$ and that $J_R = c_{zz}/m_{zz}$. From this it easily follows that (iv) and (v) are equivalent.

Clearly (vi) implies (v). The proof that (iv) implies (vi) is similar to the proof above, and we omit it. □

Example C.4.7. Let Γ and S_N be as in Example C.4.4. Define Γ_N by $P_{00}(N) = 1 - N^{-2}$, $P_{0N}(N) = N^{-2}$, and $P_{ii-1}(N) = 1$ for $1 \le i \le N$. This makes Γ_N into an irreducible MC, and it is easy to see, using reasoning similar to that in Example C.1.3, that $\pi_0(N) = 1 + N^{-1}$ which converges to π_0.

Assume that $C(i) = i$. Then $c_{00} = 0$, but

$$\begin{aligned} c_{00}(N) &= \frac{1}{N^2}\left(\frac{N(N+1)}{2}\right) \\ &= \frac{N+1}{2N} \to \frac{1}{2}. \end{aligned} \tag{C.26}$$

□

The following definition embodies the idea that the convergence is properly behaved:

Definition C.4.8. Assume that Γ is z standard. An AS is *conforming* if the following hold:

(i) There exists N^* such that Γ_N is unichain with z an element of the positive recurrent class for $N \ge N^*$.
(ii) We have $m_{iz}(N) \to m_{iz}$ and $c_{iz}(N) \to c_{iz}$ for all i. □

Here are some consequences of the notion of conformity.

Proposition C.4.9. Assume that Γ is z standard and that the AS is conforming.

(i) $\pi_i(N) \to \pi_i$ for all i.
(ii) $J(N) \to J_R$, where $J(N)$ is the constant average cost on Γ_N for $N \ge N^*$.

Proof: If $i \in U$, then the convergence in (i) follows from Proposition C.4.3(i). If $i \in R$, then it follows from Proposition C.4.6. To prove (ii), note that $J(N) = c_{zz}(N)/m_{zz}(N) \to c_{zz}/m_{zz} = J_R$. □

It is sometimes useful to have the following weaker notion of conformity:

Definition C.4.10. Assume that Γ has a positive recurrent class R with finite average cost J_R. Then an AS is *conforming on R* if, for $i \in R$, we have $\pi_i(N) \to \pi_i$ and $J(i)(N) \to J_R$. □

C.5 SUFFICIENT CONDITIONS FOR CONFORMITY

The examples in Section C.4 tell us that achieving conformity requires additional assumptions on the approximating sequence. Developing appropriate assumptions is the task of this section. These assumptions require the AS to be of a special type, namely the counterpart of the augmentation type approximating sequence introduced in Definition 2.5.3 for MDCs. For clarity we give the definition here for MCs.

Definition C.5.1. Assume that we have a MC with an approximating sequence. The AS is an *augmentation type approximating sequence* (ATAS) if the following holds: Given $i \in S_N$, for each $r \notin S_N$ there exists a probability distribution $(q_j(i, r, N))_{j \in S_N}$, called the augmentation distribution associated with (i, r, N), such that

$$P_{ij}(N) = P_{ij} + \sum_{r \in S - S_N} P_{ir} q_j(i, r, N), \qquad j \in S_N. \tag{C.27}$$

□

The idea is that the original probability associated with states in S_N is not changed, but excess probability associated with a transition to a state outside of S_N is redistributed to the elements of S_N according to some probability distribution.

The next result involves an ATAS that sends excess probability to a finite set.

Proposition C.5.2. Let Γ be a z standard MC, and let G be a finite nonempty subset of S. Any ATAS that sends excess probability to G is conforming. The weaker notion of conformity in Definition C.4.10 also holds as long as G is a subset of R.

**Proof:* We first argue that there is no loss of generality in assuming that $z \in G$, since if G does not already contain z, then we may add it in. The approximating sequence is still an ATAS that sends the excess probability to G. (There is no requirement that any excess probability be sent to z.)

Now assume that N is so large that $G \subset S_N$. We claim that $(^*)$: ${}_GP_{ik}^{(t)}(N) \le {}_GP_{ik}^{(t)}$ for $t \ge 1$, $i \in S_N$, and $k \in S_N - G$. Think about why this is intuitively clear! We prove $(^*)$ by induction. For $t = 1$ we have ${}_GP_{ik}^{(1)}(N) = P_{ik}(N) = P_{ik} = {}_GP_{ik}^{(1)}$. Now assume that the result holds for t. Then

$$
\begin{aligned}
{}_G P_{ik}^{(t+1)}(N) &= \sum_{j \in S_N - G} {}_G P_{ij}^{(t)}(N) P_{jk}(N) \\
&\le \sum_{j \in S_N - G} {}_G P_{ij}^{(t)} P_{jk} \\
&\le \sum_{j \in S - G} {}_G P_{ij}^{(t)} P_{jk} \\
&= {}_G P_{ik}^{(t+1)}.
\end{aligned} \tag{C.28}
$$

The second line follows from the induction hypothesis and what was just proved for $t = 1$. The third and fourth lines are clear. This completes the induction and hence (*) holds.

It then follows from Proposition C.1.4(i) that (**): ${}_G u_{ik}(N) \le {}_G u_{ik}$ for $k \in S_N - G$. Then

$$
\begin{aligned}
m_{iG}(N) &= I(i \in G) + \sum_{k \in S_N - G} {}_G u_{ik}(N) \\
&\le I(i \in G) + \sum_{k \in S_N - G} {}_G u_{ik} \\
&\le I(i \in G) + \sum_{k \in S - G} {}_G u_{ik} \\
&= m_{iG}.
\end{aligned} \tag{C.29}
$$

The first and last line follow from Proposition C.1.4(ii). The second line follows from (**), and the third line from the nonnegativity of the terms. From (C.29) we have (***): $m_{iG}(N) \le m_{iG}$ for $i \in S_N$. Then Proposition C.4.2(iii) implies that $m_{iG}(N) \to m_{iG}$ for all i.

Since $z \in G$ and Γ is z standard, it follows that $m_{iG} < \infty$ for all i. Hence it follows from (***) that $m_{iG}(N) < \infty$ for $i \in S_N$. This implies that G must intersect every positive recurrent class in Γ_N. Moreover it follows from Proposition C.1.4(ii) that ${}_G u_{ij}(N) < \infty$ for $i, j \in S_N$.

Consider the first equation in Proposition C.1.4(iv). It was stated for a finite subset of a positive recurrent class. It is easy to check that under the above conditions the equation holds in general for S_N. Thus

$$
\pi_j(N) = \sum_{i \in G} \pi_i(N) {}_G u_{ij}(N), \qquad j \in S_N. \tag{C.30}
$$

Now let $F(N)$ be the number of positive recurrent classes in Γ_N, and note that $F(N) \geq 1$. Then adding up the terms in (C.30) yields

$$
\begin{aligned}
F(N) &= \sum_{i \in G} \pi_i(N) m_{iG}(N) \\
&\leq \sum_{i \in G} \pi_i(N) m_{iG}.
\end{aligned} \tag{C.31}
$$

The first line follows from Proposition C.1.4(ii), and the second line follows from (***).

Given a sequence N_r, there exist a subsequence N_s and b, with $0 \leq b \leq 1$, such that $\pi_i(N_s) \to b\pi_i$ for $i \in R$. Moreover $\pi_i(N) \to 0$ for $i \in U$. This follows from Proposition C.4.3(ii). Let $H = G \cap R$, and note that H is nonempty. Using these facts, it follows from (C.31) that $1 \leq \liminf_{s \to \infty} F(N_s) \leq \limsup_{s \to \infty} F(N_s) \leq b \sum_{i \in H} \pi_i m_{iG} = b \sum_{i \in H} \pi_i m_{iH} = b$. The next to last equality follows since $m_{iG} = m_{iH}$ for $i \in H \subset R$. The last equality follows from Proposition C.1.4(iv). Hence $b = 1$. Since this holds for any sequence, it follows that $\pi_i(N) \to \pi_i$ for $i \in R$. Then Proposition C.4.6 yields that $m_{iz}(N) \to m_{iz}$ for $i \in R$.

This also proves that $F(N) \to 1$. Since $F(N)$ is an integer, we must have $F(N) = 1$ for sufficiently large N. It follows from what has been proved that z must lie in the positive recurrent class $R(N)$ for sufficiently large N, say $N \geq N^*$. This verifies Definition C.4.8(i).

Using Proposition C.2.2(i) and (**), we obtain $c_{iG}(N) \leq c_{iG}$ for $i \in S_N$. Then it follows from Proposition C.4.5 that $c_{iG}(N) \to c_{iG}$. For $N \geq N^*$, since Γ_N is unichain, we see that Proposition C.2.2(iii) may be generalized to give $J(N) = \sum_{i \in G} \pi_i(N) c_{iG}(N)$. Then this yields $J(N) \to \sum_{i \in H} \pi_i c_{iG} = \sum_{i \in H} \pi_i c_{iH} = J_R$. It follows from Proposition C.4.6 that $c_{iz}(N) \to c_{iz}$ for $i \in R$.

It remains to verify Definition C.4.8(ii) for i transient. We reason, in general, for initial state $i \neq z$. Let $T_i(N)$ be the time to first reach either z or $S - S_N$ (call this first passage 1), and let T_i be the first passage time to z (call this first passage 2). Note that both first passages take place in Γ. Let $u^1_{ik}(N)$ (respectively, u^2_{ik}) be the expected number of visits to k during first passage 1 (respectively, first passage 2). Clearly it is the case that $u^1_{ik}(N) \leq u^2_{ik}$. Summing both sides over k yields $E[T_i(N)] \leq m_{iz}$.

Now assume that $i \in S_N$, and consider $m_{iz}(N)$. Note that Γ_N operates just as Γ until either z is reached (and the first passage is completed) or until $S - S_N$ is reached. If the latter occurs, then the process is reset to an element of G according to some probability distribution and then begins anew an attempt to reach z (unless it is reset to z). Let us define $y_i(N) =: P(\Gamma$ reaches $S - S_N$ before it reaches $z | X_0 = i)$. Then we see that

$$\begin{aligned}
m_{iz}(N) &= E[T_i(N)] \\
&\quad + \sum_{r \in S - S_N} E[\text{additional time to reach } z | X_{T_i(N)} = r] P(X_{T_i(N)} = r) \\
&\le m_{iz} + y_i(N) \left(\sum_{j \in G - \{z\}} m_{jz}(N) \right) \\
&= m_{iz} + y_i(N) M(N), \qquad \text{(C.32)}
\end{aligned}$$

where $M(N)$ is defined to be the summation in brackets in the second line. Note that $M(N) < \infty$ for $N \ge N*$.

Now let us sum both sides of (C.32) over $j \in G - \{z\}$. Solving for $M(N)$ yields

$$M(N) \le \frac{\sum_{j \in G - \{z\}} m_{jz}}{1 - \sum_{j \in G - \{z\}} y_j(N)}. \qquad \text{(C.33)}$$

Then substituting (C.33) into (C.32) yields

$$m_{iz}(N) \le m_{iz} + \left(\sum_{j \in G - \{z\}} m_{jz} \right) \left\{ \frac{y_i(N)}{1 - \sum_{j \in G - \{z\}} y_j(N)} \right\}. \qquad \text{(C.34)}$$

We now prove that $y_{\_}(N) \to 0$. This will imply that $\limsup_{N \to \infty} m_{iz}(N) \le m_{iz}$, and hence by Proposition C.4.2(iii) it follows that $m_{iz}(N) \to m_{iz}$.

Observe that $y_i(N)$ is decreasing in N, and hence $\lim_{N \to \infty} y_i(N) =: y_i$ exists. Now

$$\begin{aligned}
y_i(N) &= \sum_{r \in S - S_N} P_{ir} + \sum_{k \in S_N - \{z\}} P_{ik} y_k(N) \\
&\le \sum_{r \in S - S_N} P_{ir} + \sum_{k \in S - \{z\}} P_{ik} y_k(N), \qquad i \ne z. \qquad \text{(C.35)}
\end{aligned}$$

Take the limit of both sides of (C.35) as $N \to \infty$. The first term on the right approaches 0. We may apply Corollary A.2.4 to the summation (with bounding function 1) to obtain $y_i \le \sum_{k \ne z} P_{ik} y_k$. Iterating this yields

$$
\begin{aligned}
y_i &\le \sum_{k \neq z} {}_zP_{ik}^{(n)} y_k \\
&\le \sum_{k \neq z} {}_zP_{ik}^{(n)} \\
&= P(T_i > n).
\end{aligned} \tag{C.36}
$$

The second line follows, since $y_k \le 1$. The finiteness of m_{iz} implies that $P(T_i > n) \to 0$ as $n \to \infty$. this implies that $y_i = 0$.

We now deal with the first passage costs. The above argument may be easily modified to yield the analogue of (C.34) for the expected costs. This yields $\limsup_{N \to \infty} c_{iz}(N) \le c_{iz}$, and hence by Proposition C.4.5 it follows that $c_{iz}(N) \to c_{iz}$.

It remains to prove the second statement of the proposition. To accomplish this, we may simply reduce S to the positive recurrent class R (in which case it is z standard for any $z \in R$) and apply the first statement. □

The next result utilizes an ATAS satisfying a structural property.

Proposition C.5.3. Let Γ be z standard. Assume that we have an ATAS and a nonnegative integer N^* such that the augmentation distributions satisfy

$$
\sum_{j \in S_n - \{z\}} q_j(i, r, N) m_{jz} \le m_{rz}, \qquad i \in S_N, r \notin S_N, N \ge N^*, \tag{C.37}
$$

and

$$
\sum_{j \in S_n - \{z\}} q_j(i, r, N) c_{jz} \le c_{rz}, \qquad i \in S_N, r \notin S_N, N \ge N^*. \tag{C.38}
$$

Then the ATAS is conforming.

Proof: The basic idea of (C.37) is that the convex combination of first passage times in Γ corresponding to excess probability P_{ir} cannot exceed the first passage time associated with r. This is a type of structural property. A similar comment holds for (C.38).

Observe that

$$\sum_{j \in S_N - \{z\}} P_{ij}(N)m_{jz} = \sum_{j \in S_N - \{z\}} P_{ij}m_{jz} + \sum_{r \notin S_N} P_{ir}\left(\sum_{j \in S_N - \{z\}} q_j(i,r,N)m_{jz}\right)$$
$$\leq \sum_{j \in S_N - \{z\}} P_{ij}m_{jz} + \sum_{r \notin S_N} P_{ir}m_{rz}$$
$$= m_{iz} - 1, \qquad i \in S_N. \tag{C.39}$$

The first line follows from (C.27). The second line follows from (C.37), and the last line from (C.4).

The hypotheses of Corollary C.1.6 are satisfied for Γ_N (with $y(i) = m_{iz}$ for $i \in S_N - \{z\}$ and $y(z) = 0$), and hence it follows that $m_{iz}(N) \leq m_{iz}$ for $i \in S_N - \{z\}$. Proposition C.4.2(iii) implies that $m_{iz}(N) \to m_{iz}$ for $i \neq z$.

Now

$$m_{zz}(N) = 1 + \sum_{j \in S_N - \{z\}} P_{zj}(N)m_{jz}(N). \tag{C.40}$$

We wish to apply Theorem A.2.6 with bounding function m_{jz}. The hypotheses will hold if it can be shown that

(*): $\lim_{N \to \infty} \sum_{j \in S_N - \{z\}} P_{zj}(N)m_{jz} = \sum_{j \neq z} P_{zj}m_{jz}$.

If (*) can be shown then Theorem A.2.6 yields $m_{zz}(N) \to 1 + \sum_{j \neq z} P_{zj}m_{jz} = m_{zz}$.

So let us show (*). It follows from (C.39) and Proposition A.1.8 that

$$m_{zz} - 1 \geq \limsup_{N \to \infty} \sum_{j \in S_N - \{z\}} P_{zj}(N)m_{jz}$$
$$\geq \liminf_{N \to \infty} \sum_{j \in S_N - \{z\}} P_{zj}(N)m_{jz}$$
$$\geq \sum_{j \neq z} P_{zj}m_{jz}$$
$$= m_{zz} - 1. \tag{C.41}$$

Hence all these terms are equal, and (*) holds.

The proof for the costs is similar and is omitted. □

We now explore two special results valid when $S = \{0, 1, 2, \ldots\}$. If $P_{ij} = 0$ for $i \geq 2$ and $j < i - 1$, then the transition matrix is *upper Hessenberg*. In this case the MC can transition downward at most one state at a time. If $P_{ij} = 0$ for $i \geq 0$ and $j > i + 1$, then the transition matrix is *lower Hessenberg*. In this case the MC can transition upward at most one state at a time.

Corollary C.5.4. Let Γ be a 0 standard MC with upper Hessenberg transition matrix. Assume that $S_N = \{0, 1, \ldots, N\}$ for $N \geq 1$ and that the ATAS sends the excess probability to N. Then it is conforming.

Proof: We apply Proposition C.5.3 with $N^* = 1$. In this case we have $q_N(.) \equiv 1$, and (C.37) becomes the requirement that $m_{N0} \leq m_{r0}$ for $r > N$. This is equivalent to the requirement that m_{i0} be increasing in i for $i \geq 1$. But this is clear for an upper Hessenberg matrix because $m_{i0} = m_{ii-1} + m_{i-10}$. Similar comments are true for the expected first passage costs. □

Another proof of this result is given in Sennott (1997a) and is based on Gibson and Seneta (1987). The following result for lower Hessenberg transition matrices is stated (for the steady state probabilities alone) in Gibson and Seneta (1987) with a proof in Gibson and Seneta (1986). A complete proof, based on the Gibson and Seneta proof and including the cost structure, is given in Sennott (1997a). We state the result here.

Proposition C.5.5. Let Γ be a standard MC with lower Hessenberg transition matrix that is irreducible on S. Assume that $S_N = \{0, 1, \ldots, N\}$ for $N \geq 1$. Let α_N be a probability distribution on S_N that converges to a probability distribution α on S as $N \to \infty$. Let the ATAS satisfy $q_j(N, N+1, N) = \alpha_N(j)$. Then it is conforming.

Note that there is excess probability only in state N and it is P_{NN+1}. This probability is distributed to the states of S_N according to the probability distribution α_N.

An example in Gibson and Seneta (1987) shows that conformity may fail to hold for a MC with a lower Hessenberg transition matrix and an ATAS that sends the excess probability to N.

BIBLIOGRAPHIC NOTES

Proposition C.1.5 and Corollary C.1.6 are modifications of a result due originally to Foster (1953) and generalized by Pakes (1969). For much additional material, see Tweedie (1976, 1983), and for these results and recent developments, see Meyn and Tweedie (1993).

For versions of Proposition C.2.3 and Corollary C.2.4, see Sennott (1989a) and Meyn and Tweedie (1993).

Concerning approximating sequences for Markov chains, earlier authors have restricted attention to Markov chains without costs. We have developed the subject to include costs, and most of our results and proof techniques are modifications of prior work.

To obtain Proposition C.4.3, we have generalized a result due to Wolf (1980); this paper also stimulated other results in Appendix C. Proposition C.5.2 is basically due to Gibson and Seneta (1987). Proposition C.5.3 is due to Sennott (1997b). Corollary C.5.4 is due to Gibson and Seneta (1987), and see Heyman (1991) for another proof. Our proof uses Proposition C.5.3. Proposition C.5.5 is also due to Gibson and Seneta (1987) with a proof in Gibson and Seneta (1986). Based on their proof, Sennott (1997a) gives a proof including costs.

An ATAS type approximating sequence for Markov chains is studied in Van Dijk (1991) and applied to some queueing systems.

Bibliography

E. Altman, P. Konstantopoulos, and Z. Liu. Stability, monotonicity and invariant quantities in general polling systems. *Queueing Sys.* **11,** 35–57 (1992).

T. Apostol. *Mathematical Analysis.* Addison-Wesley, Reading, MA, 1974.

A. Arapostathis, V. Borkar, E. Fernandez-Gaucherand, M. Ghosh, and S. Marcus. Discrete-time controlled Markov processes with average cost criterion: A survey. *SIAM J. Control Optim.* **31,** 282–344 (1993).

Y. Arian and Y. Levy. Algorithms for generalized round robin routing. *Op. Res. Letters* **12,** 313–319 (1992).

R. Barlow and F. Proschan. *Mathematical Theory of Reliability.* Wiley, New York, 1965.

R. Bellman. *Dynamic Programming.* Princeton University Press, Princeton, NJ, 1957.

D. Bertsekas. *Dynamic Programming, Deterministic and Stochastic Models.* Prentice-Hall, Englewood Cliffs, NJ, 1987.

D. Bertsekas. *Dynamic Programming and Optimal Control*, vols. 1 and 2. Athena, Belmont, MA, 1995.

D. Blackwell. Discounted dynamic programming. *Ann. Math. Stat.* **36,** 226–235 (1965).

V. Borkar. On minimum cost per unit time control of Markov chains. *SIAM J. Control Optim.* **22,** 965–978 (1984).

V. Borkar. Control of Markov chains with long-run average cost criterion. In *Stochastic Differential Systems, Stochastic Control Theory and Applications*, edited by W. Fleming and P. L. Lions. Springer-Verlag, New York, 1988.

V. Borkar. Control of Markov chains with long-run average cost criterion: the dynamic programming equations. *SIAM J. Control Optim.* **27,** 642–657 (1989).

V. Borkar. *Topics in Controlled Markov Chains*, Pitman Research Notes in Mathematics 240. Longman Scientific-Wiley, New York, 1991.

S. Borst. A globally gated polling system with a dormant server. *Prob. Eng. Info. Sci.* **9,** 239–254 (1995).

R. Bournas, F. Beutler, and D. Teneketzis. Time-average and asymptotically optimal flow control policies in networks with multiple transmitters. *Ann. Op. Res.* **35,** 327–355 (1992).

O. Boxma and W. Groenendijk. Pseudo-conservation laws in cyclic-service systems. *J. Appl. Prob.* **24,** 949–964 (1987).

S. Browne and U. Yechiali. Dynamic priority rules for cyclic-type queues. *Adv. Appl. Prob.* **21,** 432–450 (1989).

H. Bruneel and B. Kim. *Discrete-Time Models for Communication Systems Including ATM.* Kluwer, Boston, 1993.

H. Bruneel and I. Wuyts. Analysis of discrete-time multiserver queueing models with constant service times. *Op. Res. Letters* **15,** 231–236 (1994).

R. Cavazos-Cadena. Finite-state approximations for denumerable state discounted Markov decision processes. *Appl. Math. Optim.* **14,** 1–26 (1986).

R. Cavazos-Cadena. Weak conditions for the existence of optimal stationary policies in average Markov decision chains with unbounded costs. *Kybernetika* **25,** 145–156 (1989).

R. Cavazos-Cadena. Solution to the optimality equation in a class of Markov decision chains with the average cost criterion. *Kybernetika* **27,** 23–37 (1991a).

R. Cavazos-Cadena. A counterexample on the optimality equation in Markov decision chains with the average cost criterion. *Sys. Control Letters* **16,** 387–392 (1991b).

R. Cavazos-Cadena. Recent results on conditions for the existence of average optimal stationary policies. *Ann. Op. Res.* **28,** 3–27 (1991c).

R. Cavazos-Cadena and E. Fernandez-Gaucherand. Denumerable controlled Markov chains with strong average optimality criterion: Bounded and unbounded costs. *ZOR Math. Meth. Op. Res.* **43,** 281–300 (1996).

R. Cavazos-Cadena and L. Sennott. Comparing recent assumptions for the existence of average optimal stationary policies. *Op. Res. Letters* **11,** 33–37 (1992).

K. Chung. *Markov Chains with Stationary Transition Probabilities*, 2d ed. Springer-Verlag, New York, 1967.

E. Cinlar. *Introduction to Stochastic Processes.* Prentice-Hall, Englewood Cliffs, NJ, 1975.

R. Cooper. *Introduction to Queueing Theory*, 2d ed. North Holland, New York, 1981.

R. Cooper, S. Niu, and M. Srinivasan. A decomposition theorem for polling models: The switchover times are effectively additive. *Oper. Res.* **44,** 629–633 (1996).

E. Denardo. *Dynamic Programming, Models and Applications.* Prentice-Hall, Englewood Cliffs, NJ, 1982.

C. Derman. Denumerable state Markovian decision processes—Average cost criterion. *Ann. Math. Stat.* **37,** 1545–1553 (1966).

C. Derman. *Finite State Markovian Decision Processes.* Academic Press, New York, 1970.

C. Derman and A. Veinott, Jr. A solution to a countable system of equations arising in Markovian decision processes. *Ann. Math. Stat.* **38,** 582–584 (1967).

A. Federgruen and P. Schweitzer. A survey of asymptotic value-iteration for undiscounted Markovian decision processes. In *Recent Developments in Markov Decision Processes*, edited by R. Hartley, L. C. Thomas, and D. J. White. Academic Press, New York, 1980, pp. 73–109.

A. Federgruen and H. Tijms. The optimality equation in average cost denumerable state semi-Markov decision problems, recurrency conditions and algorithms. *J. Appl. Prob.* **15,** 356–373 (1978).

A. Federgruen and P. Zipkin. An inventory model with limited production capacity and uncertain demands I: The average cost criterion. *Math. Op. Res.* **11,** 193–207 (1986a).

A. Federgruen and P. Zipkin. An inventory model with limited production capacity and uncertain demands II: The discounted cost criterion. *Math. Op. Res.* **11,** 208–215 (1986b).

A. Federgruen, A. Hordijk, and H. Tijms. Denumerable state semi-Markov decision processes with unbounded costs, average cost criterion. *Stoc. Proc. Appl.* **9,** 223–235 (1979).

A. Federgruen, P. Schweitzer, and H. Tijms. Denumerable undiscounted semi-Markov decision processes with unbounded costs. *Math. Op. Res.* **8,** 298–313 (1983).

E. Feinberg. Non-randomized strategies in stochastic decision processes. *Ann. Op. Res.* **29,** 315–332 (1991).

E. Feinberg and H. Park. Finite state Markov decision models with average reward criteria. *Stoc. Proc. and Appl.* **49,** 159–177 (1994).

L. Fisher and S. Ross. An example in denumerable decision processes. *Ann. Math. Stat.* **39,** 674–675 (1968).

J. Flynn. Averaging vs. discounting in dynamic programming: A counterexample. *Ann. Stat.* **2,** 411–413 (1974).

F. Foster. On the stochastic matrices associated with certain queueing processes. *Ann. Math. Stat.* **24,** 355–360 (1953).

B. Fox. Finite-state approximations to denumerable-state dynamic programs. *J. Math. Anal. Appl.* **34,** 665–670 (1971).

C. Fricker and M. Jaibi. Monotonicity and stability of periodic polling models. *Queueing Sys.* **15,** 211–238 (1994).

L. Georgiadis and W. Szpankowski. Stability of token passing rings. *Queueing Sys.* **11,** 7–33 (1992).

D. Gibson and E. Seneta. Augmented truncations of infinite stochastic matrices. University of Sydney Report. Sydney, Australia, 1986.

D. Gibson and E. Seneta. Augmented truncations of infinite stochastic matrices. *J. Appl. Prob.* **24,** 600–608 (1987).

W. Grassmann, M. Taksar, and D. Heyman. Regenerative analysis and steady state distributions for Markov chains. *Op. Res.* **33,** 1107–1116 (1985).

D. Gross and C. Harris. *Fundamentals of Queueing Theory*, 3d ed. Wiley, New York, 1998.

B. Hajek. Optimal control of two interacting service stations. *IEEE Trans. Auto. Control* **AC-29,** 491–499 (1984).

B. Hajek. Extremal splitting of point processes. *Math. Oper. Res.* **10,** 543–556 (1985).

O. Hernandez-Lerma. Finite state approximations for denumerable multidimensional state discounted Markov decision processes. *J. Math. Anal. Appl.* **113,** 382–389 (1986).

O. Hernandez-Lerma. Average optimality in dynamic programming on Borel spaces—Unbounded costs and controls. *Sys. Control Letters* **17,** 237–242 (1991).

O. Hernandez-Lerma. Existence of average optimal policies in Markov control processes with strictly unbounded costs. *Kybernetika* **29,** 1–17 (1993).

O. Hernandez-Lerma and J. Lasserre. Average cost optimal policies for Markov control processes with Borel state space and unbounded costs. *Sys. Control Letters* **15,** 349–356 (1990).

O. Hernandez-Lerma and J. Lasserre. *Discrete-Time Markov Control Processes.* Springer-Verlag, New York, 1996.

D. Heyman. Approximating the stationary distribution of an infinite stochastic matrix. *J. Appl. Prob.* **28,** 96–103 (1991).

D. Heyman and M. Sobel. *Stochastic Models in Operations Research*, vol. 1. McGraw-Hill, New York, 1982.

D. Heyman and M. Sobel. *Stochastic Models in Operations Research*, vol. 2. McGraw-Hill, New York, 1984.

K. Hinderer. *Foundations of Non-Stationary Dynamic Programming with Discrete Time Parameter.* Springer-Verlag, New York, 1970.

A. Hordijk. Regenerative Markov decision models. *Math. Prog. Study* **6,** 49–72 (1976).

A. Hordijk. *Dynamic Programming and Markov Potential Theory*, 2d ed. Mathematisch Centrum Tract 51. Amsterdam, 1977.

A. Hordijk, P. Schweitzer, and H. Tijms. The asymptotic behavior of the minimal total expected cost for the denumerable state Markov decision model. *J. Appl. Prob.* **12,** 298–305 (1975).

R. Howard. *Dynamic Programming and Markov Processes.* MIT Press, Cambridge, 1960.

Q. Hu. Discounted and average Markov decision processes with unbounded rewards: New conditions. *J. Math. Anal. Appl.* **171,** 111–124 (1992).

W. Jewell. Markov-renewal programming I: Formulation, finite return models; Markov-renewal programming II: Infinite return models, example. *Op. Res.* **11,** 938–971 (1963).

N. Johnson and S. Kotz. *Discrete Distributions.* Houghton Mifflin, Boston, 1969.

N. Johnson, S. Kotz, and N. Balakrishnan. *Discrete Multivariate Distributions.* Wiley, New York, 1997.

S. Karlin. The structure of dynamic programming models. *Naval Res. Log. Quart.* **2,** 285–294 (1955).

S. Karlin and H. Taylor. *A First Course in Stochastic Processes*, 2d ed. Academic Press, New York, 1975.

E. Kim, M. Van Oyen, and M. Rieders. *General dynamic programming algorithms applied to polling systems.* Technical Report, Northwestern University, Evanston, IL (1996).

M. Kitaev and V. Rykov. *Controlled Queueing Systems.* CRC Press, Boca Raton, 1995.

L. Kleinrock. *Queueing Systems*, vol. 1. Wiley, New York, 1975.

K. Krishnan. Joining the right queue: A Markov decision rule. *Proc. IEEE Conf. Decision and Control* (1987), pp. 1863–1868.

J. Labetoulle and J. Roberts, eds. *The Fundamental Role of Teletraffic in the Evolution of Telecommunications Networks*, vol. 1a. Elsevier, Amsterdam, 1994.

H.-J. Langen. Convergence of dynamic programming models. *Math. Op. Res.* **6,** 493–512 (1981).

T. Liggett and S. Lippman. Stochastic games with perfect information and time average payoff. *SIAM Review* **11,** 604–607 (1969).

W. Lin and P. Kumar. Optimal control of a queueing system with two hetergeneous servers. *IEEE Trans. Auto. Control* **AC-29,** 696–703 (1984).

S. Lippman. On dynamic programming with unbounded rewards. *Man. Sci.* **21,** 1225–1233 (1975a).

S. Lippman. Applying a new device in the optimization of exponential queueing systems. *Op. Res.* **23,** 687–710 (1975b).

A. Makowski and A. Shwartz. On the Poisson equation for Markov chains: Existence of solutions and parameter dependence by probabilistic methods. Preprint (1994).

S. Meyn and R. Tweedie. *Markov Chains and Stochastic Stability.* Springer-Verlag, New York, 1993.

R. Milito and E. Fernandez-Gaucherand. Open-loop routing of N arrivals to M parallel queues. *IEEE Trans. Auto. Control* **AC-40,** 2108–2114 (1995).

R. Montes-de-Oca and O. Hernandez-Lerma. Conditions for average optimality in Markov control processes with unbounded costs and controls. *J. Math. Sys. Estimation and Control* **4,** 1–19 (1994).

A. Odoni. On finding the maximal gain for Markov decision processes. *Op. Res.* **17,** 857–860 (1969).

A. Pakes. Some conditions for ergodicity and recurrence of Markov chains. *Op. Res.* **17,** 1058–1061 (1969).

W. Pervin. *Foundations of General Topology.* Academic Press, New York, 1964.

M. Puterman. *Markov Decision Processes.* Wiley, New York, 1994.

R. Ritt and L. Sennott. Optimal stationary policies in general state space Markov decision chains with finite action sets. *Math. Op. Res.* **17,** 901–909 (1992).

Z. Rosberg. Deterministic routing to buffered channels. *IEEE Trans. on Comm.* **COM:34,** 504–507 (1986).

S. Ross. Non-discounted denumerable Markovian decision models. *Ann. Math. Stat.* **39,** 412–423 (1968).

S. Ross. *Applied Probability Models with Optimization Applications.* Holden-Day, San Francisco, 1970.

S. Ross. On the nonexistence of ε-optimal randomized stationary policies in average cost Markovian decision models. *Ann. Math. Stat.* **42,** 1767–1768 (1971).

S. Ross. *Introduction to Stochastic Dynamic Programming.* Academic Press, New York, 1983.

S. Ross. *Stochastic Processes*, 2d ed. Wiley, New York, 1996.

H. Royden. *Real Analysis*, 2d ed. Macmillan, New York, 1968.

H. Scarf. The optimality of (s, S) policies in the dynamic inventory problem. In *Studies in the Mathematical Theory of Inventory and Production*, edited by K. Arrow, S. Karlin, and P. Suppes. Stanford University Press, 1960.

M. Schal. Conditions for optimality in dynamic programming and for the limit of n-stage optimal policies to be optimal. *Z. Wahr. Geb.* **32,** 179–196 (1975).

M. Schal. On the optimality of (s, S) policies in dynamic inventory models with finite horizon. *SIAM J. Appl. Math.* **13,** 528–537 (1976).

M. Schal. Average optimality in dynamic programming with general state space. *Math. Op. Res.* **18,** 163–172 (1993).

P. Schweitzer. Iterative solution of the functional equations of undiscounted Markov renewal programming. *J. Math. Anal. Appl.* **34,** 495–501 (1971).

P. Schweitzer and A. Federgruen, The asymptotic behavior of undiscounted value iteration in Markov decision problems, *Math. Op. Res.* **2,** 360–381 (1978).

L. Sennott, P. Humblet, and R. Tweedie. Mean drifts and the non-ergodicity of Markov chains. *Op. Res.* **31,** 783–789 (1983).

L. Sennott. A new condition for the existence of optimum stationary policies in average cost Markov decision processes. *Op. Res. Letters* **5,** 17–23 (1986a).

L. Sennott. A new condition for the existence of optimum stationary policies in average cost Markov decision processes—Unbounded cost case, *Proc. 25th Conf. Decision and Control*, Athens, Greece (1986b), pp. 1719–1721.

L. Sennott. Average cost optimal stationary policies in infinite state Markov decision processes with unbounded costs. *Op. Res.* **37,** 626–633 (1989a).

L. Sennott. Average cost semi-Markov decision processes and the control of queueing systems. *Prob. Eng. Info. Sci.* **3,** 247–272 (1989b).

L. Sennott. Value iteration in countable state average cost Markov decision processes with unbounded costs. *Ann. Op. Res.* **28,** 261–272 (1991).

L. Sennott. The average cost optimality equation and critical number policies. *Prob. Eng. Info. Sci.* **7,** 47–67 (1993).

L. Sennott. Another set of conditions for average optimality in Markov control processes. *Sys. Control Letters* **24,** 147–151 (1995).

L. Sennott. The computation of average optimal policies in denumerable state Markov decision chains. *Adv. Appl. Prob.* **29,** 114–137 (1997a).

L. Sennott. On computing average cost optimal policies with application to routing to parallel queues. *ZOR Math. Meth. Op. Res.* **45,** 45–62 (1997b).

R. Serfozo. An equivalence between continuous and discrete time Markov decision processes. *Op. Res.* **27,** 616–620 (1979).

J. Shanthikumar and S. Xu. Asymptotically optimal routing and service rate allocation in a multi-server queueing system. *Op. Res.*, to appear.

S. Shenker and A. Weinrib. The optimal control of heterogeneous queueing systems: A paradigm for load-sharing and routing. *IEEE Trans. Compu.* **38,** 1724–1735 (1989).

F. Spieksma. Geometrically ergodic Markov chains and the optimal control of queues. Ph.D. dissertation. Leiden University, 1990.

M. Srinivasan, S. Niu, and R. Cooper. Relating polling models with zero and nonzero switchover times. *Queueing Sys.* **19,** 149–168 (1995).

S. Stidham, Jr. On the convergence of successive approximations in dynamic programming with non-zero terminal reward. *Z. Op. Res.* **25,** 57–77 (1981).

S. Stidham, Jr., and R. Weber. Monotonic and insensitive optimal policies for control of queues with undiscounted costs. *Op. Res.* **87,** 611–625 (1989).

S. Stidham, Jr., and R. Weber, A survey of Markov decision models for control of networks of queues. *Queueing Systems* **13,** 291–314 (1993).

R. Strauch. Negative dynamic programming. *Ann. Math. Stat.* **37,** 871–890 (1966).

H. Takagi. *Analysis of Polling Systems.* MIT, Cambridge, 1986.

H. Takagi. Queueing analysis of polling models: An update. In *Stochastic Analysis of Computer and Communication Systems*, edited by H. Takagi. North Holland, New York, 1990.

H. Takagi. Queueing analysis of polling models: progress in 1990–1994. In *Frontiers in Queueing*, edited by J. Dshalalow. CRC Press, Boca Raton, 1997.

H. Taylor. Markovian sequential replacement processes. *Ann. Math. Stat.* **36,** 1677–1694 (1965).

H. Taylor and S. Karlin. *An Introduction to Stochastic Modeling.* Academic Press, New York, 1984.

L. Thomas and D. Stengos. Finite state approximation algorithms for average cost denumerable state Markov decision processes. *OR Spektrum* **7,** 27–37 (1985).

H. Tijms. *Stochastic Models, An Algorithmic Approach.* Wiley, New York, 1994.

E. Titchmarsh. *Theory of Functions*, 2d ed. Oxford University Press, Oxford, 1939.

K. Tseng and M.-T. Hsiao. Optimal control of arrivals to token ring networks with exhaustive service discipline. *Op. Res.* **43,** 89–101 (1995).

R. Tweedie. Criteria for classifying general Markov chains. *Adv. Appl. Prob.* **8,** 737–771 (1976).

R. Tweedie. The existence of moments for stationary Markov chains. *J. Appl. Prob.* **20,** 191–196 (1983).

N. Van Dijk. Truncation of Markov chains with applications to queueing. *Op. Res.* **39,** 1018–1026 (1991a).

N. Van Dijk. On truncations and perturbations of Markov decision problems with an application to queueing network overflow control. *Ann. Oper. Res.* **29,** 515–536 (1991b).

A. Veinott, Jr. On the optimality of (s, S) inventory policies: New conditions and a new proof. *SIAM J. Appl. Math.* **14,** 1067–1083 (1966).

J. Walrand. *An Introduction to Queueing Networks.* Prentice-Hall, Englewood Cliffs, NJ, 1988.

R. Weber. On the optimal assignment of customers to parallel queues. *J. Appl. Prob.* **15,** 406–413 (1978).

C. White III and D. White. Markov decision processes. *Eur. J. Op. Res.* **39,** 1–16 (1989).

D. White. Dynamic programming, Markov chains, and the method of successive approximations. *J. Math. Anal. Appl.* **6,** 373–376 (1963).

D. White. Finite state approximations for denumerable state infinite horizon discounted Markov decision processes: The method of successive approximation. In *Recent Developments in Markov Decision Processes*, edited by R. Hartley, L. Thomas, and D. White. Academic Press, New York, 1980a, pp. 57–72.

D. White. Finite state approximations for denumerable state infinite horizon discounted Markov decision processes. *J. Math. Anal. Appl.* **74,** 292–295 (1980b).

D. White. Finite state approximations for denumerable state infinite horizon discounted Markov decision processes with unbounded rewards. *J. Math. Anal. Appl.* **86,** 292—306 (1982).

W. Whitt. Approximations of dynamic programs I. *Math. Op. Res.* **3,** 231–243 (1978).

W. Whitt. A priori bounds for approximations of Markov programs. *J. Math. Anal. Appl.* **71,** 297–302 (1979a).

W. Whitt. Approximations of dynamic programs II. *Math. Op. Res.* **4,** 179–185 (1979b).

D. Widder. *The Laplace Transform.* Princeton University Press, Princeton, NJ, 1941.

J. Wijngaard. Existence of average optimal strategies in Markovian decision problems with strictly unbounded costs. In *Dynamic Programming and Its Applications*, edited by M. L. Puterman. Academic Press, New York, 1978.

W. Winston. Optimality of the shortest line discipline. *J. Appl. Prob.* **14,** 181–189 (1977).

D. Wolf. Approximation of the invariant probability measure of an infinite stochastic matrix. *Adv. Appl. Prob.* **12,** 710–726 (1980).

R. Wolff. *Stochastic Modeling and the Theory of Queues.* Prentice-Hall, Englewood Cliffs, NJ, 1989.

Index

WILEY SERIES IN PROBABILITY AND STATISTICS

ESTABLISHED BY WALTER A. SHEWHART AND SAMUEL S. WILKS

Probability and Statistics Section

*ANDERSON · The Statistical Analysis of Time Series
ARNOLD, BALAKRISHNAN, and NAGARAJA · A First Course in Order Statistics
ARNOLD, BALAKRISHNAN, and NAGARAJA · Records
BACCELLI, COHEN, OLSDER, and QUADRAT · Synchronization and Linearity: An Algebra for Discrete Event Systems
BASILEVSKY · Statistical Factor Analysis and Related Methods: Theory and Applications
BERNARDO and SMITH · Bayesian Statistical Concepts and Theory
BILLINGSLEY · Convergence of Probability Measures
BOROVKOV · Asymptotic Methods in Queuing Theory
BRANDT, FRANKEN, and LISEK · Stationary Stochastic Models
CAINES · Linear Stochastic Systems
CAIROLI and DALANG · Sequential Stochastic Optimization
CONSTANTINE · Combinatorial Theory and Statistical Design
COVER and THOMAS · Elements of Information Theory
CSÖRGŐ and HORVÁTH · Weighted Approximations in Probability Statistics
CSÖRGŐ and HORVÁTH · Limit Theorems in Change Point Analysis
DETTE and STUDDEN · The Theory of Canonical Moments with Applications in Statistics, Probability, and Analysis
*DOOB · Stochastic Processes
DRYDEN and MARDIA · Statistical Analysis of Shape
DUPUIS and ELLIS · A Weak Convergence Approach to the Theory of Large Deviations
ETHIER and KURTZ · Markov Processes: Characterization and Convergence
FELLER · An Introduction to Probability Theory and Its Applications, Volume I, *Third Edition,* Revised; Volume II, *Second Edition*
FULLER · Introduction to Statistical Time Series, *Second Edition*
FULLER · Measurement Error Models
GELFAND and SMITH · Bayesian Computation
GHOSH, MUKHOPADHYAY, and SEN · Sequential Estimation
GIFI · Nonlinear Multivariate Analysis
GUTTORP · Statistical Inference for Branching Processes
HALL · Introduction to the Theory of Coverage Processes
HAMPEL · Robust Statistics: The Approach Based on Influence Functions
HANNAN and DEISTLER · The Statistical Theory of Linear Systems
HUBER · Robust Statistics
IMAN and CONOVER · A Modern Approach to Statistics
JUREK and MASON · Operator-Limit Distributions in Probability Theory
KASS and VOS · Geometrical Foundations of Asymptotic Inference
KAUFMAN and ROUSSEEUW · Finding Groups in Data: An Introduction to Cluster Analysis

*Now available in a lower priced paperback edition in the Wiley Classics Library.

Probability and Statistics (Continued)

KELLY · Probability, Statistics, and Optimization
LINDVALL · Lectures on the Coupling Method
McFADDEN · Management of Data in Clinical Trials
MANTON, WOODBURY, and TOLLEY · Statistical Applications Using Fuzzy Sets
MARDIA and JUPP · Statistics of Directional Data, *Second Edition*
MORGENTHALER and TUKEY · Configural Polysampling: A Route to Practical Robustness
MUIRHEAD · Aspects of Multivariate Statistical Theory
OLIVER and SMITH · Influence Diagrams, Belief Nets and Decision Analysis
*PARZEN · Modern Probability Theory and Its Applications
PRESS · Bayesian Statistics: Principles, Models, and Applications
PUKELSHEIM · Optimal Experimental Design
RAO · Asymptotic Theory of Statistical Inference
RAO · Linear Statistical Inference and Its Applications, *Second Edition*
*RAO and SHANBHAG · Choquet-Deny Type Functional Equations with Applications to Stochastic Models
ROBERTSON, WRIGHT, and DYKSTRA · Order Restricted Statistical Inference
ROGERS and WILLIAMS · Diffusions, Markov Processes, and Martingales, Volume I: Foundations, *Second Edition;* Volume II: Îto Calculus
RUBINSTEIN and SHAPIRO · Discrete Event Systems: Sensitivity Analysis and Stochastic Optimization by the Score Function Method
RUZSA and SZEKELY · Algebraic Probability Theory
SCHEFFE · The Analysis of Variance
SEBER · Linear Regression Analysis
SEBER · Multivariate Observations
SEBER and WILD · Nonlinear Regression
SERFLING · Approximation Theorems of Mathematical Statistics
SHORACK and WELLNER · Empirical Processes with Applications to Statistics
SMALL and McLEISH · Hilbert Space Methods in Probability and Statistical Inference
STAPLETON · Linear Statistical Models
STAUDTE and SHEATHER · Robust Estimation and Testing
STOYANOV · Counterexamples in Probability
TANAKA · Time Series Analysis: Nonstationary and Noninvertible Distribution Theory
THOMPSON and SEBER · Adaptive Sampling
WELSH · Aspects of Statistical Inference
WHITTAKER · Graphical Models in Applied Multivariate Statistics
YANG · The Construction Theory of Denumerable Markov Processes

Applied Probability and Statistics Section

ABRAHAM and LEDOLTER · Statistical Methods for Forecasting
AGRESTI · Analysis of Ordinal Categorical Data
AGRESTI · Categorical Data Analysis
ANDERSON, AUQUIER, HAUCK, OAKES, VANDAELE, and WEISBERG · Statistical Methods for Comparative Studies
ARMITAGE and DAVID (editors) · Advances in Biometry
*ARTHANARI and DODGE · Mathematical Programming in Statistics
ASMUSSEN · Applied Probability and Queues
*BAILEY · The Elements of Stochastic Processes with Applications to the Natural Sciences
BARNETT and LEWIS · Outliers in Statistical Data, *Third Edition*
BARTHOLOMEW, FORBES, and McLEAN · Statistical Techniques for Manpower Planning, *Second Edition*

*Now available in a lower priced paperback edition in the Wiley Classics Library.

Applied Probability and Statistics (Continued)

BATES and WATTS · Nonlinear Regression Analysis and Its Applications
BECHHOFER, SANTNER, and GOLDSMAN · Design and Analysis of Experiments for Statistical Selection, Screening, and Multiple Comparisons
BELSLEY · Conditioning Diagnostics: Collinearity and Weak Data in Regression
BELSLEY, KUH, and WELSCH · Regression Diagnostics: Identifying Influential Data and Sources of Collinearity
BHAT · Elements of Applied Stochastic Processes, *Second Edition*
BHATTACHARYA and WAYMIRE · Stochastic Processes with Applications
BIRKES and DODGE · Alternative Methods of Regression
BLOOMFIELD · Fourier Analysis of Time Series: An Introduction
BOLLEN · Structural Equations with Latent Variables
BOULEAU · Numerical Methods for Stochastic Processes
BOX · Bayesian Inference in Statistical Analysis
BOX and DRAPER · Empirical Model-Building and Response Surfaces
BOX and DRAPER · Evolutionary Operation: A Statistical Method for Process Improvement
BUCKLEW · Large Deviation Techniques in Decision, Simulation, and Estimation
BUNKE and BUNKE · Nonlinear Regression, Functional Relations and Robust Methods: Statistical Methods of Model Building
CHATTERJEE and HADI · Sensitivity Analysis in Linear Regression
CHOW and LIU · Design and Analysis of Clinical Trials: Concepts and Methodologies
CLARKE and DISNEY · Probability and Random Processes: A First Course with Applications, *Second Edition*
*COCHRAN and COX · Experimental Designs, *Second Edition*
CONOVER · Practical Nonparametric Statistics, *Second Edition*
CORNELL · Experiments with Mixtures, Designs, Models, and the Analysis of Mixture Data, *Second Edition*
*COX · Planning of Experiments
CRESSIE · Statistics for Spatial Data, *Revised Edition*
DANIEL · Applications of Statistics to Industrial Experimentation
DANIEL · Biostatistics: A Foundation for Analysis in the Health Sciences, *Sixth Edition*
DAVID · Order Statistics, *Second Edition*
*DEGROOT, FIENBERG, and KADANE · Statistics and the Law
DODGE · Alternative Methods of Regression
DOWDY and WEARDEN · Statistics for Research, *Second Edition*
DUNN and CLARK · Applied Statistics: Analysis of Variance and Regression, *Second Edition*
ELANDT-JOHNSON and JOHNSON · Survival Models and Data Analysis
EVANS, PEACOCK, and HASTINGS · Statistical Distributions, *Second Edition*
FLEISS · The Design and Analysis of Clinical Experiments
FLEISS · Statistical Methods for Rates and Proportions, *Second Edition*
FLEMING and HARRINGTON · Counting Processes and Survival Analysis
GALLANT · Nonlinear Statistical Models
GLASSERMAN and YAO · Monotone Structure in Discrete-Event Systems
GNANADESIKAN · Methods for Statistical Data Analysis of Multivariate Observations, *Second Edition*
GOLDSTEIN and LEWIS · Assessment: Problems, Development, and Statistical Issues
GREENWOOD and NIKULIN · A Guide to Chi-Squared Testing
*HAHN · Statistical Models in Engineering
HAHN and MEEKER · Statistical Intervals: A Guide for Practitioners
HAND · Construction and Assessment of Classification Rules
HAND · Discrimination and Classification
HEIBERGER · Computation for the Analysis of Designed Experiments

*Now available in a lower priced paperback edition in the Wiley Classics Library.

Applied Probability and Statistics (Continued)

HINKELMAN and KEMPTHORNE: · Design and Analysis of Experiments, Volume 1: Introduction to Experimental Design

HOAGLIN, MOSTELLER, and TUKEY · Exploratory Approach to Analysis of Variance

HOAGLIN, MOSTELLER, and TUKEY · Exploring Data Tables, Trends and Shapes

HOAGLIN, MOSTELLER, and TUKEY · Understanding Robust and Exploratory Data Analysis

HOCHBERG and TAMHANE · Multiple Comparison Procedures

HOCKING · Methods and Applications of Linear Models: Regression and the Analysis of Variables

HOGG and KLUGMAN · Loss Distributions

HOLLANDER and WOLFE · Nonparametric Statistical Methods

HOSMER and LEMESHOW · Applied Logistic Regression

HØYLAND and RAUSAND · System Reliability Theory: Models and Statistical Methods

HUBERTY · Applied Discriminant Analysis

JACKSON · A User's Guide to Principle Components

JOHN · Statistical Methods in Engineering and Quality Assurance

JOHNSON · Multivariate Statistical Simulation

JOHNSON and KOTZ · Distributions in Statistics
Continuous Multivariate Distributions

JOHNSON, KOTZ, and BALAKRISHNAN · Continuous Univariate Distributions, Volume 1, *Second Edition*

JOHNSON, KOTZ, and BALAKRISHNAN · Continuous Univariate Distributions, Volume 2, *Second Edition*

JOHNSON, KOTZ, and BALAKRISHNAN · Discrete Multivariate Distributions

JOHNSON, KOTZ, and KEMP · Univariate Discrete Distributions, *Second Edition*

JUREČKOVÁ and SEN · Robust Statistical Procedures: Aymptotics and Interrelations

KADANE · Bayesian Methods and Ethics in a Clinical Trial Design

KADANE AND SCHUM · A Probabilistic Analysis of the Sacco and Vanzetti Evidence

KALBFLEISCH and PRENTICE · The Statistical Analysis of Failure Time Data

KELLY · Reversability and Stochastic Networks

KHURI, MATHEW, and SINHA · Statistical Tests for Mixed Linear Models

KLUGMAN, PANJER, and WILLMOT · Loss Models: From Data to Decisions

KLUGMAN, PANJER, and WILLMOT · Solutions Manual to Accompany Loss Models: From Data to Decisions

KOVALENKO, KUZNETZOV, and PEGG · Mathematical Theory of Reliability of Time-Dependent Systems with Practical Applications

LAD · Operational Subjective Statistical Methods: A Mathematical, Philosophical, and Historical Introduction

LANGE, RYAN, BILLARD, BRILLINGER, CONQUEST, and GREENHOUSE · Case Studies in Biometry

LAWLESS · Statistical Models and Methods for Lifetime Data

LEE · Statistical Methods for Survival Data Analysis, *Second Edition*

LePAGE and BILLARD · Exploring the Limits of Bootstrap

LINHART and ZUCCHINI · Model Selection

LITTLE and RUBIN · Statistical Analysis with Missing Data

MAGNUS and NEUDECKER · Matrix Differential Calculus with Applications in Statistics and Econometrics

MALLER and ZHOU · Survival Analysis with Long Term Survivors

MANN, SCHAFER, and SINGPURWALLA · Methods for Statistical Analysis of Reliability and Life Data

McLACHLAN and KRISHNAN · The EM Algorithm and Extensions

McLACHLAN · Discriminant Analysis and Statistical Pattern Recognition

McNEIL · Epidemiological Research Methods

*Now available in a lower priced paperback edition in the Wiley Classics Library.

Applied Probability and Statistics (Continued)

MEEKER and ESCOBAR · Statistical Methods for Reliability Data
MILLER · Survival Analysis
MONTGOMERY and PECK · Introduction to Linear Regression Analysis, *Second Edition*
MYERS and MONTGOMERY · Response Surface Methodology: Process and Product in Optimization Using Designed Experiments
NELSON · Accelerated Testing, Statistical Models, Test Plans, and Data Analyses
NELSON · Applied Life Data Analysis
OCHI · Applied Probability and Stochastic Processes in Engineering and Physical Sciences
OKABE, BOOTS, and SUGIHARA · Spatial Tesselations: Concepts and Applications of Voronoi Diagrams
PANKRATZ · Forecasting with Dynamic Regression Models
PANKRATZ · Forecasting with Univariate Box-Jenkins Models: Concepts and Cases
PIANTADOSI · Clinical Trials: A Methodologic Perspective
PORT · Theoretical Probability for Applications
PUTERMAN · Markov Decision Processes: Discrete Stochastic Dynamic Programming
RACHEV · Probability Metrics and the Stability of Stochastic Models
RÉNYI · A Diary on Information Theory
RIPLEY · Spatial Statistics
RIPLEY · Stochastic Simulation
ROUSSEEUW and LEROY · Robust Regression and Outlier Detection
RUBIN · Multiple Imputation for Nonresponse in Surveys
RUBINSTEIN · Simulation and the Monte Carlo Method
RUBINSTEIN and MELAMED · Modern Simulation and Modeling
RYAN · Statistical Methods for Quality Improvement
SCHUSS · Theory and Applications of Stochastic Differential Equations
SCOTT · Multivariate Density Estimation: Theory, Practice, and Visualization
*SEARLE · Linear Models
SEARLE · Linear Models for Unbalanced Data
SEARLE, CASELLA, and McCULLOCH · Variance Components
SENNOTT · Stochastic Dynamic Programming and the Control of Queueing Systems
STOYAN, KENDALL, and MECKE · Stochastic Geometry and Its Applications, *Second Edition*
STOYAN and STOYAN · Fractals, Random Shapes and Point Fields: Methods of Geometrical Statistics
THOMPSON · Empirical Model Building
THOMPSON · Sampling
TIJMS · Stochastic Modeling and Analysis: A Computational Approach
TIJMS · Stochastic Models: An Algorithmic Approach
TITTERINGTON, SMITH, and MAKOV · Statistical Analysis of Finite Mixture Distributions
UPTON and FINGLETON · Spatial Data Analysis by Example, Volume 1: Point Pattern and Quantitative Data
UPTON and FINGLETON · Spatial Data Analysis by Example, Volume II: Categorical and Directional Data
VAN RIJCKEVORSEL and DE LEEUW · Component and Correspondence Analysis
WEISBERG · Applied Linear Regression, *Second Edition*
WESTFALL and YOUNG · Resampling-Based Multiple Testing: Examples and Methods for p-Value Adjustment
WHITTLE · Systems in Stochastic Equilibrium
WOODING · Planning Pharmaceutical Clinical Trials: Basic Statistical Principles
WOOLSON · Statistical Methods for the Analysis of Biomedical Data
*ZELLNER · An Introduction to Bayesian Inference in Econometrics

*Now available in a lower priced paperback edition in the Wiley Classics Library.

Texts and References Section

AGRESTI · An Introduction to Categorical Data Analysis
ANDERSON · An Introduction to Multivariate Statistical Analysis, *Second Edition*
ANDERSON and LOYNES · The Teaching of Practical Statistics
ARMITAGE and COLTON · Encyclopedia of Biostatistics: Volumes 1 to 6 with Index
BARTOSZYNSKI and NIEWIADOMSKA-BUGAJ · Probability and Statistical Inference
BERRY, CHALONER, and GEWEKE · Bayesian Analysis in Statistics and Econometrics: Essays in Honor of Arnold Zellner
BHATTACHARYA and JOHNSON · Statistical Concepts and Methods
BILLINGSLEY · Probability and Measure, *Second Edition*
BOX · R. A. Fisher, the Life of a Scientist
BOX, HUNTER, and HUNTER · Statistics for Experimenters: An Introduction to Design, Data Analysis, and Model Building
BOX and LUCEÑO · Statistical Control by Monitoring and Feedback Adjustment
BROWN and HOLLANDER · Statistics: A Biomedical Introduction
CHATTERJEE and PRICE · Regression Analysis by Example, *Second Edition*
COOK and WEISBERG · An Introduction to Regression Graphics
COX · A Handbook of Introductory Statistical Methods
DILLON and GOLDSTEIN · Multivariate Analysis: Methods and Applications
DODGE and ROMIG · Sampling Inspection Tables, *Second Edition*
DRAPER and SMITH · Applied Regression Analysis, *Third Edition*
DUDEWICZ and MISHRA · Modern Mathematical Statistics
DUNN · Basic Statistics: A Primer for the Biomedical Sciences, *Second Edition*
FISHER and VAN BELLE · Biostatistics: A Methodology for the Health Sciences
FREEMAN and SMITH · Aspects of Uncertainty: A Tribute to D. V. Lindley
GROSS and HARRIS · Fundamentals of Queueing Theory, *Third Edition*
HALD · A History of Probability and Statistics and their Applications Before 1750
HALD · A History of Mathematical Statistics from 1750 to 1930
HELLER · MACSYMA for Statisticians
HOEL · Introduction to Mathematical Statistics, *Fifth Edition*
JOHNSON and BALAKRISHNAN · Advances in the Theory and Practice of Statistics: A Volume in Honor of Samuel Kotz
JOHNSON and KOTZ (editors) · Leading Personalities in Statistical Sciences: From the Seventeenth Century to the Present
JUDGE, GRIFFITHS, HILL, LÜTKEPOHL, and LEE · The Theory and Practice of Econometrics, *Second Edition*
KHURI · Advanced Calculus with Applications in Statistics
KOTZ and JOHNSON (editors) · Encyclopedia of Statistical Sciences: Volumes 1 to 9 wtih Index
KOTZ and JOHNSON (editors) · Encyclopedia of Statistical Sciences: Supplement Volume
KOTZ, REED, and BANKS (editors) · Encyclopedia of Statistical Sciences: Update Volume 1
KOTZ, REED, and BANKS (editors) · Encyclopedia of Statistical Sciences: Update Volume 2
LAMPERTI · Probability: A Survey of the Mathematical Theory, *Second Edition*
LARSON · Introduction to Probability Theory and Statistical Inference, *Third Edition*
LE · Applied Categorical Data Analysis
LE · Applied Survival Analysis
MALLOWS · Design, Data, and Analysis by Some Friends of Cuthbert Daniel
MARDIA · The Art of Statistical Science: A Tribute to G. S. Watson
MASON, GUNST, and HESS · Statistical Design and Analysis of Experiments with Applications to Engineering and Science

*Now available in a lower priced paperback edition in the Wiley Classics Library.

Texts and References (Continued)

MURRAY · X-STAT 2.0 Statistical Experimentation, Design Data Analysis, and Nonlinear Optimization
PURI, VILAPLANA, and WERTZ · New Perspectives in Theoretical and Applied Statistics
RENCHER · Methods of Multivariate Analysis
RENCHER · Multivariate Statistical Inference with Applications
ROSS · Introduction to Probability and Statistics for Engineers and Scientists
ROHATGI · An Introduction to Probability Theory and Mathematical Statistics
RYAN · Modern Regression Methods
SCHOTT · Matrix Analysis for Statistics
SEARLE · Matrix Algebra Useful for Statistics
STYAN · The Collected Papers of T. W. Anderson: 1943–1985
TIERNEY · LISP-STAT: An Object-Oriented Environment for Statistical Computing and Dynamic Graphics
WONNACOTT and WONNACOTT · Econometrics, *Second Edition*

WILEY SERIES IN PROBABILITY AND STATISTICS

ESTABLISHED BY WALTER A. SHEWHART AND SAMUEL S. WILKS

Editors
Robert M. Groves, Graham Kalton, J. N. K. Rao, Norbert Schwarz, Christopher Skinner

Survey Methodology Section

BIEMER, GROVES, LYBERG, MATHIOWETZ, and SUDMAN · Measurement Errors in Surveys
COCHRAN · Sampling Techniques, *Third Edition*
COX, BINDER, CHINNAPPA, CHRISTIANSON, COLLEDGE, and KOTT (editors) · Business Survey Methods
*DEMING · Sample Design in Business Research
DILLMAN · Mail and Telephone Surveys: The Total Design Method
GROVES and COUPER · Nonresponse in Household Interview Surveys
GROVES · Survey Errors and Survey Costs
GROVES, BIEMER, LYBERG, MASSEY, NICHOLLS, and WAKSBERG · Telephone Survey Methodology
*HANSEN, HURWITZ, and MADOW · Sample Survey Methods and Theory, Volume I: Methods and Applications
*HANSEN, HURWITZ, and MADOW · Sample Survey Methods and Theory, Volume II: Theory
KASPRZYK, DUNCAN, KALTON, and SINGH · Panel Surveys
KISH · Statistical Design for Research
*KISH · Survey Sampling
LESSLER and KALSBEEK · Nonsampling Error in Surveys
LEVY and LEMESHOW · Sampling of Populations: Methods and Applications
LYBERG, BIEMER, COLLINS, de LEEUW, DIPPO, SCHWARZ, TREWIN (editors) · Survey Measurement and Process Quality
SKINNER, HOLT, and SMITH · Analysis of Complex Surveys

*Now available in a lower priced paperback edition in the Wiley Classics Library.